Pastoral practices in High Asia

Advances in Asian Human-Environmental Research

Aims and Scope

The series aims at fostering the discussion on the complex relationships between physical landscapes, natural resources, and their modification by human land use in various environments of Asia. It is widely acknowledged that human-environment-interactions become increasingly important in area studies and development research, taking into account regional differences as well as bio-physical, socio-economic and cultural particularities.

The book series seeks to explore theoretic and conceptual reflection on dynamic human-environment systems applying advanced methodology and innovative research perspectives. The main themes of the series cover urban and rural landscapes in Asia. Examples include topics such as land and forest degradation, glaciers in Asia, mountain environments, dams in Asia, medical geography, vulnerability and mitigation strategies, natural hazards and risk management concepts, environmental change, impacts studies and consequences for local communities. The relevant themes of the series are mainly focused on geographical research perspectives of area studies, however there is scope for interdisciplinary contributions.

For further volumes:
http://www.springer.com/series/8560

Hermann Kreutzmann
Editor

Pastoral practices in High Asia

Agency of 'development' effected by modernisation, resettlement and transformation

Springer

Editor
Hermann Kreutzmann
Centre for Development Studies
Geographic Sciences
Freie Universität Berlin
Malteserstr. 74-100, House K
D-12249 Berlin, Germany

ISBN 978-94-007-3845-4 ISBN 978-94-007-3846-1 (eBook)
DOI 10.1007/978-94-007-3846-1
Springer Dordrecht Heidelberg New York London

Library of Congress Control Number: 2012935644

Printed on acid-free paper

Springer is part of Springer Science+Business Media (www.springer.com)

Preface

Eckart Ehlers

If we believe the results of the synthesis report of the Millennium Ecosystem Assessment, published in 2005 under the title 'Ecosystems and Human Well-Being', only very few terrestrial biomes are still in a somewhat pristine condition. Amongst them are, not surprisingly, the desert regions of the earth, its boreal forests and tundras and, of course, the ice-covered regions of the northern and southern hemispheres. But also parts of the tropical forests and of the montane regions are so far comparatively little affected by human conversion. These – the montane grasslands and shrublands as well as the tropical and subtropical moist broadleaf forests – will, however, experience dramatic changes until the midst of the twenty-first century, especially due to human interferences and impacts on their natural environments.

High Asia – defined as a region between the Hindukush in the West and the fringes of the Tibetan Plateau in the East, between the Altay and its Mongolian promontories in the North and the Himalayas in the South – is the most significant part of this montane milieu on a global scale. Covering a highly dynamic region of recent state formations, of fundamental political changes and of remarkable socio-economic developments, ongoing and future transformation processes have dramatic impacts on all spheres of life – and will continue to have so even more in the future. They will affect the fragile and sensitive natural environments of the montane milieus, because their inhabitants' traditional lifestyles, land uses and land management practices are under rapidly increasing pressures from population growth and modernization processes.

These developments set the frame for the contents of this book. '*Pastoral Practices in High Asia*' gives insights into the aforementioned processes of change. Based on a number of preparatory workshops in Germany and in the regions under review, on field visits and discussions with local stakeholders and incorporating the experiences of local experts and their intimate knowledge of the problems at stake, the book contains a broad set of articles, in which the wide range of crucial coping and adaptation strategies of pastoral nomads and mountain farmers and their struggles with change, both natural and structural, are presented and discussed. A remarkable feature of all contributions is their focused approach to those aspects of herding practices that are crucial to its sustainable future in the light of modernization and globalization. Due to the fact that all presented case studies are located in similar ecological environments, on the one hand, in different political and socio-economic settings, on the other, however, each of the case studies contains an element of comparability and transferability. It may therefore be argued with good reasons that the 16 regionally and/or thematically different analyses of pastoral practices and experiences in High Asia are in fact a comprehensive survey of adaptation and transformation processes in a comparative view – an approach which enables researchers and practitioners to reflect on best practices and to consider lessons to be learnt from each other.

Specific mention must be made of the both careful and focused editorship of this book. Guided by an almost all-embracing introduction by the editor of this volume, all authors are more or less embedded in a structural frame in which they were expected 'to elaborate on the varied expressions of pastoral practices, frame conditions and performances'. Arguing that agencies, that is, institutions and actors on local to global levels, cause developments which not seldom lead to *endisms* (e.g. the end of nomadism), poses an intellectual challenge and hypothesis which authors had to respond to. The editor takes up this challenge in his final conclusions. Analysis and critical evaluation of the presented case studies lead him to basic reflections on Hardin's 'tragedy of the commons'. An important result of this book is the proposal to complement this tragedy by a 'tragedy of responsibility'. Arguing that 'vital interests of rural people and communities are at stake and grossly neglected' opens a new dimension of research on the future role and potentials, that is, aspects of good governance and responsible decision-making processes as indispensable preconditions of a long-term sustainable use and preservation of the montane grass- and shrublands of High Asia. As a matter of fact, most contributions to this collection of articles give testimony to obvious negligence of pastoral interests, to a lack of understanding of the ecological, economic and social potentials of pastoralism and a corresponding irresponsibility of governmental policies. Such a finding is the more deplorable as pastoralists, and mountain farmers have accumulated knowledge systems that have enabled them again and again to overcome short-term natural catastrophes, to adapt to risks and hazards of their mountainous environments and to cope with longer-term changes of nature and society. Pastoralists and mountain farmers with their specific forms of animal husbandry have been the real protectors of one of the last nature reserves and stewards of their sustainable uses in the past. This book pays tribute to their achievements – and shows pathways of how to preserve and apply their accumulated knowledge and experiences in a modernizing and globalizing world.

In summarizing the findings of the various case studies, the editor rightly concludes that while 'pastoral activities have been shrinking further since the competition between combined mountain farmers and pastoralists increased the demand for grazing lands', the latter have proven their adaptive capacities again and again. Thus, it is easy to follow his conclusion according to which the presented transformations of pastoral practices are not necessarily part of those endisms of high-mountain pastoralism, but signifiers and indicators of its flexibility to cope with changing survival conditions.

Our introductory reference to the findings of the Millennium Ecosystem Assessment and our specific reference to the extremely pressurized montane grass- and shrubland biomes are alarmingly underpinned by the extensive and partly generalizable results of the presented case studies in this volume. I dare say that this book can claim the status of a work of reference – at least for the manifold aspects of pastoral practices in High Asia, their challenges and responses. It is a major contribution to a specific aspect of our planet under pressure. And it is to be hoped that this profound collection of articles will instigate similar research not only on pastoral lifestyles in other endangered and vulnerable mountain, desert and steppe environments, but also in tropical and subtropical forest biomes and beyond! This extremely thorough, thoughtful and thought-provoking book will hopefully reach policy-makers and practitioners not only in pastoral environments in High Asia and beyond, but everywhere where our fragile natural environments are at stake and need careful and responsible human stewardship.

Contents

Contributors

Christoph Bergmann Department of Social Anthropology, South Asia Institute, University of Heidelberg, Im Neuenheimer Feld 330, D-69120 Heidelberg, Germany

Ted Callahan Boston University Arts & Sciences Anthropology, Boston, MA 02215, USA

Andrei Dörre Human Geography, Department of Earth Sciences, Centre for Development Studies (ZELF), Freie Universität Berlin, Berlin, Germany

Eckart Ehlers ZEF, Bonn University, Bonn, Germany

Fazlur-Rahman Department of Geography, Urban & Regional Planning, University of Peshawar, Peshawar, Pakistan

Martin Gerwin Department of Geography, South Asia Institute, University of Heidelberg, Heidelberg, Germany

Melvyn C. Goldstein Center for Research on Tibet, Case Western Reserve University, Cleveland, OH, USA

Andreas Gruschke Institute of Oriental Studies, University of Leipzig, Leipzig, Germany

Arnd Holdschlag Institute of Geography, Hamburg University, Hamburg, Germany

Toni Huber Central Asian Seminar, Institute for Asian and African Studies, Humboldt University, Berlin, Germany

Muhammad Ismail Rangeland Resources Management, International Centre for Integrated Mountain Development, Kathmandu, Nepal

Tobias Kraudzun Department of Earth Sciences, Centre for Development Studies (ZELF), Geographic Sciences, Freie Universität Berlin, Berlin, Germany

Hermann Kreutzmann Human Geography, Department of Earth Sciences, Centre for Development Studies (ZELF), Freie Universität Berlin, Berlin, Germany

Wu Ning Chengdu Institute of Biology, Chinese Academy of Sciences, Chengdu, Sichuan, China

Marcus Nüsser Department of Geography, South Asia Institute, University of Heidelberg, Heidelberg, Germany

Jarmila Ptackova Central Asian Seminar, Institute for Asian and African Studies, Humboldt University, Berlin, Germany

William S. Sax Department of Social Anthropology, South Asia Institute, University of Heidelberg, Im Neuenheimer Feld 330, D-69120 Heidelberg, Germany

Stefan Schütte Department of Earth Sciences, Centre for Development Studies (ZELF), Freie Universität Berlin, Berlin, Germany

Yi Shaoliang NRM (Land & Water), Aga Khan Foundation, P.O Box 5753, House No.2, Qala-e-Fatullah, Kabul, Afghanistan

Chetan Singh Department of History, Himachal Pradesh University, Shimla, India

Bernd Steimann Human Geography, Department of Geography, University of Zurich, Zürich, Switzerland

Emilia Róża Sułek Central Asian Seminar, Institute of Asian and African Studies, Humboldt University, Berlin, Germany

Lu Tao Chengdu Institute of Biology, Chinese Academy of Sciences, Chengdu, Sichuan, China

Yan Zhaoli Chengdu Institute of Biology, Chinese Academy of Sciences, Chengdu, Sichuan, China

List of Abbreviations and Acronyms

ADB	Asian Development Bank
ADGM	Administrative documents concerning grassland management
AKRSP	Aga Khan Rural Support Programme
AN KSSR	Academy of Sciences of the Kyrgyz Socialist Soviet Republic
art.	Article in a legal document
asl	Above sea level
CCP	Chinese Communist Party
CCS	Chitral Conservation Strategy
CIA	Central Intelligence Agency
CPC	Communist Party of China
CPI	Consumer price index
CR	Cultural Revolution
CSG	Chamba State, 1904. 1910. Punjab States Gazetteers. Vol. XXII A. Lahore: The Civil and Military Press
est.	Estimated
ETP	Eastern Tibetan Plateau
FES	Foundation for Ecological Security
GAOZh	State Archive of the Zhalalabad Oblast'
GDP	Gross domestic product
GIPROZEM	State Design Institute for Land Management
GJ	*Guomin jingji he shehui fazhan tongji ziliao huibian 2007* [Statistical Yearbook on People's Economic and Social Development 2007]
GK KSSR IPK	State Committee of the Kyrgyz Socialist Soviet Republic for publishing, print and book trade
GLSKR and GUL	State Forestry Service of the Kyrgyz Republic and Main Department of Forest Regulation
GoNWFP	Government of North-West Frontier Province
GOP	Decree of the Head of the rayon administration Bazar Korgon 'Borders of remote pastures'
GoP	Government of Pakistan

GOSREGISTR	State Property Registry
GoU	Government of Uttarakhand
GTH	*Guoluo zhou tuimu huancao gongcheng jingshi chengxiao xianzhe* [Guoluo Prefecture's Tuimu huancao Construction Program is a Remarkable Success]
GUGK	Head office for geodesy and cartography
ha	Hectare
IBRD	International Bank for Reconstruction and Development
ICG	International Crisis Group
IITM	Indian Institute of Tropical Meteorology
INR	Indian Rupee
IOR	India Office Library Records
IUCN	International Union for the Conservation of Nature
KDG 1918	Kangra District, Parts II, III, and IV, 1917. 1918. Punjab District Gazetteers, Vol. XXX A. Lahore: Superintendent Government Printing, Punjab
KDG 1926	Kangra District, 1924–1925. 1926. Punjab District Gazetteers, Vol. VII. Part A. Lahore: Superintendent Government Printing, Punjab
kg	Kilogram
KGS	Kyrgyzstan Som (national currency)
KIRGIZGIPROZEM	Kyrgyz State Design Institute for Land Management of the Kyrgyz Socialist Soviet Republic
KKH	Karakoram Highway
km	Kilometre
km^2	Square kilometre
KR	Kyrgyz Republic
KSSR	Kyrgyz Socialist Soviet Republic
LARC	Legal Assistance to Rural Citizens
LKKR	Forestry Code of the Kyrgyz Republic
MAWPRI	Ministry of Agriculture, Water Resources and Processing Industry of the Kyrgyz Republic
MOA	Ministry of Agriculture
MSG	Mandi State, 1904. 1908. Punjab States Gazetteers, Vol. XII A. Lahore: Civil and Military Press, Punjab Government
MQ	*Maqin xian zhi* [Annals of Maqin County]
MT	Metric ton
N.S.	New series
NATO	North Atlantic Treaty Organisation
NDC	National Documentation Centre
NSKKR	National Statistical Committee of the Kyrgyz Republic
OSCE	Organization for Security and Co-Operation in Europe
par.	Paragraph
PNV	Pastoral new village

POPKR	List of remote pastures of the Kyrgyz Republic. Confirmed by decree of the Government of the Kyrgyz Republic
PPPAIP	Resolution 'On pasture land lease and use'
PR	People's Republic
PRC	People's Republic of China
PVC	Polyvinyl chloride
QD	Qinghai Daily
QTP	Qinghai Tibet Plateau
RMB	Renminbi, Chinese currency Yuan
RRMP	Regional Rangeland Management Programme
SAEPFUGKR	State Agency on Environment Protection and Forestry under the Government of the Kyrgyz Republic, shortly State Forestry Agency
Sanjiangyuan	Sanjiangyuan guojia ji ziran baohu qu (Three Rivers' Headwaters National Nature Reserve)
SHSG	Shimla Hill States, 1910. 1911. Punjab States Gazetteers, Vol. VII. Lahore: The Civil and Military Gazette Press, Punjab Government
SKBK	Schematic map of the Bazar Kurgan Rayon of the Zhalalabad Oblast' of the Kyrgyz Socialist Soviet Republic
SSR	Soviet Socialist Republic
SU	Sheep unit
TAP	Tibetan Autonomous Prefecture
TAR	Tibetan Autonomous Region
TCA	The Times of Central Asia
TJS	Tajikistan Somoni, Tajikistan's currency
UNDP	United Nations Development Program
UNDP RBECIS	United Nations Development Program Regional Bureau for Europe and the Commonwealth of Independent States
UNDPKR	United Nations Development Program in the Kyrgyz Republic
UNEP	United Nations Environment Program
UPKR-MNRGPZAR	Decree of the President of the Kyrgyz Republic "On measures for further development and state support to the land and agrarian reform in the Kyrgyz Republic"
USD	US dollar
VDC	Village Development Committee
WFP	World Food Programme
YSB	Yushu TAP Statistical Bureau
YZZ	Yushu TAP Local Archives Compilation Committee
ZKKR	Land Code of the Kyrgyz Republic
ZKKR-UZSN	Law of the Kyrgyz Republic 'On management of agricultural lands'
ZOP	Law of the Kyrgyz Republic 'On pastures'

Chapter 1
Pastoral Practices in Transition: Animal Husbandry in High Asian Contexts

Hermann Kreutzmann

Abstract Vast tracts of High Asia are utilised for pastoral strategies of survival, and the mountainous areas provide livelihoods to herders and their households. Locally adopted and adapted pastoral practices reflect politico-historical and socio-economic changes that are often the result of external intervention. Pastoral practices in the mountain periphery seem to be a vital indicator of change. Two regions will receive special attention – the Pamirian Knot and the Tibetan Plateau – in 16 case studies grounded in the wider framework of. External and internal boundary-making and quite distinct path-dependent developments are reflected in the typology given here. The focus of the case studies is directed towards the variation of experiences in a wider angle, drawing attention to marginalised groups in the mountainous periphery of High Asia.

Keywords Modernisation • Development • Pastoral adaptation strategies • Hindukush-Karakoram-Himalaya • Pamir • Tibetan Plateau

1.1 Introduction to Pastoral Practices in Central Asia and on the Tibetan Plateau

The position of Central Asian deserts and oases between the densely populated regions of Asia and Europe and their respective centres of gravity has strongly influenced economic exchange, territorial power games and communicative curiosity

H. Kreutzmann (✉)
Human Geography, Department of Earth Sciences, Centre for Development Studies (ZELF) and Institute of Geographical Sciences, Freie Universität Berlin, Malteserstrasse 74-100, D-12249 Berlin, Germany
e-mail: h.kreutzmann@fu-berlin.de

H. Kreutzmann (ed.), *Pastoral practices in High Asia*, Advances in Asian Human-Environmental Research, DOI 10.1007/978-94-007-3846-1_1,

directed towards sparsely inhabited and marginally utilised areas. This region would be underestimated if it were only reduced to a corridor of traverse and a link between seats of major powers. High Asia – as prominently defined and for the first time ever perceived as a complex entity by Robert von Schlagintweit (1865) and Hermann von Schlagintweit-Sakünlünski (1869–1880, 1870) – is on the one hand characterised by its relationship to outside influences and imperial forces that have shaped the boundaries, fate and destiny of principalities, kingdoms, states and regions until today. In various contexts, the connotation of High Asia reappears and contributes to a debate about communication, experiences, practices and shared commonalities that focuses on borderlands, boundaries and territoriality. Most spatial references though are made to the eastern Himalayas and the Tibetan Plateau (cf. Blackburn 2007; van Driem 2001; Klieger 2006; Shneiderman 2010). While Willem van Schendel (2002) attributes a certain centrality to the region, others like James Scott (2009) primarily perceive its peripheral location and status or highlight the potential for escape from state intervention and dominance. On the other hand, the natural wealth, internal power games and competition over resources and people created microcosms within the macro-system that have changed in space and time. Pastoralists have played an important role in shaping relationships, connecting regions, exchanging goods and valuable information.

Throughout long historical periods, Central Asia and the Tibetan Plateau became the focus of knowledge-seeking explorers, elaborate expeditions and expansionist imperial conquests that depended on services provided by oasis dwellers and pastoralists alike. During the *Great Game* and thereafter, the territorial division of High Asia resulted in effective boundary-making that has significantly affected local livelihoods. Borders and fences restrict movements, define territorially applicable legal systems, rules and regulations, and identify spaces of mobility and exchange. Diachronic enquiries into politics and society provide insights for the interpretation of history and economy as they affect pastoral environments and livelihoods. Consequently, ecological properties are creating the arena in which socio-economic struggles for survival and political power games are taking place. All three parameters need to be taken into account when exploring the position of pastoralism in High Asia.

The symbiosis of highly productive and spatially concentrated oases in a wide-ranging environment with extensively utilised rangelands in deserts and steppe regions is modified by the third dimension represented in the verticality of Central Asian high mountains and the Tibetan Plateau (Fig. 1.1). The vertical dimension is often connected with the prevalence of yaks and their hybrids when it comes to pastoral practices. They form a significant part of the livestock kept by pastoralists and have the reputation of enduring harsh environments and high-altitude conditions (Photo 1.1).

In evaluating and assessing the environmental potential, the vast area under consideration requires a fine-tuned approach based on latitudinal and longitudinal position, but in a mountainous environment, regional and micro-scale variations also need to be accounted for. Mountains provide a higher degree of ecological variability in a clear-cut spatial segment than any other eco-zone. In an initial approximation,

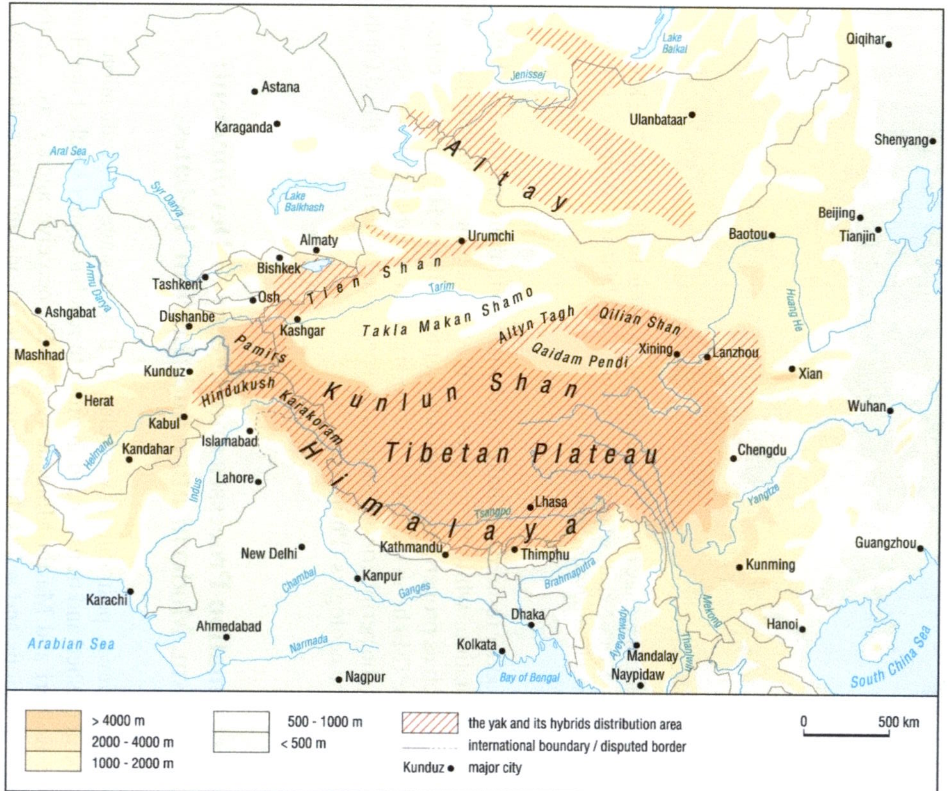

Design: H.Kreutzmann Cartography: B.Hilberer

Fig. 1.1 High Asia stretching from the Central Asian mountain ranges to the Tibetan Plateau and from the Altay to the Himalaya. The distribution of the yak and its hybrids constitutes a roughly contiguous area that serves as a common structural indicator for the scope of this book

Photo 1.1 Yaks are put on pasture close to the Sherpa village of in Beding (3,692 m; Rolwaling Valley, Gauri Shankar VDC, Nepal) where ample pasture is available in a shrinking community (Photograph © Hermann Kreutzmann, September 27, 2011)

the availability of fodder resources is linked to thermal conditions and the distribution of water and vegetation, while their accessibility is based on environmental and societal criteria such as property rights and entitlements. In the High Asian context, aridity and altitude (Miehe et al. 2001) as two significant limiting parameters for human activities at the peripheries of settlement space need to be highlighted for the perception of steppe ecologies (Photo 1.2).

1.2 Transformation Processes and Agency of Development: Pastoralism, Modernisation and 'Endism' Debates

In conventional views, pastoralism was classified as a stage of civilisation that needed to be abolished and transcended in order to reach a higher level of development. Uma Kothari and Martin Minogue (2002, 13) perceive agency as: '… the network of institutions and actors that through their actions and interactions "produce" development. The analysis of agency is crucial because it allows us to capture the complexities of the process by which ideas are mediated into objectives and translated into practice'. By looking at agency and actors, we might gain insights into the scope of transformation and development, understand better localised forms of empowerment and participation (Natarajan 2005) and shall be able to bridge the gap between globalised phenomena, national responses and regional effects. In this context, global approaches to modernising a rural society have been ubiquitous

Photo 1.2 The Tibetan Plateau combines available waters in different aggregate states. Glaciers and lakes dominate the physical landscape where pastures and agricultural settlements are interspersed (Photograph © Hermann Kreutzmann, September 12, 2000)

phenomena independent of ideological and regional contexts (Dyson-Hudson and Dyson-Hudson 1980; Montero et al. 2009; Salzman and Galaty 1990). The twentieth century experienced a variety of concepts to settle nomads and to adapt their lifestyles to modern expectations and perceptions. 'When nomads settle' (Salzman 1980), then obviously the 'future of pastoral peoples' (Galaty et al. 1981) has to come into focus. Is sedentarisation the result of an inevitable modernisation process or an adaptation to changed frame conditions? Does the settlement in itself form a crisis of pastoralism, or is this just another approach to cope with societal and economic challenges? Permanent settlements have often been the vivid expression of an ideology-driven approach that aimed '... at reducing flexibility in favour of concentration and rootedness. Modernisation theory translated into development practice captured all elements of pastoral life and tried to optimise breeding techniques, pasture utilisation, transport of animals and products, and related processing concepts to increase the value of livestock products' (Kreutzmann and Schütte 2011, 104). The aspect of higher requirements for inputs tended to be neglected when the modernisation of animal husbandry was at stake.

New insights into other aspects of pastoralism, such as its role as an *adaptive strategy to use marginal resources* in remote locations with difficult access (Ehlers and Kreutzmann 2000), its function as *high reliability pastoralism* (Roe et al. 1998), the distinction of the *sediment of nomadism* (Kaufmann 2009) in its puristic relation to hybrid forms of inessentials, the objective of governance expansion as a tool of *spatial appropriation* (Kreutzmann 2011a, b), the impact of the civilisation project as a *strategy for dominance and exploitation* (Scott 2009) and the potential of *globalising scapes of mobility and insecurity* (Gertel and Breuer 2007), could only be understood as a critique of external interventions by powerful actors and stakeholders as well as capitalist and communist concepts of modernisation. The rejection of input-dominated theories that triggered the enhancement of outputs but neglected ecological considerations regarding sustainability opened up a new field for research combining ecology, economy and society. This perception might gain further importance when mitigation strategies coping with climate change and societal challenges are debated.

Pastoral practices can be perceived as flexible strategies to adapt to changing survival conditions, rather than transitory stages on the path to modern development. A variety of pastoral practices were adopted by people when opportunities arose, when it was economically sound, and when the challenges posed by ecological and socio-political environments could be managed. Consequently, our emphasis on pastoralism studies provides us with an important tool to understand society in general and human-environmental relations in particular.

Nevertheless, an *endism* debate is accompanying such observations and thoughts. The 'end of nomadism' was rightly discussed by Caroline Humphrey and David Sneath (1999) when analysing fundamental transformations that had taken place in now post-communist societies of Central Asia. Within the twentieth century, structural and reformist interventions had resulted in two phases of modernisation: (1) collectivisation processes after the respective revolutions in the Russian and Chinese empires and (2) privatisation and deregulation after the dissolution of the Soviet Union and reform movements in the People's Republic of China and Mongolia. Both interventions have led researchers to question whether the resulting pastoral practices could or should be termed nomadic any longer. From a more structural and classificatory point of view, other authors debated alternative scenarios (Barfield 1993; Karmyševa 1981; Weissleder 1978) and identified substantial changes that made them refrain from using a concept of *nomadism* in a classical manner and promoted the theses of the *last nomads* (Benson and Svanberg 1998), *changing nomads* (Ginat and Khazanov 1998), the *demise of traditional nomadic pastoralism* (Miller 2000), *former nomads* (Gruschke 2008) and/or *nomadism in decline* (Scholz 2008). Most authors agree on significant changes in pastoral practices that mainly follow the direction shown by the modernisation paradigm (Brower and Johnston 2007; Goldstein and Beall 1991; Kreutzmann et al. 2011a, b; Montero et al. 2009; Sheehy et al. 2006). Nevertheless, there are a few exceptions worth mentioning. Pastoral practices in less regulated societies such as Afghanistan seem to be even further away from any *endism* debate (Barfield 2008; Glatzer 1981). India and Pakistan experience little reflection about such classification and structural aspects. Recent fieldwork has produced evidence that pastoral practices are used in a flexible manner, mainly when an investment is expected to be a profitable one, and when institutional obstacles in the socio-political environment can be tackled (Alden Wiley 2004, 2009; Inam-ur-Rahim and Amin Beg 2011; Dangwal 2009; Davies and Hatfield 2007; Ehlers and Kreutzmann 2000; Ferdinand 2006; Finke 2005; Kreutzmann 2004; Kreutzmann and Schütte 2011; Li and Huntsinger 2011; Manderscheid 2001; Nüsser and Gerwin 2008; Rao and Casimir 2003; Tapper 2008). The cases from South Asia have shown a significant dynamism over time. No single trend has been identified as pastoral practitioners probably base their decision-making on a different set of parameters than, for example, advocates of a one-directional modernisation process. Development as a phenomenon is repeatedly challenged, obviously path-dependent and contradicted by the adaptive potential of actors and flexibility of certain stakeholders. One of the major challenges for the authors contributing to this volume was to elaborate on the varied expressions of pastoral practices, frame conditions and performances in the study area.

1.3 Structure and Practice in Diagrams

In a diagrammatic approach, structural aspects of pastoral practices are introduced as a reference point for positioning the individual case studies and the related transformation processes (cf. for an earlier version Kreutzmann 2011, 205–211). 'Classical' practices in combined mountain agriculture and nomadism have generated 'modern' expressions that reflect on the one hand strategies adopted by pastoralists and on the other hand pinpoint strong external interventions in the livestock sector and the utilisation of high pastures (Fig. 1.2).

1. Combined mountain agriculture represents the pastoral practice operating from a settlement that is the base for agricultural activities and in most cases the residential centre of its practitioners for most of the year. It has the advantage of simultaneous fodder production in the permanent homesteads for herds which are grazed in the high-lying (and rarely low-lying) pastures during the summers (Photo 1.3).

   In his recent book on the 'third dimension', Jon Mathieu (2011, 101–114) has shown how the transformation processes in mountain mobility systems are linked to the combination of crop-farming and pastoralism. The limiting factor here is that feed has to be provided for up to 9 months, and it has to be produced on

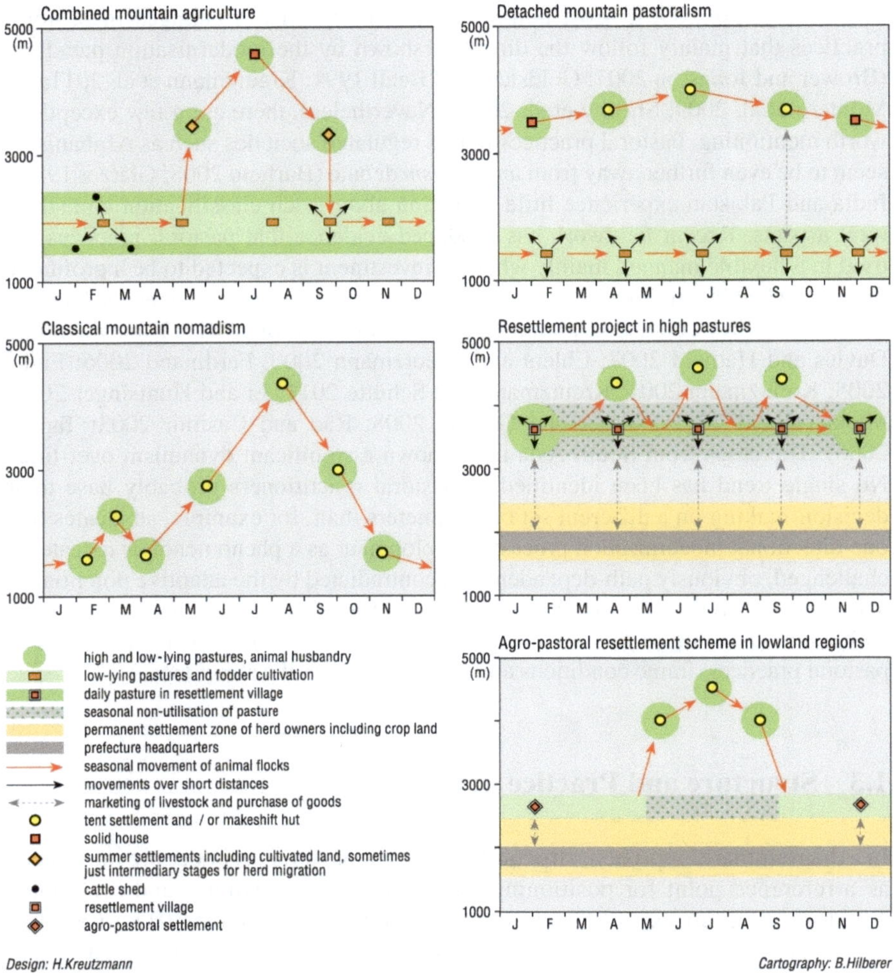

Fig. 1.2 Pastoral strategies in High Asian mountain regions

private or common property village lands (Photo 1.4). In recent years, the importance of the settled operational base has grown in most communities, while animal husbandry's contribution is shrinking. In the context of our study area in general terms, the mountain regions of India, Pakistan and Tajikistan are cases in point, whereas the same cannot be stated for Afghanistan.

In the context of this book, the term 'transhumance' is generally avoided because of its eurocentric connotations. Nevertheless, described about a century ago as a regional pastoral practice in Southern France, the term transhumance has been applied to pastoral practices in the circum-Mediterranean. It describes pastoral practices with an emphasis on proprietary rights in flocks and the

Photo 1.3 The summer settlement of combined mountain farmers from the Ishkoman Valley in Hindis (2,750 m) en route to Panji Pass (4,450 m). Their makeshift huts consist of piled-up stone walls on which juniper branches are erected to form a conus-shaped dome to protect the shepherds and their household members from wind and rains. In between the shelters, there are single-cropping fields that are only cultivated during the pasture season while other household members dwell on the double-cropping village lands in the winter settlement (Photograph © Hermann Kreutzmann, September 3, 1990)

relationship between crop production and a detached form of pastoralism (Beuermann 1967; Blache 1934; Ehlers and Kreutzmann 2000; Johnson 1969; Jones 2005; Mathieu 2011, 103–106; Rinschede 1979). Transhumance involves seasonal migrations of herds (sheep and goats, cattle) between summer pastures in the mountains and winter pastures in the lowlands. In contrast to prevalent perceptions of mountain nomadism, the shepherds of a migrating team are not necessarily that strongly affiliated with each other to form a group of relatives managing their own resources. The shepherds serve as wage labourers hired by the livestock proprietors on a permanent basis. As a rule, the shepherds are neither related to them, nor do they have livestock of their own. The proprietors of the flocks can be farmers or non-agrarian entrepreneurs. Management-wise the year-round migration between suitable grazing grounds is independent from other economic activities of the proprietors. Nevertheless, sometimes proprietor farmers provide shelter and grazing on their fields after harvest or on meadows. Usually common property pastures are utilised in the mountains while customary rights or contracts with residents in the lowlands establish the winter grazing conditions. Pastoral practices resembling a transhumance of this kind seem to be found in mountainous regions of all continents (cf. Rinschede 1988, 99–100).

Photo 1.4 Wakhi farmers of Sarhad-e Wakhan (3,600 m) have brought their yaks from the summer pastures for ploughing and threshing practising combined mountain agriculture (Photograph © Hermann Kreutzmann, October 14, 1999)

For contributions to the scholarly discussion of transhumance in the High Asian context (cf. Alirol 1979; Bahsin 1996; Banjade and Paudel 2008; Chakravarty-Kaul 1998; Ehlers and Kreutzmann 2000; Jettmar 1960; Kreutzmann 2004; Nautiyal et al. 2003; Tucker 1986; Snoy 1993; Uhlig 1976, 1995). Transhumance gained global application by some authors (e.g. Guillet 1983; Messerschmidt 1976; Stevens 1993) in Anglo-American publications. Stevens (1993, especially chapter two) uses it in a wider and ambiguous sense synonymous for pastoralism as a comprehensive concept and all kinds of pastoral activities.

2. At least traditionally, nomadic groups were able to exploit natural resources at dispersed locations. Great distances in the order of several hundreds of kilometres separated economically valuable mountain pastures in summers from winter camp sites of less economic interest lying in between. Sometimes spring and/or autumn pastures were frequented when suitable forage was provided. Projected on the studied region, a 'classical mountain nomadism' with functional migration cycles could be established in a historical context. Nevertheless, socio-political pressure was one of the major driving forces to amend and change mobility patterns. State evasion and escape from domination were coping strategies in an ever shrinking arena of pastoral activities. James Scott (2009) has presented his interpretation of mountain mobilities and escape strategies for the Southeast Asian highlands, while M. Nazif Shahrani (1979) contributed the focal text for nomadism as an 'adaptation to closed frontiers' (Photo 1.5). The debate

Photo 1.5 Kirghiz pastoralists from Little Pamir ride their yaks to a food distribution point near Ghundjibai (3,900 m), their pastures are located in the side valleys of the main plateau (Photograph © Hermann Kreutzmann, June 11, 2000)

on nomadism that is going over and above structural aspects of habitations and social organisation is strongly inspired by analysing power relations, state interference and 'modernisation' in the disguise of 'development' (Bauer 2006; Ferdinand 2006; Kreutzmann et al. 2011a, b; Salzman 2004; Scholz 2002; Tapper 2008). In the context of this book, case studies will be mainly presented from areas where transformation is analysed as transition from a system that was denominated as 'nomadic' earlier on.

3. Detached mountain pastoralism is a more recent strategy reflecting societal transformations, collectivisation and forced sedentarisation and settlement in high-lying grazing grounds. It is, as well, an adaptation to new borders, the borders of administration under collective regimes. Pastoral brigades and herding collectives were made part of entities such as people's communes (*gongshe*), collective economies (*kolchoz*) and state farms (*sovchoz*) that were confined in their pastoral regimes to an assigned territory consisting mainly of former summer pastures in the periphery of their respective states. This entailed a significant adaptation process in conjunction with completely new exchange and supply patterns (Kreutzmann 2009; Qonunov 2011; Robinson and Whitton 2010; Robinson et al. 2010). Basically, the long-distance migrations ceased to exist and were replaced by short-distance migrations at a rather high altitude of permanent abodes, usually above 3,000 m. The former winter grazing grounds have become rural settlements with crop cultivation and their own livestock regime. Both areas are connected by annual exchanges of goods. Livestock products are exported from the high pastures and exchanged for all kinds of necessities needed for life in the Pamirs. Prominent examples for these enforced adaptations are to be found

in Gorno-Badakhshan (Tajikistan), Xinjiang (PR of China) as well as in other high mountain pastoral areas that were part of the collectivisation processes within the Soviet Union and PR of China.

In recent years, two new forms of organising pastoralists have been tested and implemented that can be regarded as a further step in 'developing' and 'modernising' peripheral communities and finding their visual expression in fencing pasture areas. Under the heading of 'resettlement', two approaches are followed:

4. Resettlement in high pastures. In furthering the modernisation attempts of previous interventions in pastoral communities, schemes have been designed that bring features of urbanisation to pastures and their inhabitants. A concentration of pastoralists' habitations in newly built townships envisages economies of scale in terms of infrastructure provision, health and educational institutions, agricultural extension services, marketing of livestock products and supply of basic goods. Township development in remote locations and enormous investment in infrastructure development are based on external subsidies that affect the livestock sector by creating fenced spaces at a hitherto unknown scale. A strong relationship between the pastoral counties and their newly built townships, on the one hand, and the cities functioning as prefecture seats, on the other, is the basis for a modern network of communication and exchange. This model has been tested in the PR of China over some years now and has been implemented on broad scale in the pastoral provinces and regions of Inner Mongolia, Qinghai, Tibet and Xinjiang as well as in adjacent prefectures and counties of other provinces (Goldstein and Beall 2002; Ho 2000; Ptackova 2011; Sheehy et al. 2006; Tao Lu et al. 2009; Zhizhong and Wen 2008; Yeh 2005; Zhao 2011). The idea is well in tune with conventional regional planning inspired by modernisation theories, thus creating a web of settlements that is integrated into a system of central places at higher levels. As a welcome side effect, pastoralists are not only concentrated in townships but also reduced in number. The planning departments welcome this shift and even support an exodus across borders (Cerny 2010) by emphasising the ecological challenges and the urgent need to reduce grazing pressure and ecological degradation.
5. Agro-pastoral resettlement schemes in lowland locations. Within the framework of prevalent resettlement strategies, a second option is going a step further. In contrast to establishing resettlement townships within the mountains, this second form of 'modern adaptation and transformation' directs the pastoralists to leave the mountains. It is inspired by the notion that development does not take place in remote mountain areas. Consequently, mountain dwellers are resettled in lowland regions close to urban areas where infrastructure is available and easily accessed, where high-quality fodder production is possible and markets are close. The price is the abandonment of winter settlements and the relocation of pastoralists far from their summer pastures close to the consumer markets. The greater distances can be covered only if additional support is provided for herd migration or rather herd transport on trucks to their respective summer pastures. Agro-pastoral resettlement is envisaged for those mountain areas that are located at the fringes of the Tibetan Plateau or that are close to low-lying areas where distances

are manageable. Experience with these strategies was gathered in Xinjiang where highly productive oases are close to the mountain abodes (An et al. 2011). Such approaches need substantial investments and will contribute to a significant reduction of mountain pastoralists and their settlements. Finally, the pastoralists who were confined to mountain abodes during the collectivisation periods are experiencing a kind of reverse reform, the resettlement in the lowland oases where their ancestors grazed their herds during winters.

All approaches can result in competition for natural resources at the same location and have frequently been discussed from that perspective. The ecological aspect has gained momentum as planning institutions have adopted nature protection as an additional measure for implementing their modernisation packages. The debate needs to be expanded to include conflicting economic strategies and perceptions of development. Sometimes the political dimension of power and influence, grazing taxes and the levying of them, threat and security has been neglected in historical contexts. In recent times, the debate needs to be shifted to the challenges between external planners and local user groups, between perceptions of modernity and preferences of lifestyles, between subsidised interventions and stakeholder participation. Pastoral practices in the mountain periphery seem to be a vital indicator for societal change. In the framework of this volume, two regions will receive special attention: the Pamirian Knot and the Tibetan Plateau. Both regions are located at the interface of political entities with different socio-political backgrounds and quite distinct path-dependent developments that are reflected in the typology given above. Our focus is directed towards the variation of experiences in a wider angle, thus incorporating developments at the fringes of the focal areas as well.

1.4 Variegated Pastoral Practices in the Pamirian Knot

Pastoralism has played a major role in Central Asia since times immemorial. Nomads functioned as transporters and communicators between oasis settlements, as powerful actors controlling passages and providing security to trade caravans. The Silk Road exchange over vast tracts of deserts, steppes and mountain environments became feasible because of pastoralists covering huge distances with transport animals and valuable loads (Christian 2000; Cosmo 1999; Khazanov 1984, 2005; Seaman 1989). Mountain passes functioned as thoroughfares for the Inner Asian traverses, but especially for the connection between the Tibetan Plateau and South Asian rim lands across the Himalayas as well as between the Central Asian oases along the Silk Road and the transmountain areas beyond the Hindukush, Karakoram, Kun Lun Shan and Pamir. Vital passages were controlled by herding communities who, in addition to animal husbandry and livestock breeding, engaged in transport services across difficult passages and functioned as guides and guards for trade caravans. These activities hint at three important aspects of pastoralism in Central

Asian mountain regions. First, pastoralism needs to be interpreted as embedded in the overall economic exchange of goods between farmers and bazaar traders on the one hand and pastoral meat and milk producers on the other. Second, an understanding of pastoralism alone would produce an isolated picture of livelihood generation, while other important components would be omitted. Third, the socio-political framework needs to be taken into consideration as pastoral strategies incorporate answers to challenges by powerful actors and stakeholders. State interference has often resulted in state evasion; consequently, pastoralists have tried to avoid payment of heavy grazing taxes, to escape from forced conscription and bureaucratic domination and to elude slavery, forced labour and religious persecution or proselytism. Pastoral strategies of resource utilisation in remote locations are an example of survival strategies in the periphery. Attempts at modernisation in the aftermath of the Russian and Chinese revolutions incorporated settlement programmes, the introduction of hitherto unknown forms of social organisation as well as transformations of production strategies and consumption patterns.

The complexity of socio-cultural problems and the manifestation of transformation processes in societies with economies based on pastoralism vary from region to region. The cases presented here emphasise the rearrangements in the livestock sector initiated by external intervention. Our concern with pastoral practices in Central Asian mountain regions requires adopting different perspectives in order to understand the environmental setting for mountain pastoralism (Marcus Nüsser, Arnd Holdschlag and Fazlur Rahman), to investigate the diachronic dimension of change in pastoral settings, to assess power and politics as factors for multi-fold insecurities (Stefan Schütte), to identify survival options for marginalised communities (Ted Callahan) and to evaluate attempts at forced modernisation and adherence to sustainable development.

The developments in post-socialist societies as a result of two major transformations need to be investigated in terms of 'new livestock breeders' (Tobias Kraudzun), changing institutional and legal arrangements (Andrei Dörre) and contested spaces (Bernd Steimann). Here a broad spectrum of challenging topics is expanded that surpasses conventional dealings with aspects of pastoral practices and that is strongly linked to processes of modernisation in a broad sense. Worshippers of modernisation made it a vital argument to perceive progress as a transition from mobile economies to settled farming and entrepreneurship. The co-existence of both in Central Asian mountain regions reflects the complementarities and interdependence involved: nomadism/pastoralism is not feasible without exchange relations with farmers and markets. Nevertheless, the altitudinal limits of habitations and the utilisation of marginal lands have significantly shifted towards high-lying and arid areas. The extensive utilisation of marginal resources has been superseded by intensification and increasing external inputs. Thus, it is not surprising that mountain farmers and pastoralists have been a prime target for 'development', which aims to reduce subsistence levels by integrating people from the periphery into the mainstream of nation states. Thus, 'traditional' lifestyles and locally developed economic strategies are endangered and transformed.

1.5 Recent Transformations on the Tibetan Plateau

The High Asian vast mountain tracts constitute an area that is characterised by the High Himalayas in the south, with their deeply incised river valleys and gorges. The Himalayas of Pakistan, India, Nepal and Bhutan and additional contested and uncontested mountain areas in the east form the periphery of the Tibetan Plateau with verticality as its dominant feature (cf. Fig. 1.1). Further northwards, the Changtang covers an extensive area and is dominated by horizontal mobility at high altitudinal levels.

The Tibetan Plateau is a huge ecological area perfectly predisposed for mountain pastoralism. Debates about pastoral practices, common property regimes and rangeland management have gained pace in recent years since the environmental challenges and economic returns have been discussed in the framework of climate and global change (Harris 2010). Initially, the remote lifestyles of Tibetan nomads in peripheral regions were highlighted on account of sustenance from their own produce and subsistence economies. Even then mountain pastoralists were embedded in a network of mutual exchange relations that enabled them to survive in remote mountain plateaux and valleys. Their command of yak breeding, their abilities to adapt to harsh environmental conditions and to cover huge distances between extensive natural pastures and market centres brought admiration from outside observers. Their 'traditional lifestyle' seemed to be the perfect adaptation to environmental conditions. Conventional thinking attributed more importance to ecological factors than to politics, economy and society. Thus, the focus was directed to local production and less to animal husbandry as one component of a mountain-based household production system. The twentieth century proved to be one of socio-economic change in all regions and for all communities. The Tibetan Plateau is no exception to this. Collectivisation in the aftermath of the Chinese Revolution introduced forms of social organisation that have been modified several times since. The shift from peoples' communes to the production responsibility system, from strict state governance and command economy to lenient forms of private ownership and entrepreneurship and from pasture laws to regulating rangeland management has introduced new concepts, policies and management tools that are significantly changing the institutional structure and organisational setup (Banks et al. 2003; Foggin 2008; Miller 2008; Yamaguchi 2011). The challenges are environmental degradation processes triggered by social and climate change, contested commons and their boundaries, external development and modernisation strategies versus local and regional perceptions and participation demands in decision-making processes. Present-day practices are the result of adaptive mechanisms as a response to all kinds of challenges and external reforms, infrastructure development and provision of subsidies. Mountain pastoral economies and societies of our time are strongly linked with neighbouring markets and interrelated socio-economic structures. This especially holds true for pastoralists of the Tibetan Plateau who are significantly embedded in government programmes and poverty alleviation strategies and who respond to rangeland management packages and pasture laws by adapting to change.

Their sources of income are highly diversified, drawing on animal husbandry and agriculture, and are augmented by government salaries, proceeds from wage labour and incomes from trade and entrepreneurship. Today's mountain pastoralism is, on the one hand, a livestock-based activity and related to different localities connected through mobility. On the other hand, significant incomes are derived from other local and non-local sources, often surpassing the amount generated from pastoralism. In future, part of pastoralists' income may be derived from payment for ecological services that are provided by pastoralists as landscape managers for the maintenance of fragile environments and rewarding sustainable stocking practices. Pastoral practices reflect adaptive strategies that respond to the ecological, socio-economic and political environment over time, thus giving insights into path-dependent developments in remote mountain areas.

China controls a major share of these pastures and rangelands. Three-quarters of its rangelands are located in the semi-arid, sparsely settled areas of the north and west of the PR of China. Out of China's 400 million hectares of rangeland, 140 million hectares are to be found in mountainous regions of the Tibetan Plateau; in addition, there are 57 million hectares of natural pasture in Xinjiang. Both areas comprise about one-fifth of China's land area. Close to 40 million people live in 260 predominantly pastoral counties (Miller 2002, 22; Tashi et al. 2010, 54–55). The livelihoods of their less affluent citizens are strongly related to the natural resources of the rangelands.

About five million pastoralists and combined mountain farmers make a living on the Tibetan Plateau by keeping 12 million yaks – three-quarters of all yaks worldwide – and 30 million sheep and goats. In an ecological definition, the Tibetan steppe covers 165 million hectares equalling more than two-fifths of China's grazing areas (Sheehy et al. 2006, 143). China has assessed the potential of its mountain pastoralism in great detail and initiated a number of programmes that are now being implemented (Photo 1.6).

Giving an example of the aim, objectives and sophistication of governmental planning and implementation, a series of quotes from a recent presentation at the Lhasa workshop on pastoralism by Zhang Younian, Deputy Secretary-General, Tibet Autonomous Region People's Government, are presented in the following:

1. *On progress in rangeland restoration and resettlement*

> We have made solid progress on rangeland restoration, pastoralist resettlement and other major development projects. Since 2004, 3273 *mu* of rangelands have been restored, with the project area's vegetation coverage increased to more than 55% and fodder yield up by 25–30%. The grass production capacity of these rangelands has been effectively restored. We have adhered to the development and restructuring of pastoralism, especially pastoralism in farming areas, creating an increasingly integrated pastoral and agricultural development model. Livestock raising is now seeing a strong momentum in the rural and suburban areas. The process of rangeland livestock breeding shows new changes. Meat output in 2009 in TAR reached 25 million tons, an increase of 63% over 2000. Milk output amounted 30 million tons, up 45% compared to 2000 (Zhang Younian 2011, 267).

2. *On housing projects within the resettlement programme*

> Since 2006, the TAR party committee and government have made the housing projects in the pastoral area a starting point of building the new countryside. We have spared no efforts

Photo 1.6 Pastoral settlement on the Tibetan Plateau (Photograph © Hermann Kreutzmann, October 18, 2010)

to promote the pastoralist resettlement programme, which also involves poverty alleviation and reconstruction of rural housing, greatly improving the living and housing conditions for farmers and pastoralists. By the end of 2009, a total 230,000 households of 1.2 million farmers and herdsmen have moved into their new and more comfortable houses. At the same time, we have strengthened construction of supporting facilities in these housing projects, putting in place an integrated network of water, electricity, roads, telecommunications, gas, etc. The rural landscape is moving from local improvement to overall advancement. The past 10 years have witnessed 85% of the masses getting safe drinking water, 114,000 households using biogas and 395,000 farmers and herdsmen families being provided with solar cookers by the government. All towns and 80% of villages have become accessible by road. 70% of the population in agricultural and pastoral areas has access to electricity. 85% of villages are connected by phone lines and 80% of towns by post. We have basically realized the development goals of 'cable to the county and telephone/fax to the town' (Zhang Younian 2011, 269).

3. *On the scope of investment*

Since 2001, the central government has made a total investment of more than ten billion yuan in Tibetan agriculture and pastoralism. It has implemented some hefty development projects such as the pastoralist resettlement programme, the rangeland restoration programme, the high-quality grain and oil production base, and the niche agricultural and pastoral industries. Especially since the beginning of the 'Eleventh-Five-Year-plan' period, the relevant ministries of the State Council has increased their support for infrastructure building in agriculture and pastoralism, giving a strong impetus to the rapid development of agricultural and pastoral economies (Zhang Younian 2011, 270–271).

4. *On the goals to narrow the socio-economic gap*

> ... when the Fifth Working Conference on Tibet was held, the central authorities set out some ambitious goals for Tibet. By 2015, the gap between the per capita net income of Tibetan farmers and herdsmen and the national average is to be significantly reduced, with basic public services much improved, the ecological environment better conserved and the living and working conditions of farmers and herdsmen enhanced (Zhang Younian 2011, 273–274).

5. *On poverty alleviation*

> We shall concentrate on poverty reduction for people whose per-capita annual net income is less than 1,700 yuan. We shall proceed by lifting a whole town out of poverty and promote the anti-poverty work in the border areas, minority areas with small populations and areas with harsh natural conditions. We shall spare efforts to improve the production and living conditions for low-income population, and accelerate their pace to get rich. Specifically, we shall focus on eight tasks. We shall complete the housing projects for 22,000 poor farmers and herdsmen. We shall promote poverty alleviation for 200 rural villages and towns. We shall focus on supporting the 206 border villages, minority villages with small populations (Zhang Younian 2011, 278–279).

6. *On overgrazing and degradation*

> To deal with overgrazing and the degradation, desertification and salinization of rangelands, we shall strengthen law enforcement and supervision. To maintain the balance between livestock and rangelands, we shall increase livestock slaughtering, provide more incentives for rangeland ecological protection, and strengthen rangeland restoration and other major conservation projects. We shall enhance rangeland resource monitoring and establish an early warning system of ecological environment (Zhang Younian 2011, 279).

The quoted statements by a leading decision-maker show the importance attached to comprehensive change in the pastoral communities and the significant investments undertaken. The pastoral resettlement programme as a central strategy to change life on the Tibetan Plateau will be addressed in several case studies in this volume. The stages of 'development' will be reflected in practices related to hunting and wildlife utilisation by pastoralists (Toni Huber), in the overall effects of the resettlement programme (Jarmila Ptackova) and its related activities in enclosures and fencing (Wu Ning, Yan Zhaoli and Lu Tao), and its changes in housing (Emilia Roza Sulek). Change and continuity (Melwyn Goldstein) over time and their linkages to political legacies and political interference (Andreas Gruschke) will highlight the dimensions and scope of transformation in order to assess the importance and effect of what is going on today in the pastoral sector in the PR of China.

1.6 Recent Transformations in Himalayan Mountain Pastoralism

A similar attention reflected in the impact on and scope of interventions in the pastoral sector cannot be ascertained for adjacent areas of the Hindukush-Karakoram-Himalaya. High Asian pastoralism in the neighbouring countries of the South is

Photo 1.7 Cattle herders in Rolpa District of Nepal. Here cows are kept for milk production only, while Tamang people refrain from slaughtering cattle for human consumption (Photograph © Hermann Kreutzmann, September 3, 2010)

limited in the context of this book to a contiguous area in the Himalayan belt. It is difficult to retrieve reliable statistical data or even trustworthy estimates. In an overview publication, the livestock population in the Indian Himalaya was estimated at 50 million domestic animals, whereby a large number of livestock is kept in systems of combined mountain agriculture (cf. Fig. 1.2), while the upper levels of grazing are occupied by mobile pastoral communities (Tulachan 2001, 30). Other publications hint on the fact that nobody knows the size of animal husbandry in the Indian Himalayas, for Himachal Pradesh a total livestock of app. five million heads was returned in the 1992 Livestock Census (Sharma et al. 2003, 39). Despite the size of animal husbandry in mountainous regions in India, an explicit pastoral policy seems to be absent: 'There are no official pastoral development policies; in fact both the Ministry of Agriculture and the Ministry of Environment and Forest are remarkable for their stance against pastoralists' (Sharma et al. 2003, iii). Similar forms are to be found in Nepal where nearly half of the animal feed is provided by crop residues. The livestock figures for Nepal (Photo 1.7) were estimated at about 15 million (Tulachan 2001, 36). The latest figures on Bhutan's livestock range at about 400,000 for cattle, yak and horses, while goats and sheep play a minor role; yak numbers have been rising since the 1990s, while others show a decrease (Roder et al. 2001; Joshi and Gurung 2009, 4–5). Less than 1,400 yak herding households with less than 50,000 yaks contribute about 3% to the livestock products generated in Bhutan (Derville and Bonnemaire 2010, 2). For the households mainly depending on animal husbandry, the share of subsistence production is quite significant.

Photo 1.8 Wakhi woman in Mulung Kir (3,600 m), Gojal (Gilgit-Baltistan, Pakistan) drying on the elevated platform (*ulina*) dehydrated whey (*qurut*) for storage and transport to her village Morkhun (Photograph © Hermann Kreutzmann, July 8, 1990)

For Pakistan, statistical evidence has been even more intangible. Most studies that address pastoralism in Pakistan state that there is a significant pressure on grazing lands as well as an increase in livestock numbers and pastoral households, but fail to even estimate the importance and size of pastoral practices in the mountainous regions (cf. Azhar-Hewitt 1999; Dost 2003; Gura 2006; Omer et al. 2006; Rahman et al. 2008). Statistical evidence provided by the Pakistan Agricultural Research Council (PARC) identifies 5.85 million hectares as high mountain pastures and grazing lands in the northern mountain ranges. On the basis of a recent estimate, Inam-ur-Rahim and Amin Beg (2011) state that about 37% of the gross farm income in Northern Pakistan is derived from livestock production. In every respect, these data can only hint at a dimension of pastoral activities that is difficult for planners and development actors to grasp (Photos 1.8 and 1.9).

The Himalayan mountain regions form the narrow, elongated belt where mountain pastoralism plays a significant role and where millions of pastoralists are engaged in utilising the natural pasture resources. In their respective regions, Bakrewals, Gaddi (Photo 1.10), Gujur and Bhotia – to name just a few – are well known for their adaptive strategies in combining different ecological belts in seasonally varying mobility patterns (Agrawal and Saberwal 2004; Axelby 2007; Chakravarty-Kaul 1998; Dangwal 2009; Gura 2006; Rao and Casimir 2003; Saberwal 1999; Sharma et al. 2003; Shashi 1979). Besides the adaptive potential of pastoral strategies, the growing exclusion of valuable pastures through administrative acts of nature protection contributes to shrinking spaces: 'Today Himalayan pastoralism is perceived by decision-makers and politicians as an environmental threat to the Himalayas and the local pastoral groups are incessantly blamed for overgrazing and livestock increase' (Sharma et al. 2003, 29).

Photo 1.9 Wakhi shepherd women in their pasture hut in Boiber (3,400 m), Gojal (Gilgit-Baltistan, Pakistan) preparing their evening meal (Photograph © Hermann Kreutzmann, July 6, 1990)

Photo 1.10 Gaddi shepherds returning from their summer pastures near lake Chandra Tal (4,270 m) drive their sheep and goats down the Chandrabagha valley, India (Photograph © Hermann Kreutzmann, August 20, 2009)

Consequently, the topical evidence presented in this volume will range around different aspects of pastoral practices. A historical perspective is provided on the Western Himalayas where pre-colonial practices and their transformations are traced (Chetan Singh). The role of the Gaddis of Chamba as part of an exchange economy for mutual benefit, participation in long-distance exchange of the trading pastoralists of Kinnaur as well as the conflicts between grazing and forest management in Kulu contribute to an understanding of social hierarchies and their internal and external relationships. A perspective on the interrelationship between state policies and local expressions of pastoral practices is provided from the Kumaon Himalayas (Christoph Bergmann, Martin Gerwin, Marcus Nüsser and William Sax). Finally, a more comparative view on herders' adaptive capacities and strategies is contributed (Muhammad Ismail and Yi Shaoliang), stressing the importance of human-environmental relations in the light of the global debate on climate change and its impact on pastoral systems. The Chinese borderlands with Pakistan and Nepal provide the arena in which the case studies are located.

1.7 Grounded Practices: Case Studies from High Asia

In our collection of case studies, societal and political changes play a major role for pastoralists and their adaptive capacities. The case studies are based on fieldwork by all authors and address quite a diverse spatial spectrum from the Pamirs, Tien Shan, Hindukush and Karakorum to the Himalayan arc that forms the southern boundary of the Tibetan Plateau.

High Asia constitutes an elevated and unique arena for shedding some light on the spectrum of mountain pastoralism and rangeland management (Fig. 1.3). In a similar ecological environment of mainly desert-steppe conditions – the only exceptions being some more humid areas under monsoonal conditions – the cases of different communities and localities are presented in order to illustrate various paths of socio-economic and politico-historical developments that are the result of inner-communal dynamics and external interventions. Societal and political transformations during the twentieth century significantly modified the economic frame conditions, possibilities of political participation in decision-making processes and cross-border exchange relations. Former commonalities among the studied communities have been transformed by inner-societal processes and by external linkages in response to decoupled exchange options.

The common ground of all case studies is the similarity in ecological challenges and risks exemplified in harsh climatic conditions, seasonal variability and probabilities in precipitation, thermal conditions and hazardous events that affect fodder availability in well-established pastures of varying quality and quantity. The important economic, political and socio-cultural aspects are embedded in these environmental arenas and are a reflection of local competition, power structures and external interferences.

The case studies that have been invited for this book cover a range of topics and localities. A number of case studies address Central Asian challenges that are strongly linked with post-socialist transformations in the livestock sector of Kyrgyzstan (Andrei Dörre, Bernd Steimann) and Tajikistan (Tobias Kraudzun). Across the Amu Darya in Afghanistan, the living conditions significantly change.

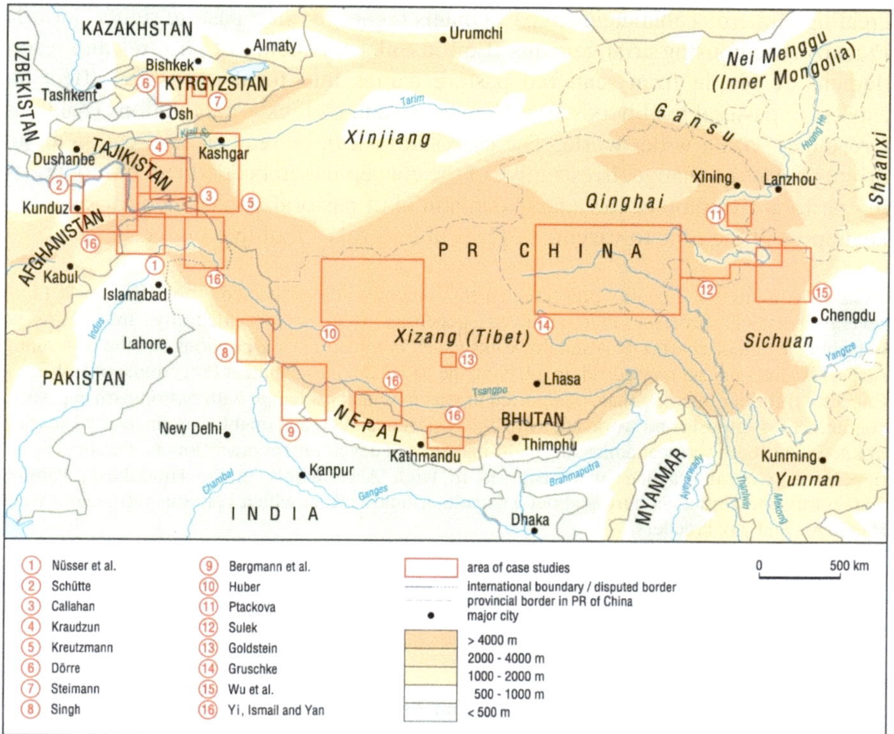

Design: H.Kreutzmann

Cartography: B.Hilberer

Fig. 1.3 The location of presented case studies in High Asia

Patterns of insecurity caused by one generation-long political turmoil and war are the guiding principles for pastoralists in the remote valleys of the Afghan Pamirs (Ted Callahan) or on the Shewa pastures of Badakhshan (Stefan Schütte). In neighbouring Pakistan, different patterns of pastoralism are observed (Marcus Nüsser, Arnd Holdschlag and Fazlur-Rahman; M. Ismail and Yi Shaoliang); its overall importance has weakened in recent decades, although certain communities continue to depend on its proceeds. In India, a comparatively long history of competition between forest managers and pastoralists (Chetan Singh) can be documented as well as the impact of pastoral services for cross-boundary exchange (Christoph Bergmann et al.). Here administrative interventions and geopolitically important conflicts causing border closures and military interference contributed to a modified arena for pastoralists. Seven case studies are devoted to the PR of China between the Western Kun Lun Shan (Hermann Kreutzmann) and the Tibetan Plateau. In all case studies, external historical developments are prominent features of change. They affected all walks of life including the hunting sector (Toni Huber), which did not escape a number of reforms such as the organisation of animal husbandry in the Changtang (Melwyn Goldstein) or Qinghai. The recent implementation and propagation of resettlement schemes are creating numerous challenges to participants regarding their pasture rights (Jarmila Ptackova) or housing arrangements (Emilia Sułek). Fencing of pastures and resettlement of people from scattered pasture camps into focal townships affect all aspects of pastoral livelihoods (Wu Ning et al.) and their external relations (Andreas Gruschke) with a wider world. The case studies can be read as a reflection about driving forces of change. Irrespective of the conceptual embeddedness of modernising pastoral livelihoods, the adaptive potential of pastoralists is called for, as it has been most of the time to varying degrees in a world of multiple insecurities.

Acknowledgements Insights presented in this chapter were collected from 1984 to 2011 in several research projects that received generous funding provided by the Deutsche Forschungsgemeinschaft (DFG) and the Volkswagen Foundation. The previous and the continuing support is gratefully acknowledged. In addition, the Mountain Development Programme of InWEnt – Capacity Building International – supported the meeting and exchange with pastoralism experts in conferences, symposia and workshops in High Asia since 2000 and enabled an intensified sharing of knowledge between academics, decision-makers and development practitioners. Personal experiences of pastoral practices were enjoyed in High Asia mainly in the Hindukush, Pamirs, Karakoram and Himalaya during about a decade of fieldwork in which I significantly learnt from shepherds and yak breeders.

References

Agrawal A, Saberwal VK (2004) Whither South Asian pastoralism? An introduction. Nomadic Peoples 8:36–53

Alden Wily L (2004) Looking for peace on the pastures. Rural land relations in Afghanistan. Afghanistan Research and Evaluation Unit, Kabul

Alden Wily L (2009) Recommended strategy for conflict resolution of competing high pasture claims of settled and nomadic communities in Afghanistan. UNEP, Kabul

Alirol P (1979) Transhuming animal husbandry systems in the Kalinchowk region (Central Nepal): a comprehensive study of animal husbandry on the southern slopes of the Himalayas. SDC, Bern

An S-Z, Jian D, Zhaohui L, Yonggang M, Xiao-Yuan L (2011) Pastoralism in China's Xinjiang Kizilsu Kirghiz autonomous prefecture. In: Kreutzmann H, Abdulalishoev K, Zhaohui L, Richter J (eds) Regional workshop in Khorog and Kashgar. Pastoralism and rangeland management in mountain areas in the context of climate and global change. GIZ, Bonn, pp 119–131

Axelby R (2007) 'It takes two hands to clap': how Gaddi shepherds in the Indian Himalayas negotiate access to grazing. J Agrarian Change 7(1):35–75

Azhar-Hewitt F (1999) Women of the high pastures and the global economy: reflections on the impacts of modernization in the Hushe valley of the Karakorum, Northern Pakistan. Mt Res Dev 19(2):141–151

Banjade MR, Paudel NS (2008) Mobile pastoralism in crisis: challenges, conflicts and status of pasture tenure in Nepal mountains. J Livelihood 7(1):49–57

Banks T, Richard C, Ping L, Zhaoli Y (2003) Community-based grassland management in western China: rationale, pilot project experience, and policy implications. Mt Res Dev 23(2):132–140

Barfield T (1993) The nomadic alternative. Prentice Hall, Englewoods Cliff

Barfield T (2008) Nomadic pastoralists in Afghanistan. Reconstruction of the pastoral economy. In: Janzen J, Enkhtuvshin B (eds) Present state and perspectives of nomadism in a globalizing world. Admon, Ulaanbaatar, pp 189–204

Bauer K (2006) Common property and power: insights from a spatial analysis of historical and contemporary pasture boundaries among pastoralists in Central Tibet. J Political Ecol 13:24–47
Benson L, Svanberg I (1998) China's last nomads: the history and culture of China's Kazaks. M. E. Sharpe, Armonk
Beuermann A (1967) Fernweidewirtschaft in Südosteuropa. Westermann, Braunschweig
Bhasin V (1996) Transhumants of Himalayas. Changpas of Ladakh, Gaddis of Himachal Pradesh and Bhutias of Sikkim. Kamla-Raj Enterprises, Delhi
Blache J (1934) L'homme et la montagne. Géographie humaine 3. Gallimard, Paris
Blackburn S (2007) Oral stories and culture areas: from northeast India to southwest China. South Asia 30(3):423–424
Brower B, Johnston BR (eds) (2007) Disappearing peoples? Indigenous groups and ethnic minorities in South and Central Asia. Left Coast Press, Walnut Creek
Cerny A (2010) Going where the grass is greener: China Kazaks and the Oralman immigration policy in Kazakhstan. Pastoralism 1(2):218–247
Chakravarty-Kaul M (1998) Transhumance and customary pastoral rights in Himachal Pradesh: claiming the high pastures for Gaddis. Mt Res Dev 18(1):5–17
Christian D (2000) Silk roads or steppe roads? The silk roads in world history. J World Hist 11(1):1–26
Dangwal DD (2009) The lost mobility. Pastoralism and modernity in Uttarakhand Himalaya (India). Nomadic Peoples 13(2):84–101
Davies J, Hatfield R (2007) The economics of mobile pastoralism: a global summary. Nomadic Peoples 11(1):91–116
Derville M, Bonnemaire J (2010) Marginalisation of yak herders in Bhutan: can public policy generate new stabilities that can support the transformation of their skills and organisations? Innovation and Sustainable Development in Agriculture and Food. ISDA 2010, Montpellier, June 28–30, 2010, pp 1–10. http://hal.archives-ouvertes.fr/docs/00/52/20/45/PDF/DervilleBonnemaire_Marginalisation_of_yak.pdf. Accessed 8 Aug 2011
di Cosmo N (1999) State formation and periodization in inner Asian history. J World Hist 10(1):1–40
Dost M (2003) Pastoral resource profile for Pakistan. http://www.fao.org/ag/AGP/AGPC/doc/Counprof/Pakistan.htm
Dyson-Hudson R, Dyson-Hudson N (1980) Nomadic pastoralism. Annu Rev Anthropol 9:15–61
Ehlers E, Kreutzmann H (eds) (2000) High mountain pastoralism in Northern Pakistan. Steiner, Stuttgart
Ferdinand K (2006) Afghan nomads. Caravans, conflicts and trade in Afghanistan and British India 1800–1980. Rhodos, Copenhagen
Finke P (2005) Contemporary pastoralism in Central Asia. In: Rasuly-Paleczek G, Katschnig J (eds) Central Asia on display. Proceedings of the VII. Conference of the European Society for Central Asian. LIT, Vienna, pp 397–410
Foggin MJ (2008) Depopulating the Tibetan grasslands. National policies and perspectives for the future of Tibetan herders in Qinghai Province, China. Mt Res Dev 28(1):26–31
Galaty JG, Aronson D, Salzman PC, Chouinard A (eds) (1981) The future of pastoral peoples. International Development Research Centre, Ottawa
Gertel J, Breuer I (eds) (2007) Pastoral Morocco. Globalizing scapes of mobility and insecurity. Reichert, Wiesbaden
Ginat J, Khazanov AM (eds) (1998) Changing nomads in a changing world. Sussex Academic Press, Brighton
Glatzer B (1981) Processes of nomadization in West Afghanistan. In: Salzman PC (ed) Contemporary nomadic and pastoral peoples: Asia and the North. Studies in third world societies 18. Department of Anthropology, Williamsburg, pp 61–86
Goldstein M, Beall C (1991) Change and continuity in nomadic pastoralism on the Western Tibetan Plateau. Nomadic Peoples 28:105–122

Goldstein M, Beall C (2002) Changing patterns of Tibetan nomadic pastoralism. In: Leonard WR, Crawford MH (eds) Human biology of pastoral populations. Cambridge University Press, Cambridge, pp 131–150

Gruschke A (2008) Nomads without pastures? Globalization, regionalization, and livelihood security of nomads and former nomads in Northern Khams. J Int Assoc Tibetan Stud 4:1–40

Guillet D (1983) Toward a cultural ecology of mountains. The Central Andes and the Himalaya compared. Curr Anthropol 24(5):561–574

Gura S (2006) Review of the literature on pastoral economics and marketing: Afghanistan, India, Iran, Iraq, Israel, Jordan, Pakistan, Palestine, Syria, and Turkey. IUCN, World Initiative for Sustainable Pastoralism, Nairobi

Harris RB (2010) Rangeland degradation on the Qinghai-Tibetan plateau: a review of the evidence of its magnitude and causes. J Arid Environ 74:1–12

Ho P (2000) The clash over state and collective property: the making of the rangeland law. China Q 16(1):240–263

Humphrey C, Sneath D (1999) The end of nomadism? Society, state and the environment in inner Asia. Duke University Press, Durham

Inam-ur-Rahim, Amin Beg (2011) Production and marketing of livestock products in the Hindu Kush-Karakoram-Himalaya. In: Kreutzmann H, Yong Y, Richter J (eds) Regional workshop in Lhasa 2010 – pastoralism and rangeland management on the Tibetan Plateau in the context of climate and global change. GIZ, Bonn, pp 177–199

Jettmar K (1960) Soziale und wirtschaftliche Dynamik bei asiatischen Gebirgsbauern. Sociologus, N.S. 10:120–138

Johnson DC (1969) The nature of nomadism: a comparative study of pastoral migrations in Southwestern Asia and Northern Africa. University Press, Chicago

Jones S (2005) Transhumance re-examined. J R Anthropol Inst 11(2):357–359

Joshi SR, Gurung BR (2009) Value chain analysis of dairy in Merak and Sakteng. Ministry of Agriculture & Forests, Thimphu

Karmyševa B (1981) Versuch einer Typologisierung der traditionellen Formen der Viehwirtschaft Mittelasiens und Kasachstans am Ende des 19. – Anfang des 20. Jh. In: Krusche R (ed) Die Nomaden in Geschichte und Gegenwart, Veröffentlichungen des Museums für Völkerkunde zu Leipzig 33. Akademie-Verlag, Berlin, pp 91–96

Kaufmann JC (2009) The sediment of nomadism. Hist Afr 36:235–264

Khazanov AM (1984) Nomads and the outside world. Cambridge University Press, Cambridge (second edition Madison: University of Wisconsin Press 1994)

Khazanov AM (2005) Nomads and cities in the Eurasian Steppe region and adjacent countries: a historical overview. In: Leder S, Streck B (eds) Shifts and drifts in nomad-sedentary relations. Reichert, Wiesbaden, pp 163–178

Klieger PC (ed) (2006) Tibetan borderlands. Brill, Leiden

Kothari U, Minogue M (eds) (2002) Development theory and practice. Critical perspectives. Palgrave, Hampshire

Kreutzmann H (2004) Pastoral practices and their transformation in the North-Western Karakoram. Nomadic Peoples 8(2):54–88

Kreutzmann H (2009) Transformations of high mountain pastoral strategies in the Pamirian knot. Nomadic Peoples 13(2):102–123

Kreutzmann H (2011) Pastoral practices on the move. Recent transformations in mountain pastoralism on the Tibetan Plateau. In: Kreutzmann H, Yong Y, Richter J (eds) Regional workshop in Lhasa 2010. Pastoralism and rangeland management on the Tibetan Plateau in the context of climate and global change. GIZ, Bonn, pp 200–224

Kreutzmann H, Schütte S (2011) Contested commons – multiple insecurities of pastoralists in North-Eastern Afghanistan. Erdkunde 65(2):99–119

Kreutzmann H, Abdulalishoev K, Zhaohui L, Richter J (eds) (2011a) Regional workshop in Khorog and Kashgar. Pastoralism and rangeland management in mountain areas in the context of climate and global change. GIZ, Bonn

Kreutzmann H, Yong Y, Richter J (eds) (2011b) Regional workshop in Lhasa 2010. Pastoralism and rangeland management on the Tibetan Plateau in the context of climate and global change. GIZ, Bonn

Li W, Huntsinger L (2011) China's grassland contract policy and its impacts on herder ability to benefit in Inner Mongolia: tragic feedbacks. Ecol Soc 16(2):1. [online] URL: http://www.ecologyandsociety.org/vol16/iss2/art1/

Lu T, Wu N, Luo P (2009) Sedentarization of Tibetan nomads. Conserv Biol 23(5):1074

Manderscheid A (2001) Decline and re-emergence of Nomadism: Tibetan pastoralists revive a nomadic way of life and production. GeoJournal 53:173–182

Mathieu J (2011) The third dimension. A comparative history of mountains in the modern Era. The White Horse Press, Cambridge

Messerschmidt DC (1976) Ecological change and adaptation among the Gurungs of the Nepal Himalaya. Hum Ecol 4(2):167–185

Miehe G et al (2001) The climatic diagram map of High Asia. Purpose and concepts. Erdkunde 55:94–97

Miller DJ (2000) Tough times for Tibetan nomads in Western China: snowstorms, settling down, fences, and the demise of traditional nomadic pastoralism. Nomadic Peoples 4(1):83–109

Miller DJ (2002) The importance of China's nomads. The sustainable future development of China's rangelands depends on integrating nomads' indigenous knowledge. Rangelands 24(1):22–24

Miller DJ (2008) Drokpa: nomads of the Tibetan Plateau and Himalaya. Vajra Books, Kathmandu

Montero RG, Mathieu J, Singh C (2009) Mountain pastoralism 1500–2000: an introduction. Nomadic Peoples 13(2):1–16

Natarajan T (2005) Agency of development and agents of change: localization, resistance, and empowerment. J Econ Issues 39(2):409–418

Nautiyal S, Rao KS, Maiykhuri RK, Saxena KG (2003) Transhumant pastoralism in the Nanda Devi Biosphere Reserve, India. A case study in the buffer zone. Mt Res Dev 23(3):255–262

Nüsser M, Gerwin M (2008) Diversity, complexity and dynamics: land use patterns in the Central Himalayas of Kumaon, Northern India. In: Löffler J, Stadelbauer J (eds) Diversity in mountain systems, Colloquium geographicum 31. Asgard, Sankt Augustin, pp 107–119

Omer RM, Hester AJ, Gordon IJ, Swaine MD, Raffique SM (2006) Seasonal changes in pasture biomass, production and offtake under the transhumance system in northern Pakistan. J Arid Environ 67:641–660

Ptackova J (2011) Sedentarisation of Tibetan nomads in China: implementation of the nomadic settlement project in the Tibetan Amdo area; Qinghai and Sichuan Provinces. Pastoralism Res, Policy Pract 1:4. http://www.pastoralismjournal.com/content/1/1/4

Qonunov Y (2011) Recent changes in pastoral systems. Case study on Tajikistan. In: Kreutzmann H, Abdulalishoev K, Zhaohui L, Richter J (eds) Regional workshop in Khorog and Kashgar. Pastoralism and rangeland management in mountain areas in the context of climate and global change. GIZ, Bonn, pp 82–101

Rahman A, Duncan AJ, Miller DW, Clemens J, Frutos P, Gordon IJ, Rehman A-ur, Baig A, Ali F, Wright IA (2008) Livestock feed resources, production and management in the agro-pastoral system of the Hindu Kush – Karakoram – Himalayan region of Pakistan: the effect of accessibility. Agr Syst 96:26–36

Rao A, Casimir MJ (eds) (2003) Nomadism in South Asia. Oxford University Press, New Delhi

Rinschede G (1979) Die Transhumance in den französischen Alpen und in den Pyrenäen, Westfälische Geographische Studien 32. Im Selbstverlag der Geographischen Kommission für Westfalen, Münster

Rinschede G (1988) Transhumance in European and American Mountains. In: Allan NJR, Knapp G, Stadel C (eds) Human impact on mountains. Rowman & Littlefield, Totowa, pp 96–108

Robinson S, Whitton M (2010) Pasture in Gorno-Badakhshan, Tajikistan: common resource or private property? Pastoralism 1(2):198–217

Robinson S, Whitton M, Biber-Klemm S, Muzofirshoev N (2010) The impact of land reform legislation on pasture tenure in Gorno-Badakhshan. Mt Res Dev 30(1):4–13

Roder W, Wangdi K, Gyamtsho P, Dorji K (2001) Feeding the herds. Improving fodder resources in Bhutan. Icimod, Kathmandu
Roe E, Huntsinger L, Labnow K (1998) High reliability pastoralism. J Arid Environ 39:39–55
Saberwal VK (1999) Pastoral politics: shepherds, bureaucrats, and conservation in the Western Himalayas. Oxford University Press, Delhi/New York
Salzman PC (ed) (1980) When nomads settle: processes of sedentarization as adaptation and response. Praeger, New York
Salzman PC (2004) Pastoralists. Equality, hierarchy, and the state. Westview Press, Boulder
Salzman PC, Galaty JG (1990) Nomads in a changing world: issues and problems. In: Salzman PC, Galaty JG (eds) Nomads in a changing world. Instituto Universitario Orientale, Naples, pp 3–21
Scholz F (2002) Nomadism and colonialism. A hundred years of Baluchistan 1872–1972. Oxford University Press, Karachi
Scholz F (2008) Nomadism. A socioecological mode of culture. International Institute for the Study of Nomadic Civilizations, Ulaanbaatar
Scott JC (2009) The art of not being governed. An anarchist history of upland Southeast Asia. Yale University Press, New Haven/London
Seaman G (ed) (1989) Ecology and empire. Nomads in the cultural evolution of the old world. Ethnographics/USC, Los Angeles
Shahrani MN (1979) The Kirghiz and Wakhi of Afghanistan. Adaptation to closed frontiers. University of Washington Press, Seattle/London
Sharma VP, Köhler-Rollefson I, Morton J (2003) Pastoralism in India. A scoping study. DFID, New Delhi
Shashi SS (1979) The nomads of the Himalayas. Sundeep Prakashan, Delhi
Sheehy DP, Miller D, Johnson DA (2006) Transformation of traditional pastoral livestock systems on the Tibetan steppe. Sécheresse 17(1–2):142–151
Shneiderman S (2010) Are the Central Himalayas in Zomia? Some scholarly and political considerations across time and space. J Glob Hist 5:289–312
Snoy P (1993) Alpwirtschaft in Hindukusch und Karakorum. In: Schweinfuhrt U (ed) Neue Forschungen im Himalaya, Erdkundliches Wissen 112. Steiner, Stuttgart, pp 49–73
Stevens SF (1993) Claiming the high ground: sherpas, subsistence, and environmental change in the highest Himalaya. University of California Press, Berkeley/Los Angeles/London
Tapper R (2008) Who are the Kuchi? Nomad self-identities in Afghanistan. J R Anthropol Inst 14:97–116
Tashi Y, Chamard PC, Courel M-F, Tiyip T, Tuerxun Y, Drake S (2010) The recent evolution of the oasis environment in the Taklimakan Desert, China. In: Schneier-Madanes G, Courel M-F (eds) Water and sustainability in arid regions. Springer, Berlin/Heidelberg, pp 51–74
Tucker RP (1986) The evolution of transhumant grazing in the Punjab Himalaya. Mt Res Dev 6(1):17–28
Tulachan PM (2001) State of mountain agriculture in the Hindukush-Himalayas. A regional comparative analysis. Icimod, Kathmandu
Uhlig H (1976) Bergbauern und Hirten im Himalaya. In: Uhlig H, Ehlers E (eds) Tagungsbericht und wissenschaftliche Abhandlungen. 40. Deutscher Geographentag Innsbruck (19.-25.5.1975). Steiner, Wiesbaden, pp 549–586
Uhlig H (1995) Persistence and change in high mountain agricultural systems. Mt Res Dev 15(3):199–212
van Driem G (2001) Languages of the Himalayas: an ethnolinguistic handbook of the greater Himalayan region. Brill, Leiden
van Schendel W (2002) Geographies of knowing, geographies of ignorance: jumping scale in Southeast Asia. Environ Plan D Soc Space 20(6):647–668
von Schlagintweit R (1865) Physikalisch-geographische Schilderung von Hochasien. Petermanns Geographische Mittheilungen 11:361–377
von Schlagintweit-Sakünlünski H (1869–1880) Reisen in Indien und Hochasien. Eine Darstellung der Landschaft, der Cultur und Sitten der Bewohner, in Verbindung mit klimatischen und

geologischen Verhältnissen. Basirt auf die Resultate der wissenschaftlichen Mission von Hermann, Adolph und Robert von Schlagintweit. Erster Band 1896: Indien. Zweiter Band 1871: Hochasien. I: Der Himálaya von Bhután bis Kashmir. Dritter Band 1872: Hochasien. II: Tibet zwischen der Himálaya- und der Karakorum-Kette. Vierter Band 1880: Hochasien. III: Ost-Turkistán und Umgebungen. Nebst wissenschaftlichen Zusammenstellungen über die Höhengebiete und über die thermischen Verhältnisse. Hermann Costenoble, Jena

von Schlagintweit-Sakünlünski H (1870) Erläuterung der Gebiete Hochasiens. Sitzungsberichte Akademie München 2:313–327

Weissleder W (1978) The nomadic alternative. Modes and models of interaction in the African-Asian deserts and steppes. Mouton, The Hague/Paris

Xinchun Z (2011) Pastoralism in Xinjiang: implementing the pastoralist settlement programme to promote sustainable development of pastoralism in Xinjiang. In: Kreutzmann H, Abdulalishoev K, Zhaohui L, Richter J (eds) Regional workshop in Khorog and Kashgar. Pastoralism and rangeland management in mountain areas in the context of climate and global change. GIZ, Bonn, pp 182–188

Yamaguchi T (2011) Transition of mountain pastoralism: an agrodiversity analysis of the livestock population and herding strategies in Southeast Tibet, China. Hum Ecol 39:141–154

Yeh ET (2005) Green governmentality and pastoralism in Western China: 'converting pastures to grasslands'. Nomadic Peoples 9(1–2):9–29

Younian Z (2011) Achievements and visions of agricultural and pastoral development in the Tibet autonomous region. In: Kreutzmann H, Yong Y, Richter J (eds) Production and marketing of livestock products in the Hindu Kush-Karakoram-Himalaya. Regional workshop in Lhasa 2010 – pastoralism and rangeland management on the Tibetan Plateau in the context of climate and global change. GIZ, Bonn, pp 265–280

Zhizhong Wu, Wen Du (2008) Pastoral nomad rights in inner Mongolia. Nomadic Peoples 12(2):13–33

Chapter 2
Herding on High Grounds: Diversity and Typology of Pastoral Systems in the Eastern Hindukush (Chitral, Northwest Pakistan)

Marcus Nüsser, Arnd Holdschlag, and Fazlur-Rahman

Abstract This chapter analyses pastoral migration patterns as strategies for utilising the grazing resources at the marginal belts of human habitation in Chitral, an area located in the eastern Hindukush of northern Pakistan. Beyond the common features of combined mountain agriculture, pastoral utilisation strategies vary between different tributary valleys in the region. Based on six case studies from northern, eastern and southern Chitral, similarities and differences between pastoral resource utilisation are presented and analysed here. Although the influence of heterogeneous environmental settings needs to be considered, differences in resource utilisation mainly stem from distinct settlement processes and territorial rights of access and utilisation, which in turn evolved from ethnic and social segregation between two dominant actor groups: the Kho mountain farmers and formerly nomadic Gujur. Hence, a better understanding of the complexity, diversity and dynamics involved in pastoral management systems must be based on a historically informed study of these spatial and social patterns.

Keywords Human ecology • Pastoralism • Diversity • Chitral • Pakistan

---

M. Nüsser (✉)
Department of Geography, South Asia Institute, University of Heidelberg, Im Neuenheimer Feld 330, D-69120 Heidelberg, Germany
e-mail: marcus.nuesser@uni-heidelberg.de

A. Holdschlag
Institute of Geography, Hamburg University, Bundesstr. 55 (Geomatikum), D-20146 Hamburg, Germany
e-mail: holdschlag@geowiss.uni-hamburg.de

Fazlur-Rahman
Department of Geography, Urban & Regional Planning, University of Peshawar, Peshawar, Pakistan
e-mail: fazal367@hotmail.com

H. Kreutzmann (ed.), *Pastoral practices in High Asia*, Advances in Asian Human-Environmental Research, DOI 10.1007/978-94-007-3846-1_2,

2.1 Introduction

Pastoral migration patterns in high mountain regions have predominantly been analysed as local adaptation strategies within specific socio-economic conditions and natural resource potentials. The spheres of environmental, especially biological resources, grazing management practices and subsequent land use/land cover change together constitute a complex and dynamic human-environmental setting (Nüsser 1998, 2003). A better understanding of the interactions between these various spheres is calling for an integration of the diverse perspectives of both natural and social sciences that emphasises the recognition of plural explanations (Forsyth 1998).

A deeper knowledge of pastoral land use dynamics and the concomitant changes in both production patterns and livelihood strategies strongly depend on the integration of the territoriality of land tenure (spatial dimension), historical developments (temporal dimension) and external influences (power relations). Especially in the context of communal property systems and community-based institutions (cf. Ostrom et al. 1999), the interaction of local knowledge and norms, formal and informal regulations and external development interventions all must be taken into consideration (Ives 2004; Kreutzmann 2004a).

This chapter analyses pastoral mobility patterns in different tributary valleys of Chitral in northern Pakistan. Much like in other parts of the Hindukush, Karakoram and Western Himalayas, animal husbandry in this mountain region generally takes place within a context of combined mountain agriculture (Ehlers and Kreutzmann 2000), relying on irrigated crop cultivation and mobile livestock-holding across different altitudinal belts. Despite this general pattern, livelihood strategies in the individual valleys of Chitral vary amongst themselves (Holdschlag 2011). Based on six case studies, this chapter draws attention to aspects of spatial access rights and territoriality, mobility patterns, herding arrangements and specific pasture-ecological settings. A similar comparative approach has been established for the Nanga Parbat area (Clemens and Nüsser 2008). In order to cover the diversity and dynamics of pastoral systems in Chitral, we present a typology according to historical and socio-political constellations and development paths rooted in factors of ethnic and social segregation of local and regional actors. Based on the comparative and classificatory assessment, we aim to identify specific forces that have shaped the evolution of pastoral systems. Against this background, we argue that it is fruitful to analyse various cases according to the bipolarity of persistence and change in order to better understand the impact of external (non-place-based) interventions and specific local (place-based) response.

Research has been carried out through socio-economic surveys drawing on quantitative and extensive qualitative interviews covering local household economies and livelihood strategies in a historical dimension. Empirical data collection has been organised from 1997 until 2006.

2.2 Regional Contextuality of Animal Husbandry

2.2.1 Environmental Settings

Chitral (ca. 35–37°N, 71–74°E) constitutes the northernmost district of Pakistan's province Khyber Pakhtunkhwa (until 2010 known as the North-West Frontier Province), bordering the Afghan provinces Kunar and Badakhshan to the west and north (Fig. 2.1). The valley floors reveal arid conditions, and the vegetation is characterised by open dwarf-scrub communities and sharply delimited forests and shrublands along tributary streams and irrigation channels. Whereas montane coniferous forests (*Cedrus deodara, Pinus wallichiana, Picea smithiana, Abies pindrow*) can only be found in southern Chitral, the inner valleys of northern Chitral are substantially treeless (Nüsser and Dickoré 2002).

The altitudinal distribution of cultivated areas ranges from the colline to the lower montane belt. Generally, the production of wheat (*Triticum durum, T. aestivum* up to approximately 3,300 m), maize (*Zea mays* up to 2,900 m), barley (*Hordeum vulgare* up to 3,500 m) and rice (*Oryza sativa* up to 2,300 m) depends on the availability of glacial and snowmelt water for irrigation (Israr-ud-Din 1996). The general development of cultivated areas in Chitral is characterised by an extension of irrigated areas and an increasing number of fruit and fodder trees in the village environs (Nüsser 2001). Double cropping is possible with an upper limit at approximately 2,500 m. Crop residues, hay, planted fodder trees (*Salix* spp., *Populus* spp.) along the new water channels and hygrophilous bushes (*Hippophae rhamnoides, Rosa webbiana*) form the major source for the livestock's winter supplies within the cultivated areas, supplemented by the cultivation of lucerne (*Medicago sativa, M. X varia*). The fodder situation for stall-feeding during the long winter season is generally more severe in the single-cropping areas of the upper tributaries (Haserodt 1989). On the other hand, the economic importance of animal husbandry is more pronounced in the single-cropping areas. Consequently, the poor condition of all animals in early spring is symptomatic of important nutrition problems and is responsible for diseases and low productivity. Herd size and composition generally varies with local fodder potentials, which in turn depend on the biological productivity of accessible pasture land, cultivation of fodder plants and cropping patterns. Cows, for example, need more feed throughout the year than oxen, as the latter are mostly sent to the alpine pastures for free grazing in summer. Goats are predominantly kept in those areas of southern Chitral where evergreen oak woodlands (*Quercus baloot*) are found, providing a broad basis for lopping and winter grazing.

The multifunctionality of biological resources in the pastoral economy is not limited to the aspect of fodder. Especially in the vicinity of seasonally inhabited pastoral settlements at higher elevations, thorn cushions (*Acantholimon, Astragalus*) and other dwarf-scrubs (*Artemisia, Ephedra gerardiana*) are utilised for fuel and thatched roofs. Insufficient fuel wood is often the reason for the abandonment of pasture settlements in favour of the utilisation of higher alpine grazing grounds (*ghari*).

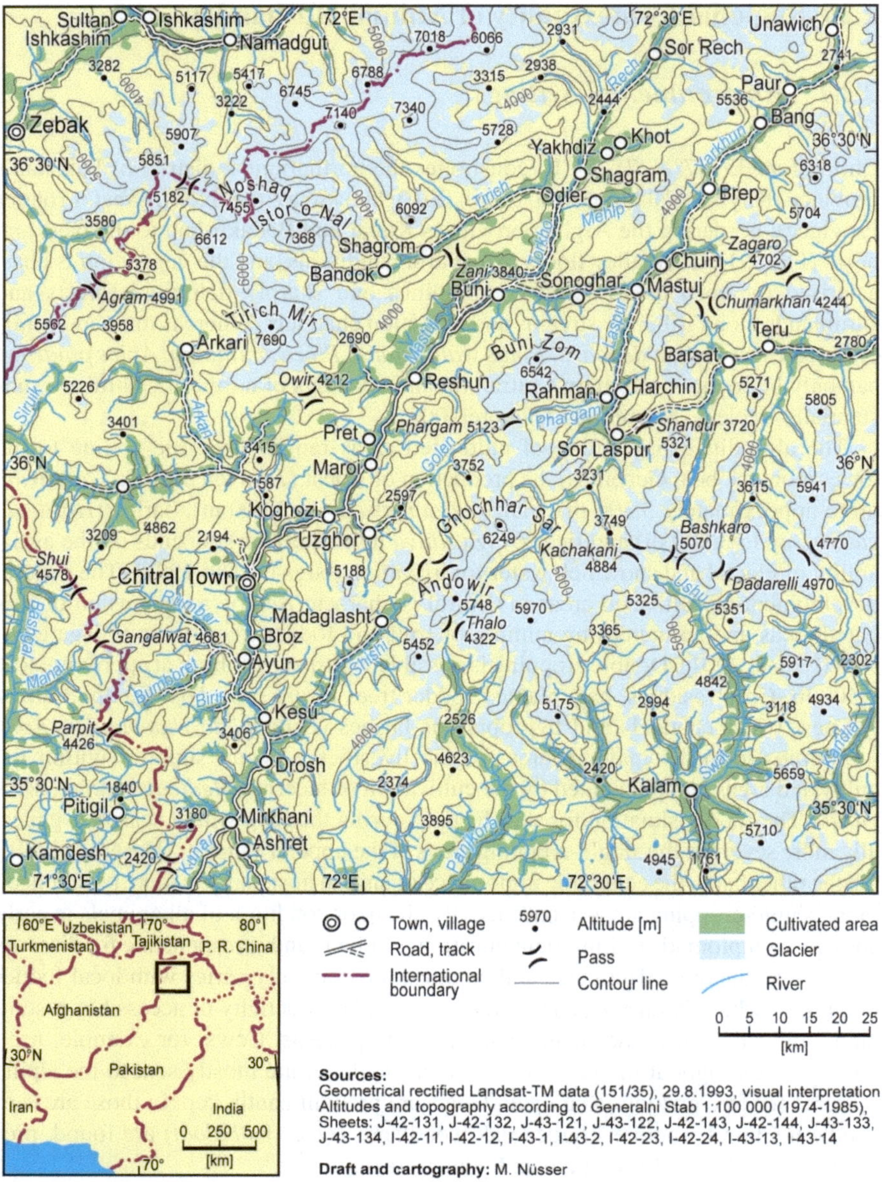

Fig. 2.1 Map of Chitral, eastern Hindukush, northwest Pakistan

2.2.2 Importance of Animal Husbandry for Livelihood Strategies

Animal husbandry has always been a critical economic activity in Chitral's combined mountain agriculture. During pre-colonial and colonial times, animal husbandry was the only possible long-term investment for ensuring the security of sustenance in such a harsh environment, one in which the local rulers exclusively owned all the land and other resources. Furthermore, the productivity of and entitlement to arable land was closely connected with livestock rearing, for example, the annual sheep tax (*thangi*) on arable land was in practice for a long period of time in the feudal state of Chitral (General Staff India 1928; Barth 1956; IUCN Pakistan 2004). According to historical sources, 800 sheep, 8 yaks, 20 horses and 400 *bati* (ca. 1,000 kg) of *ghee* (clarified butter) were collected from Torkho tahsil, for example, in one single year at the end of the nineteenth century (IOR/2/1077/235/11826/214–213). Moreover, animals were the only source of draught power and manure in the cultivated fields and the only available means of exchange and barter in the absence of a monetary system.

Although other sources of non-agrarian income and investment have recently become more attractive to the villagers, animal husbandry is still of crucial economic importance for the mountain dwellers in Chitral. This becomes obvious if one compares specific monetary values. A 4-year-old male yak, for example, costs about 10,000 Pakistani Rupees (Rs) in the local market. In comparison, the total earnings of a seasonal labourer hardly exceed 5,000 Rs for the whole winter season (Fazlur-Rahman 2007, 140). A rural household usually keeps a variety of animals to enrich the animal produce whilst optimising security against animal diseases. This diversity in animal species is one of the mountain farmers' basic strategies that ensure their subsistence livelihood by utilising all available natural resources. A herd may be composed of cattle, yaks, sheep and goats, donkeys and horses, as well as poultry. Livestock provides food for the family, raw material for domestic handicrafts and manure for the fields and can serve as draught or pack animals. Meat, eggs, milk and other dairy products still play an important role in the socio-economic as well as in the cultural spheres of the inhabitants. Although these products are imported into Chitral from the lowlands, they are not generally available in the markets of the remote villages. Instead, their availability is usually restricted to the central places, where they command higher prices and are thus unaffordable for the majority of villagers. Therefore, most of the households keep a few animals to secure the availability of the aforementioned products. Animals also serve as an investment for emergency situations and are an asset that can be easily liquidated into cash at any time.

Apart from that, varying socio-cultural dimensions of animal husbandry can be found. Animals are the main source of mutual exchange and reciprocity at the village and neighbourhood levels and between in-laws. Since a large number of households do not have enough animals to fulfil their own needs, villagers commonly practise seasonal and temporary lending and borrowing of oxen, donkeys and threshing teams, without any expectation of financial compensation. Although this trend is

Table 2.1 Composition and changes of livestock in Chitral District (1972–2006)

Animal species	1972	[%]	1986	[%]	1996	[%]	2006	[%]
Cattle[a]	71,580	14.03	100,083	12.18	173,262	13.38	174,842	15.45
Sheep	97,310	19.08	113,627	13.82	188,822	14.58	181,146	16.01
Goats	169,389	33.21	221,070	26.90	335,780	25.94	347,977	30.75
Other[b]	2,469	0.48	4,383	0.53	6,760	0.52	3,910	0.35
Poultry	169,360	33.20	382,801	46.57	590,022	45.57	423,749	37.45

Sources: GoNWFP (1997, 201), GoP (1998, 1, 2006, 1), Beg (2011, 141), Khan and Khan (n.d., 32)
[a]Includes yaks
[b]Includes horses, asses, mules, buffalos and camels

changing with current socio-economic transformations, livestock still plays a major role in maintaining relationships and exchange system throughout the region.

The main driving forces of the overall increase of livestock up until the end of the last millennium (Table 2.1) can be linked to both a pattern of constant population growth between 1951 and 1998 with an average annual growth rate of 2.5% (GoP 1999) and an increasing domestic need for animal products. The developments of the last decade (1996–2006), however, seem to signal the termination of this trend. Stocks of sheep, as well as of the traditional pack animals (horses, asses and mules), are already on the decline. According to one sample survey (Aga Khan Rural Support Programme 2000, 32), the average number of livestock is 4 head of cattle and 12 sheep/goats per Chitral household. Comparative livestock figures are given for the early 1960s based on a survey of 37 villages (Israr-ud-Din 1965, 148). He assessed on average 4–5 cattle, 14 goats and 13 sheep per household.

2.2.3 Diverse Actors: Historical Politics and Social Heterogeneity

In addition to the physical characteristics that determine the availability and productivity of a local community's natural resources, political institutions usually define the entitlement to and utilisation of pastoral resources. Access rights and ownership systems are susceptible to modifications with related transformations in the political organisation. Historically, the study area of Chitral was a mountain kingdom ruled by a hereditary ruler (*mehtar*). It consisted furthermore of various districts controlled by powerful hierarchical or conical clans (*qaum*). Social stratification was based on administrative categories which were established for state military duties of the elite on the one hand and the collection of revenues and *corvée* labour services on the other hand (Eggert 1990; Parkes 2001). Therefore, the society could be regarded as dichotomised, split into the rigid categories of 'noble' (*adamzada*) and 'ignoble' (*ghalamus*) groups (Ghulam Murtaza 1962, 168–169). All lands belonged at least nominally to the *mehtar*, and it was his prerogative to make grants to his relatives and influential clans at will. Local political regulations determined the way in which pasture resources were distinctly utilised by different user groups and social classes.

As a region affected by constant immigration, Chitral is characterised by diverse ethno-linguistic settings dominated by the Kho, but comprising about 12 different languages in all (Biddulph 1880; Strand 1997–2011; Kreutzmann 2005). Diversity is also revealed through the existence of complex religious structures (Cacopardo and Cacopardo 2001; Marsden 2005). During the feudal periods, the levy of taxes on minority groups was a common source of state income, for example, to raise a grazing tax (*qalang*) on Gujur nomads. These groups of Gujri-speakers, who were probably formerly shepherds settled in the Doabs of Punjab, were pushed towards the northern mountain belt in the course of large-scale expansion of irrigation projects since the middle of the nineteenth century onwards (Fautz 1963; Edelberg and Jones 1979). Coming from Dir, Swat and Hazara Kohistan Gujur nomads migrated to Chitral at the end of the nineteenth and the beginning of the twentieth century (Israr-ud-Din 1969, 55). According to Robertson (1896, 298) and contemporary informants, it was a part of the *mehtar*'s policy to attract Gujur herdsmen in order to secure experienced workforce for herding the large flocks of the local elite.

Gujri-speakers have established a livelihood based on combined mountain agriculture in the southern valleys of the region. Colonial sources described sedentary Gujur families, commenting: '...and even in their cultivation, chiefly maize, their first thought is for their animals' (Lawrence 1908, 35). Herdsmen were especially employed by the powerful landlords of lower Chitral, in particular by the family of the ruling Katoor dynasty (Katoorey clan). In the 1920s, the British colonial administration, for example, documented a Gujur community living in Shishi Valley and paying a tax of 60 *maunds* (ca. 2,240 kg) of *ghee* (purified butter) annually (General Staff India 1928, 199). As one of the main suppliers, the Gujur also provided 400–500 *maunds* (15–19 t) of *ghee* per annum to the regional bazaars of Chitral (town) and Drosh (ibid., 50–51). With the passage of time, the Gujur shepherds have extended their economic activities towards the north and west of Chitral.

2.3 Case Studies of Pastoral Systems

In order to display the diversity and dynamics of pastoral systems beyond common features, the results from six empirical case studies are presented. The historically informed analysis includes aspects of the economic and political power constellations and development paths. Our six case studies are re-grouped into three distinct types to make them more comparable. For this comparative approach, we have developed a framework that encompasses such aspects as territorial and social features of access to grazing grounds, herd size and animal composition, horizontal and vertical migration patterns, herding arrangements and regional differences in pasture and fodder resources (Table 2.2).

Table 2.2 Principal components and specificities of pastoral land use systems in Chitral

Principal component	Range (bipolarity)
Access rights and spatial territoriality	Agreed – contested Individual access – village commons
Herd size and animal composition	Large – small Homogeneous – heterogeneous
Horizontal and vertical mobility pattern	Short distance – long distance
Stages of vertical utilisation	Settlement pattern with cultivation – without cultivation
Herding arrangements	Internal – external
Fodder and pasture resources	Rich/sufficient – degraded

2.3.1 Laspur and Khot: Communally Controlled Systems of Combined Mountain Agriculture

The valleys of Laspur and Khot, with the village of Yakhdiz in particular, may serve as examples of communally controlled pastoral systems in the context of mountain agriculture. Laspur Valley (northeastern Chitral) is connected via the Shandur Pass (3,720 m) with Gilgit-Baltistan, an important transport route to Gilgit, especially since colonial times (cf. Lockhart and Woodthorpe 1889, 292–294; General Staff India 1928, 289–291). Historically, *adamzada* had no principal influence in Laspur (NDC 826), resulting in a comparatively egalitarian social structure.

The peripheral village of Yakhdiz is located on a northernly exposed slope of Khot Valley (Torkho) and comprises a population of about 500 people of 11 clans in 57 households. The average holdings of livestock are three head of cattle, nine sheep and six goats per household (for 1999, based on full survey $n=57$). In addition, four households keep a total of 15 yaks. The possession of animals reflects a relatively balanced socio-economic segmentation of this community resulting from the relatively low influence of ancient local rulers. Yakhdiz predominantly consists of former lower-class clans.

Both example communities have de facto communal pasturage rights to use the nearby valley slopes (*mal/pai rochini*) and the high pastures (*ghari*) of adjacent tributaries. These pastoral mobility patterns cover various altitudinal belts in a seasonally differentiated cycle. As a common adaptive practice and risk-mitigation strategy, numerous families have arable land and houses in vertically dispersed locations and move their entire households between winter and summer settlements. In Laspur, for example, only 10 of the 40 households reside in village Phargam (3,120 m) for the entire year, whereas the majority of households spend the winter months in the lower-lying permanent settlements of Rahman or Harchin. The migration distances between permanent and seasonal settlements are comparatively short in both cases examined, enabling a daily movement and, therefore, an effective coordination of labour necessities at all localities. The altitudinal distances between the homesteads of Sor Laspur, the summer field settlement Shapir Mali and the summer pasture settlement Laspur Ghari as well as between Yakhdiz and the seasonal settlements of the affiliated summer pasture Wozg reach approximately 600 m.

In the annual cycle of pastoralism, the villagers' flocks of cattle, sheep and goats are kept in stalls attached to the permanent farmsteads in the time period from January to March. Yaks are allowed to graze freely depending on weather conditions. In spring, the animals begin their grazing cycle on pastures in close proximity to the settlements. In the village of Yakhdiz, the yaks are sent to a commonly shared summer pasture at the end of Khot valley in late March. By the end of April, cattle, sheep and goats begin their movement to separate grazing grounds. Cattle, with the exception of young and lactating animals, are brought up twice a day to the irrigated parcelled grasslands, indigenously termed *adrakh*, which stretches from about 2,870 m upwards.

In contrast to this manner of family-based animal care, the labour-intensive herding of sheep and goats is managed communally in Yakhdiz. This system is based on usufructuary communal grazing rights which are valid for all lands above the highest irrigation channel. Following the receding snow line, the village's entire population of sheep and goats are driven to the surrounding *Artemisia brevifolia* steppes for daily grazing. The tending is organised by individual households on a rotational basis. This system, locally called *sotsiri*, is practised until the beginning of the winter snowfalls, with the exception of two summer months when the flock grazes on the alpine pasture. Similar *sotsiri* systems also predicated on communally arranged turns exist throughout the whole Laspur Valley, all based on either kinship or neighbourhood. Such rota systems are an effective strategy for reducing pressure on household labour resources. These regulations of shared responsibility are considered as crucial components of a sustainable management system (Masoodul Mulk 1991, 45).

During the summer months, livestock is sent to the highest available grazing grounds. The Laspur people drive their flocks to the pastures of Phargam Valley where young livestock is concentrated. In the upper section of this side valley, the remains of a former pasture settlement, abandoned due to the lack of fuel wood, can be found, indicating a more intensive pastoral utilisation of this valley in the past. Nonetheless, fire wood for the summer settlements is still collected. The majority of animals in Laspur Valley are driven to the wide and flat pastures of Shandur Pass, where permanent villages hold pasture settlements individually and daily herding is organised by local shepherds. The pasturage rights exceed the mountain pass further to the east. The summer huts set up by people from the Laspur Valley in the Shandur Region negate the claim of the people of Gilgit-Baltistan. According to local sources, the number of huts there has risen from 344 in 1959 to 479 in 2007 (Inayatullah Faizi, www.chitralnews.com/LN923.htm). About 5,000 and 6,000 sheep and goats as well as 1,500 bovines are sent for grazing at Shandur Pass between July and September. In the remaining seasons, only the free pasturing of yaks, crossbreeds, cattle and donkeys is practised. The productive pasture vegetation at Shandur Pass consists of herb-rich steppe meadows and dwarf-scrub steppes indicating intense grazing pressure.

In Yakhdiz, grazing on high grounds is carried out during July and August. Formerly, each clan could use distinct pasture areas with fixed boundaries in three different locations. Today, only the pasture of Wozg is utilised on a community basis.

The grazing area spreads 600 m along the mountain flanks between the two pasture settlements Dok-a-Shal and Touq-a-Shal (3,470 m) and the ridge (4,070 m) towards the neighbouring Mehlp Valley. These two boundaries as well as those towards west and east mark a traditionally fixed, uncontested territory.

In an annual rotational cycle, it is the responsibility of 1 of the 11 hamlets' residents to send or recruit shepherds for the whole village community's flock of sheep and goats. According to local rules, these villagers are subject to a fine of 7,000–8,000 Rs if they fail to do so. Cattle are sent uphill according to individual arrangements. In recent times, the duty of the *ghari* management has been delegated to two families. Following traditional regulations, two groups of users, utilising Dok-a-Shal and Touq-a-Shal separately, are formed, and all livestock is distributed on a hamlet-to-hamlet basis. Livestock holders and the number of animals are recorded in detail. In summer 1999, in Dok-a-Shal, 359 sheep and 203 goats of 33 households of Yakhdiz and, additionally, of six households in neighbouring villages were kept. In Touq-a-Shal, only 170 sheep and 172 goats of 22 households of Yakhdiz were tended. According to the herdsmen, the total number of livestock grazing in Wozg declined in the 1990s.

Historically, the lower segments of the society just as the Gujur in many places of Chitral were forced to run the herding of the noble clans' flocks on the higher grounds. Nevertheless, these 2 months of work prove to be quite lucrative for the 'Dok-a-Shal family': after returning from Wozg, it receives 4 kg of wheat flour per pair sheep or goats (*balach* – traditional livestock unit; see also Fazlur-Rahman 2007, 150) tended, 20 kg of wheat flour per head of cattle tended, profits from the sale of dung and the milk produced in the second month – all in return for its exertions. The shared production of the first month is handed over to the livestock holders.

In summary, the cases of Laspur and Khot Valleys demonstrate that pastoral systems of peripheral village communities are smoothly managed by means of informal institutions. Common property systems including equal access and utilisation rights as well as share-tending arrangements contribute to the efficiency and stability of the villages' arrangements. Moreover, it can be seen that these systems have been subject to various transformation processes within the last decades, for example, the abandonment of pasture settlements or the change from a clan-managed to a community-managed system of pastoralism. Though the size of herds and the role of animal husbandry in the single household economy are on the decline, as, for example, in Yakhdiz, the samples show that a majority of families still derive large-scale benefit from domestic and common livestock production.

2.3.2 *Tirich and Mehlp: Co-occurrence of Sedentary Mountain Farmers and Gujur Herders*

Tirich Valley drains the central part of the main Hindukush Range between Tirich Mir (7,690 m) and Noshaq (7,455 m) towards the east. The valley is connected with the main valley via the Zani Pass (3,840 m). The longitudinal cross section of the large tributary valley comprises semi-arid to arid steppes of the submontane and montane belts. The high pastures around Zani Pass consist of thorn cushions on the slopes and scattered individual plants on scree.

Extended irrigation systems sustain the relatively high importance of crop cultivation in this valley (Nüsser 1999). Whereas wheat is the main crop up to the highest permanently inhabited village Shagrom (2,900 m), barley dominates the fields in the summer settlement Bandok (3,140 m). Depending on the snow melt, field preparation in Bandok is carried out during May and the harvest of barley ends in September. Around the middle of August, irrigated meadows are cut, and hay transports to Shagrom are carried out. As only 8 km separate Shagrom and Bandok, all necessary irrigation work can be coordinated and organised in both cultivated areas. This example shows a remarkable continuity, a continuity reported in the same manner by Schomberg (1936, 302, 304) and Haserodt (1989) and illustrated by a bitemporal comparison (Photos 2.1 and 2.2).

The pastoral arrangements in Tirich Valley are characterised by coexistent utilisation patterns of sedentary Kho mountain farmers and individual Gujur groups practising a more pronounced form of pastoral mobility. Whereas the Kho farmers enjoy territorial privileges of access and resource utilisation based on traditional village rights, single Gujur households are allowed to occupy small and limited ecological niches in the vicinity of the Zani Pass left partly vacant for herders without customary grazing rights. In order to obtain such seasonal access rights, the Gujur commit to herding the indigenous Kho farmers' sheep and goats on the high pastures. As a means of compensation, they enjoy specific seasonal grazing rights for their own flocks on selected high pastures.

This mutual arrangement between the Kho population and the Gujur constitutes the main reason for the specific differentiation of pastoral mobility patterns in the valley. As the summer grazing of small animals has largely been delegated to individual Gujur groups, the Kho farmers are able to concentrate their economic activities on irrigated crop cultivation and stall-feeding of lactating cows. This leads to a spatial differentiation within the valley, as the longitudinal transect between Shagrom and Bandok shows a distinct concentration of stall-feeding, eventually sustained by additional grazing on the adjoining slopes. The mountain farmers of Shagrom drive their lactating cows to the summer settlement Bandok between early August and end of September. During the early summer months, utilisation of the upper valley section is limited to daily grazing from the village Shagrom.

The non-lactating cattle are sent with relatives to the upper valley sections, where the animals are left free to graze. The labour-intensive sheep and goats, which are not herded by Gujur, are driven for daily grazing from Bandok, based on the aforementioned rotational sharing between participating households (*sotsiri*).

In contrast, one can find small, seasonally inhabited grazing settlements such as Margali Shal (3,440 m) and Kathgaz Shal (3,600 m) in the vicinity of the Zani Pass. These settlements are used by individual Gujur households between May and the end of September. In both grazing settlements, some Gujur stay with 40–60 of their own sheep and goats and with pension animals from about 50 households of Tirich Valley. Together they herd between 1,200 and 1,300 small stock with a preponderance

Photo 2.1 The summer settlement of Bandok in the Tirich Valley in 1966 (Photograph © Klaus Haserodt, July 12, 1966)

of goats, preferred for the production of *ghee* and cheese. The seasonal mobility pattern of these Gujur groups is characterised by long-distance migrations between their winter locations in the colline belt of southern Chitral and their summer pastures near Zani Pass. As their winter pastures are located in the vicinity of the small bazaar town Drosh, they migrate a distance of approximately 120 km along the road.

A similar example of coexistent pastoral utilisation patterns of sedentary mountain farmers and Gujur herders can be found in the Mehlp Valley, located between the Khot-Shagram spur and the Torkho-Mastuj divide. The habitable place in the valley stretches between 2,800 and 3,700 m, and the cultivated fields of the three villages are part of the single-cropping zone of northern Chitral. The selected case village Odier consists of 120 households belonging to six clan groups who use the three summer settlements Romolasht (53 households), Nashtani (25 households) and Lashtodok (4 households) located between 3,200 and 3,300 m. The total population was 1,100 in the year 2001. Almost every household in Odier owns milk cows and a few sheep. At present, only three households have yaks. According to the collected data, there are 6.8 sheep, 3.7 goats, 1.9 milk cows and 3.6 cattle per household. It is important to keep in mind, however, that there is a wide disparity amongst the households and clan groups. In reality, more than 60% of the total households do not own goats, and 47% do not own a single bull. An important precondition for keeping yaks is to own cultivated land in one of the summer settlements (Romolasht), in order to provide winter fodder. The village's yak herd was traditionally tended

Photo 2.2 The summer settlement of Bandok in the Tirich Valley in 1997 (Photograph © Marcus Nüsser, July 5, 1997)

there, and in the early 1980s, there were 26 yak-owning households in Odier, with a total of 65 head of yaks.

The villagers' right of livestock grazing and fodder collection is complex. They have de facto right of grazing in 13 high pastures and fodder collection from only five pastures in the village environs. Generally, the usage patterns vary accordingly and lead to two different forms of organisation during the summer season. The villagers either send most of the sheep and goats to the high pastures for about 3–4 months, or else they keep the sheep and goats with them in the summer settlements with cultivation, and take them to pasture on a daily basis. In Odier, the latter pattern is practised, and more than 80% of the households change their residences from the winter to the summer houses in altitudes ranging from 3,000 to 3,300 m. Besides free grazing of the non-lactating animals and yaks in the high pastures, turn-based daily grazing (*sotsiri*) is practised for sheep and goats from mid-June to the end of September. For this purpose, households residing in nearby neighbourhoods of the village are grouped together to make a single grazing group, locally known as *roam*. Throughout the entire outdoor grazing season, women keep the lactating cows at home and feed them on irrigated grass patches in the vicinity of the dwellings. Thus, during the summer season, livestock belonging to Odier village is distributed over a wide range of elevations between 2,900 and 4,000 m.

Up until the recent past, the rotational grazing of sheep and goats was conducted through reciprocal exchange of household labour (*hoyou*) and conventional helpers on request (*yardoyee*). All household members participated in pastoral management and regularly contributed their share of work. Since the late 1980s, the increasing numbers of households participating in the seasonal migration and children attending school have also stressed the availability of labour for keeping yak and goat grazing in Odier. Consequently, the number of yak-owning household decreased drastically from 26 to 2, and the number of households without goats increased from 7 to 68. One goat-grazing group (*roam*) in the upper part of the village completely ceased. Since 1999, the villagers have changed the rotational grazing arrangements for sheep and goats. They have employed two Gujur for sheep and goat herding. On the occasion of collective seasonal movement, the households of the two groups take their own Gujur with them to the upper part of the village and summer settlement. Three small huts have been constructed for the Gujur as residences, one each in the lower and upper parts of the village and the summer settlement. The households benefiting from the Gujur's services are paying 40 Rs per household per month, together with four *bati* (about 10 kg) of wheat flour during the summer season. For the winter season, two bundles of firewood and four bundles of grass or lucerne are added to the usual honorarium. Like in the Tirich Valley, the Gujur are allowed to graze their herd on the village's communal pastures and collect firewood from the reserved pastures of the village. Both case studies illustrate the various modes of change and the principle of co-occurrence of mountain farmers and Gujur herders in combined systems.

2.3.3 Kesu, Shishi and Golen: Connected and Contested Pastoral Systems of the Gujur

The last case studies presented here focus on the current pastoral system of Gujur clans (*goth*) in southern Chitral. Their traditional activity space is Shishi Valley, which stretches about 50 km towards the southwest of the main valley. In lower Shishi, oak forests (*Quercus baloot*) reach from about 1,600 m up to 2,500 m, providing winter fodder resources especially for goats (Scheibe 1937, 127; Staley 1982, 180). The higher valley sections accommodate humid coniferous forests. In Shishi Valley, several pastures are the de facto property of influential landlords of the formerly royal Katoorey clan living in Kesu Village located near the valley mouth, a few kilometres north of Drosh. For this reason, we present a combined analysis of both the highly differentiated pastoral practices in the village community and the closely interconnected long-distance system maintained by Gujur herders.

Additionally, Kesu Village, a multilingual community composed of Khowar-, Pashto- and Gujri-speakers as well as Afghan refugees (speakers of Pashto and Persian), may serve as an example of the many local communities in southern Chitral that have experienced a definite decline of the livestock economy in recent times. About 30% of the households surveyed in Kesu (random sample $n=100$) have reduced or eliminated their flocks in the last one to two decades. Currently, the possession of sheep and goats is reduced to a few families and specialised livestock holders.

Only 22 of 100 households keep sheep or goats, with a distinct preference for goats. Amongst these, four households own comparatively large flocks of 120, 40, 30 and 20 animals, respectively, six households own between 2 and 9 animals, whereas 12 households keep one goat for milk production only. The majority of the sample households keep one or two dairy cows.

In the annual migration cycle, all livestock is wintered in stalls at the permanent settlement. Dairy goats and cows are grazed near the farmsteads the whole year and are driven to the stubble fields immediately after the harvest. Larger flocks of goats and non-lactating cattle are sent to the high pastures. The pastoral practice shows diverse and dynamic systems and patterns at the local level that are highly dependent on social parameters. Territorial dimensions of the grazing areas, access and utilisation rights have changed over time. In the past, livestock owners from Kesu had access rights to Angarba, Doumouk and Poshiden pastures in the Shishi Valley. After disputes between several user groups, the grazing territories came under the control of other communities.

At the present time, different local user groups can be identified. A household's pasturage right or system of exclusion reflects the traditional structure of local power. Although all non-arable lands are the de jure property of the Government of Pakistan, the powerful landowners continue to exercise authority over the pasture areas beyond the irrigated inner fields. The various forest and other grazing areas associated with the village are the de facto property of a few members of the Katoorey clan. The utilisation arrangements of the pastoral land owned by the two most influential landlords of Kesu will now be elaborated, demonstrating the peculiarity and complexity of these systems.

In one case, a landlord has divided his forest into two parts. The western territory can be utilised only by the Kho inhabitants of one specific area of the village (Kesudeh). The area is further divided into single parts for each clan. The eastern portion is only accessible to six Gujur families who tend both the landlord's flocks as well as their own herds. They are also the exclusive tenants of the landlord's *ghari* (Kesu Gol) where they have established permanent farmsteads. A spring provides irrigation water for the cultivation of maize; the crop residues have a favourable winter fodder value. Half of the cereal production as well as dairy products and meat are delivered to the landlord as rent.

A second powerful Katoorey household has de facto ownership over several forest and pasture areas at different places. The largest territories are located near Madaglasht in Shishi Valley. There, about 15 Gujur households utilise certain forest parts and the Andowir *ghari* at the valley end, an already overgrazed and flood-prone pasture, between July and October. The landlord's flock and their own herds are grazed there. The Gujur shepherds pay a fluctuating amount of money in cash and provide livestock products annually to the landlord. Moreover, members of Afghan refugee families, who settled in Kesu after 1979, are often hired as herdsmen.

Beyond these arrangements, different livestock owners from Kesu send their animals to further high pastures – usually between early June and late October. Access and utilisation rights to Phoushkari *ghari* are traditionally restricted to Kho clans also of the village part Kesudeh. The few owners of livestock recruit herdsmen locally or hire Gujur shepherds. With the exception of the Katoorey clan, livestock proprietors of the remaining parts of the village do not possess any local pasturage rights at all. Therefore, those excluded groups cannot follow a pastoral migration routine but are compelled instead to negotiate annual and spatially flexible arrangements with external pasture user groups from neighbouring villages, or with Gujur tenders.

The complex pastoral system of Kesu reflects the contemporary social heterogeneity of the village and the unequal balance of power between the different groups. A historically privileged minority is able to keep the usufruct landholdings and control over the grazing resources. Furthermore, the example illustrates the declining relevance of the livestock sector in the double-cropping area of Chitral – a development that is strongly related to a considerable over-exploitation and subsequent degradation of pastures in that part of the region.

Moreover, this outlined practice in Kesu Village demonstrates what sort of role that Gujur shepherds play in connection with a definite village of southern Chitral. Subsequently, the spatial and seasonal pattern of Shishi's Gujur community will be analysed to emphasise the persistence of nomadic and transhumant forms of animal husbandry.

For over a century, the Gujur community has been the principal actor in animal husbandry in the Shishi Valley (General Staff India 1928, 199, 297; Israr-ud-Din 1969, 55; Haserodt 1989, 129). It is a common practice for the mobile herders to utilise partly vacant pastoral niches which the sedentary mountain farmers have not made full use of. This is arranged through the payment of a tax (*qalang*) to local individuals or communities. However, in the course of the past decades, a number of Gujur families have established permanent settlements with cultivation in the valley and the surrounding areas. These settlements specialise in maize production, as was demonstrated in the case of Kesu Village. Nonetheless, the long distances required for pastoral mobility between the winter and the summer pastures, as well as the comparatively large flocks dominated by goats, still remain a distinctive feature. A survey amongst Shishi's Gujur community (Khan 2000) shows an average holding of 4.2 cattle, 68.7 goats and 2.1 sheep per household (authors' own calculations).

During the winter months, the majority of these mobile shepherds remain in the surroundings of Drosh (1,465 m), where the terraces and the lower side slopes of the main Chitral Valley are utilised as winter pastures. In the multistage migratory process from Drosh towards the upper Shishi Valley, the herders pass its largest village Madaglasht (2,450–2,670 m) in which Tajiks from Badakhshan settled more than 200 years ago. There, the Persian-speaking inhabitants maintain a completely contrasting pastoral system, keeping small herds of predominantly milk cows concentrating on stall-feeding and the utilisation of nearby summer pastures.

As indicated above, the upper portion of the Shishi Valley is allotted on rent to the Gujur by the Katoorey landlords. During the summer months, the herdsmen live in various pasture settlements. The Gujur of the permanent settlement of Kutik stay in Ruagol Ghari (3,100 m) located in the highly degraded coniferous forest, as well as in Ghochhar Sar (3,500 m) above the tree line, at the top of the valley. As the alpine turfs that are found there are insufficient for the flocks, the Gujur of the permanent villages of Chin Nissar and Kawash cross the high Dok (4,219 m) and Lohigal (4,487 m) Passes into the tributaries of the upper Golen Valley to utilise the additional pastures there. In summary, the seasonal migration distances covered by the Gujur herders reach about 70 km with a difference in height of more than 3,000 m in certain cases.

Between the middle of June and the beginning of September, ten households of Chin Nissar are tending in total about 1,200 goats, 200 sheep and a few cattle, using the pasture settlement Krui Uts (3,580 m) in the upper Lohigal area. Extended and level turfs are located there at altitudes of 3,900 and 4,000 m. However, these grazing grounds are contested because of the high pass crossing practice of the Gujur. The Kho pastoralists of Golen, who exclusively utilise the upper portion of the main Golen Valley, claim ownership of these pastures, arguing that the ridge and the high passes form a natural barrier. Upon these grounds, they plead for a ban on grazing by the Gujur's flocks. Yet, Haserodt (1989, 130) observed a reduction of the Gujur's practice of trans-pass pasturing between different valleys during the 1980s. Contrary to this trend, the lack of sufficient pasture areas put the herders under pressure to continue to maintain this kind of mobile grazing, making future disputes about valuable pasture resources highly probable.

2.4 Discussion and Conclusion

The diverse pastoral systems in Chitral have undergone multiple changes and modifications over the past decades. Although contrary examples can be found, it can be stated as a general trend that the economic role of animal husbandry at the household level has lost importance due to internal as well as extra-regional socio-economic factors. Population growth, land fragmentation and the division of herds as a result of the traditional law of inheritance are important driving forces. Political, institutional and legal changes have occurred, especially after the full administrative integration of Chitral into the nation-state of Pakistan in 1969, at which point new laws on state land affected customary access regulations (Faizi 1999). Law and legal pluralism and uncertainty have resulted in contested spatial territoriality and free grazing leading to pasture degradation which, in turn, has reduced the forage base. This is especially prevalent in the submontane and montane forest belts of southern Chitral, as it has been shown in the case of Kesu/Shishi.

Moreover, both the increased accessibility resulting from improved road networks constructed since colonial times (cf. Kreutzmann 1998, 2004b) and further infrastructural and agrarian improvements have contributed to the transformation of socio-economic conditions, attitudes and behaviours of the inhabitants, land use patterns and livelihood strategies. In particular, the role of labour is constantly being reassessed. Keeping livestock is regarded as demanding physical work which is no longer very profitable as compared with new off-farm income opportunities in the towns, lowland cities or the Gulf countries. In addition, the rural workforce is scarce due to the fact that a growing number of children, who have traditionally played a major role in animal tending, have increasingly begun attending school since the early 1990s (Holdschlag 2011).

Another contributing factor is the decline in the demand of sheep wool, goat hair and skins as more textiles, shoes and fabric goods at relatively cheap rates are being imported, a shift which results from the intensified supra-regional exchange patterns. The introduction of chemical fertilisers, tractors and threshing machines has also undermined animal husbandry. The number of cattle in many households has decreased because of these mechanisation processes. Nevertheless, oxen are still needed for ploughing the fields in disadvantaged locations, as seasonal conditions and topography do not favour tractor use there.

The past decades have also witnessed significant changes in the Gujur's mobility pattern, prevalent until the late 1970s. Since it became possible to claim usufructuary land rights after 1969, many families in individual areas of southern Chitral have become more sedentary with their own cultivation, particularly in regard to their maize crop, in the process adapting to a combined mountain agriculture. Despite their more sedentary way of life, however, the seasonal migration patterns of Gujur herdsmen are still characterised by higher mobility, longer distances and the crossing of high passes. In many cases, Gujur have become shepherds of the 'autochthonous' Kho population. Apart from cultivation and pastoral activities, Gujri-speakers often take work as rangers and in the timber and transportation businesses of southern Chitral.

We may arrive to the conclusion that on the local level, that is, in the community context, the transformation of pastoral systems is the rule whereas persistence is the exception. However, in spite of the various dimensions and factors of transformation analysed here, animal husbandry still plays a vital role in the economy of a large number of households. Its changing structures and functions reflect the overall socio-economic conditions of the region as well as supra-regional developments. Pastoral mobility patterns and livelihood strategies of diverse groups and individuals shall be differentiated not only as for environmental factors but according to such factors as ethnic and socio-cultural affiliation and territorial rights of resource access and utilisation, reflecting the political setting. A better understanding of the diversity, the dynamics and the complexities of pastoral systems requires identifying and evaluating the importance of these key elements, which are embedded in contexts of various temporal and spatial scales.

Regarding the six cases examined, we have identified three present types of corresponding pastoral land use systems in Chitral:

- Communally controlled systems of combined mountain agriculture (case studies 2.3.1): arranged spatial territoriality and common access rights, heterogeneous flocks, short migratory distances and internal herding arrangements
- Co-occurrence of sedentary mountain farmers and Gujur herders (case studies 2.3.2): arranged combination of traditional and seasonally negotiated access

rights, heterogeneous flocks, short as well as long migratory distances, and internal as well as external herding arrangements
- Connected and contested pastoral systems of the Gujur (case studies 2.3.3): contested territoriality, individual access rights, large homogeneous flocks, long migratory distances and internal herding arrangements

Although the fodder situation and the variability of pasture degradation between different valleys and altitudinal belts is certainly high in both summer and winter grazing areas, these differences cannot be reduced solely to differences in pastoral systems and herding practices. Clearly, a better understanding of the pasture-ecological situation and related fodder availability strongly depends on the integration of environmental diversity and distribution features of vegetation types.

Nevertheless, the stability of the identified types of pastoral systems strongly depends on political and socio-economic factors. Besides disputes between different clans and village communities, the potential for conflict is relatively high in the zones where contact with the dominant Khowar-speaking population is increasing, keeping sheep and goats together with cattle in the form of combined mountain agriculture, with Gujur groups, mainly relying on the mobile maintenance of sizeable herds of goats. Gujur herders, who are often placed on the lower end of the social strata by the Kho, have long been accused of utilising the natural resource base in an unsustainable way (Edelberg and Jones 1979, 100; Masoodul Mulk 1991, 46). Notably in the oak forest areas of southern Chitral, there are differences of resource management practices between Kho farmers (for the Kalasha in Rumbur, Bumboret and Birir Valleys, cf. Parkes 1987, 1992), who are establishing protective measures against the over-exploitation of certain plant species (Haserodt 1989, 126), and the Gujur shepherds. As the number of Afghans who are employed as herdsmen in certain locations of the region has been increasing for the last couple of years, they create a new factor in pastoral management, further contributing to the complexity of the system, for example, in terms of behavioural norms.

Improving accessibility resulting in advanced mobility, increased availability of consumer goods, as well as agricultural inputs and services will significantly transform the existing livestock sector and the pastoral systems. However, the future impact of factors such as population growth and migration, market incorporation and changing attitudes and values cannot be predicted. Until the present day, the overall regional animal husbandry system has proven to bear remarkable persistence under changing political and socio-economic conditions and, therefore, can be regarded as a long-term flexible and adaptive livelihood strategy in Chitral.

Acknowledgements This research has been substantially funded by the German Research Council (DFG) within the framework of the Pakistan-German joint research programme 'Culture Area Karakorum' (CAK). We thank Adam Knowles for proofreading.

References

Aga Khan Rural Support Programme (2000) An assessment of socio-economic trends and impact in northern Pakistan (1991–1997). AKRSP, Gilgit

Barth F (1956) Indus and Swat Kohistan. An ethnographic survey. Studies honouring the Centennial of Universitetets Etnografiske Museum, Oslo 1857–1957, 2. Forenede Trykkerierm, Oslo

Beg GA (2011) Current status of pastoral systems in Gilgit-Baltistan and Chitral. In: Kreutzmann H, Abdulalishoev K, Zhaohui L, Richter J (eds) Pastoralism and rangeland management in the context of climate and global change. GIZ, Bonn, pp 131–146

Biddulph J (1880) Tribes of the Hindoo Koosh. The Superintendent of Government Printing, Calcutta

Cacopardo AM, Cacopardo AS (2001) Gates of Peristan. History, religion and society in the Hindu Kush. Reports and Memoirs, Series Minor, Centro Scavi e Ricerche Archeologiche 5. Istituto Italiano per l'Africa e l'Oriente, Rome

Clemens J, Nüsser M (2008) Animal husbandry and utilization of alpine pastures in the Nanga Parbat region of northern Pakistan. Comparison of Raikot and Rupal Valleys. In: Israr-ud-Din (ed) Proceedings of the third international Hindu Kush cultural conference. Oxford University Press, Karachi, pp 71–81

Edelberg L, Jones S (1979) Nuristan. Akademische Druck- und Verlagsanstalt, Graz

Eggert P (1990) Die frühere Sozialordnung Moolkhos und Turkhos (Chitral), Beiträge zur Südasienforschung 134. Steiner, Stuttgart

Ehlers E, Kreutzmann H (2000) High mountain ecology and economy. Potential and constraints. In: Ehlers E, Kreutzmann H (eds) High mountain pastoralism in northern Pakistan, Erdkundliches Wissen 132. Steiner, Stuttgart, pp 9–36

Faizi I (1999) Indigenous resource management in Chitral. IUCN-CCS Study. IUCN Pakistan, Chitral

Fautz B (1963) Sozialstruktur und Bodennutzung in der Kulturlandschaft des Swat (Nordwesthimalaya), Gießener Geographische Schriften 3. Schmitz, Gießen

Fazlur-Rahman (2007) Persistence and transformation in the eastern Hindu Kush. A study of resource management systems in Mehlp Valley, Chitral, North Pakistan, Bonner Geographische Abhandlungen 118. Asgard, Sankt Augustin

Forsyth T (1998) Mountain myths revisited. Integrating natural and social environmental science. Mt Res Dev 18:107–116

General Staff India (1928) Military report and gazetteer on Chitral. Government of India Press, Calcutta

Ghulam Murtaza M (ed) (1962) Nai Tarikh-i Chitral. [New History of Chitral] [Urdu]. Authors: Mohammad Ghufran M et al. Public Art Press, Peshawar (English translation, typescript)

Government of N.W.F.P. (GoNWFP) (1997) NWFP development statistics 1997. Bureau of Statistics, Planning, Environment and Development Department, Peshawar

Government of Pakistan (GoP) (1998) NWFP Livestock Census 1996, vol. II, part I. Agricultural Census Organization Statistics Division, Lahore

Government of Pakistan (GoP) (1999) 1998 district census report of Chitral, census publication 20. Population Census Organization, Islamabad

Government of Pakistan (GoP) (2006) NWFP livestock census 2006. Agricultural Census Organization Statistics Division, Lahore

Haserodt K (1989) Chitral (Pakistanischer Hindukusch). Strukturen, Wandel und Probleme eines Lebensraumes zwischen Gletschern und Wüste. In: Haserodt K (ed) Hochgebirgsräume Nordpakistans im Hindukusch, Karakorum und Westhimalaya, Beiträge und Materialien zur Regionalen Geographie 2. TU Berlin, Berlin, pp 43–180

Holdschlag A (2011) Siedlungsgemeinschaften in Chitral, pakistanischer Hindu Kush: Sozioökonomische Organisation und Transformation in montaner Umwelt, Bonner Geographische Abhandlungen 126. Ferger, Bergisch Gladbach

India Office Records [IOR] Chitral Affairs file No. 33-C of 1886 Vols. II & III. Revenue, IOR/2/1077/235/11826/214–213
Israr-ud-Din (1965) A social geography of Chitral State. M.Sc.-thesis, King's College, University of London (2 vols)
Israr-ud-Din (1969) The people of Chitral. A survey of their ethnic diversity. Pakistan Geogr Rev 24:45–57
Israr-ud-Din (1996) Irrigation and society in Chitral District. In: Bashir E, Israr-ud-Din (eds) Proceedings of the second international Hindukush cultural conference. Hindukush and Karakoram Studies 1. Oxford University Press, Karachi, pp 19–42
IUCN Pakistan (2004) Chitral. A study in statecraft (1320–1969). IUCN Pakistan, Karachi
Ives JD (2004) Himalayan perceptions. Environmental change and the well-being of mountain peoples. Routledge, London/New York
Khan AS (2000) Migratory livestock industry among nomadic herdsmen in Chitral. IUCN-CCS Study, Chitral
Khan YM, Khan M (n.d.) Integrated land resources survey report of Chitral sub-project (Chitral District). North-West Frontier Province Forestry Pre-investment Centre Peshawar, Series 1. Peshawar
Kreutzmann H (1998) The Chitral triangle. Rise and decline of trans-montane central Asian trade, 1895–1935. Asien-Afrika-Lateinamerika 26:289–327
Kreutzmann H (2004a) Pastoral practices and their transformation in the north-western Karakoram. Nomadic Peoples (N.S.) 8:54–88
Kreutzmann H (2004b) Accessibility for High Asia. Comparative perspectives on northern Pakistan's traffic infrastructure and linkages with neighbours in the Hindukush-Karakoram-Himalaya. J Mt Sci 1:193–210
Kreutzmann H (2005) Linguistic diversity in space and time. A survey in the eastern Hindukush and Karakoram. Himal Linguist 4:1–24
Lawrence W (ed) (1908) The imperial gazetteer of India. Kashmir and Jammu, provincial series 13. Clarendon Press, Oxford (Reprint: Lahore 1983)
Lockhart WSA, Woodthorpe RG (1889) Confidential report of the Gilgit Mission 1885–86. Eyre and Spottiswoode, London
Marsden M (2005) Living Islam. Muslim religious experience in Pakistan's north-west frontier. Cambridge University Press, Cambridge
Mulk M (1991) A microcosm of farmers' strategies in Chitral. Aga Khan Rural Support Programme Occasional Paper 1. Gilgit
National Documentation Centre (National Archives of Pakistan), Acc. No. 826 (NDC 826) "Proposed transfer of Mastuj district to the Mehtar of Chitral, 1909–12"
Nüsser M (1998) Nanga Parbat (NW-Himalaya). Naturräumliche Ressourcenausstattung und humanökologische Gefügemuster der Landnutzung, Bonner Geographische Abhandlungen 97. Dümmler, Bonn
Nüsser M (1999) Mobile Tierhaltung in Chitral: Hochweidenutzung und Existenzsicherung im pakistanischen Hindukusch. In: Janzen J (ed) Räumliche Mobilität und Existenzsicherung, Abhandlungen – Anthropogeographie 60. Reimer, Berlin, pp 105–131
Nüsser M (2001) Understanding cultural landscape transformation. A re-photographic survey in Chitral, eastern Hindukush, Pakistan. Landscape Urban Plan 57:241–255
Nüsser M (2003) Ressourcennutzung und Umweltdegradation. Mensch-Umwelt-Beziehungen in peripheren Gebirgsräumen. In: Meusburger P, Schwan T (eds) Humanökologie. Ansätze zur Überwindung der Natur-Kultur-Dichotomie, Erdkundliches Wissen 135. Steiner, Stuttgart, pp 327–342
Nüsser M, Dickoré W-B (2002) A tangle in the triangle. Vegetation map of the eastern Hindukush (Chitral, northern Pakistan). Erdkunde 56:37–59
Ostrom E, Burger J, Field CB, Norgaard RB, Policansky D (1999) Revisiting the commons: local lessons, global challenges. Science 284:278–282
Parkes P (1987) Livestock symbolism and pastoral ideology among the Kafirs of the Hindu Kush. Man (N.S.) 22:637–660

Parkes P (1992) Reciprocity and redistribution in Kalasha prestige feasts. Anthropozoologica 16:37–46

Parkes P (2001) Alternative social structures and foster relations in the Hindu Kush. Milk kinship allegiance in former mountain kingdoms of northern Pakistan. Comp Stud Soc Hist 43:4–36

Robertson GS (1896) The Kafirs of the Hindukush. Lawrence and Bullen, London

Scheibe A (1937) Die Landbauverhältnisse in Nuristan. In: Scheibe A (ed) Deutsche im Hindukusch. Bericht der deutschen Hindukusch-Expedition 1935 der Deutschen Forschungsgemeinschaft, Deutsche Forschung 1. Siegismund, Berlin

Schomberg RCF (1936) Derdi and Chapursan Valleys. Mountains of N.W. Chitral. Alpine J 48:295–310

Staley J (1982) Words for my brother. Travels between the Hindu Kush and the Himalayas. Oxford University Press, Karachi

Strand RF (1997–2011). Nuristan. Hidden land of the Hindu-Kush. http://users.sedona.net/~strand/. Accessed 30 May 2011

Chapter 3
Pastoralism, Power and Politics: Access to Pastures in Northern Afghanistan

Stefan Schütte

Abstract This chapter explores the practice of nomadic pastoralism in contemporary Afghanistan and looks at how the geographies of access to pastures are shaped by asymmetric power relations and high degrees of tenure insecurity. The discussion is based on empirical fieldwork amongst Pashtun pastoralists based in the Chahar District of Kunduz Province, who seasonally migrate to the high pastures of Badakhshan. Their social and spatial practices are taken as examples of the constraints and opportunities that constitute pastoralism in Afghanistan today. The pastoral groups studied have shown enormous capacity to continuously adapt their mobility strategies in response to changing power structures, increased societal pressure and fluctuating economic opportunities. However, after 32 years of continuous warfare, the traumatic experiences of conflict and war, of power struggles and changes of authority, of insecurity and threats to survival, pastoralists today still strife for their rights to mobility and secure pasture access. Faced with changing rule systems and legal pluralism governing both the winter and spring pastures in Kunduz and the high pastures of Badakhshan around the environs of Lake Shewa, pastoralism in Afghanistan continues to be a highly insecure endeavour. The current situation of pastoral tenure insecurity is traced by reconstructing pastoral practices and mobility strategies as they are executed today, by looking at the governance structures that shape both mobility and pasture access and by examining the historical geographies of nomadic pastoralism as experienced by the group under study.

Keywords Afghanistan • Human security • Mobility • Pasture Conflict • Resources • Territoriality

---

S. Schütte (✉)
Department of Earth Sciences, Centre for Development Studies (ZELF),
Freie Universität Berlin, Malteserstrasse 74-100, D-12249 Berlin, Germany
e-mail: stefan.schuette@fu-berlin.de

H. Kreutzmann (ed.), *Pastoral practices in High Asia*, Advances in Asian Human-Environmental Research, DOI 10.1007/978-94-007-3846-1_3,

3.1 Introduction: Framing the Issues

The nexus of pastoralism and mobility has long been a focus of geographic and anthropological enquiry. Pastoralism in its multiple forms and as a livelihood strategy continues to be dynamic and flexible, with changing pastoral strategies reflecting societal pressure, shifting power structures and economic opportunities (Schlee 2005). Nowhere seems this clearer than in Afghanistan, where protracted war conditions forced pastoralists to continuously adapt and change their mobility patterns, and where pastoral livelihood systems are increasingly characterised by heightened insecurity and vulnerability.

In this chapter, the current practice of pastoralism in Afghanistan is examined through a case study of a Pashtun community that seasonally resides in the volatile District of Chahar Dara in Kunduz Province of Northern Afghanistan (Fig. 3.1). In the warm season, this community migrates to the fertile Shewa pastures in

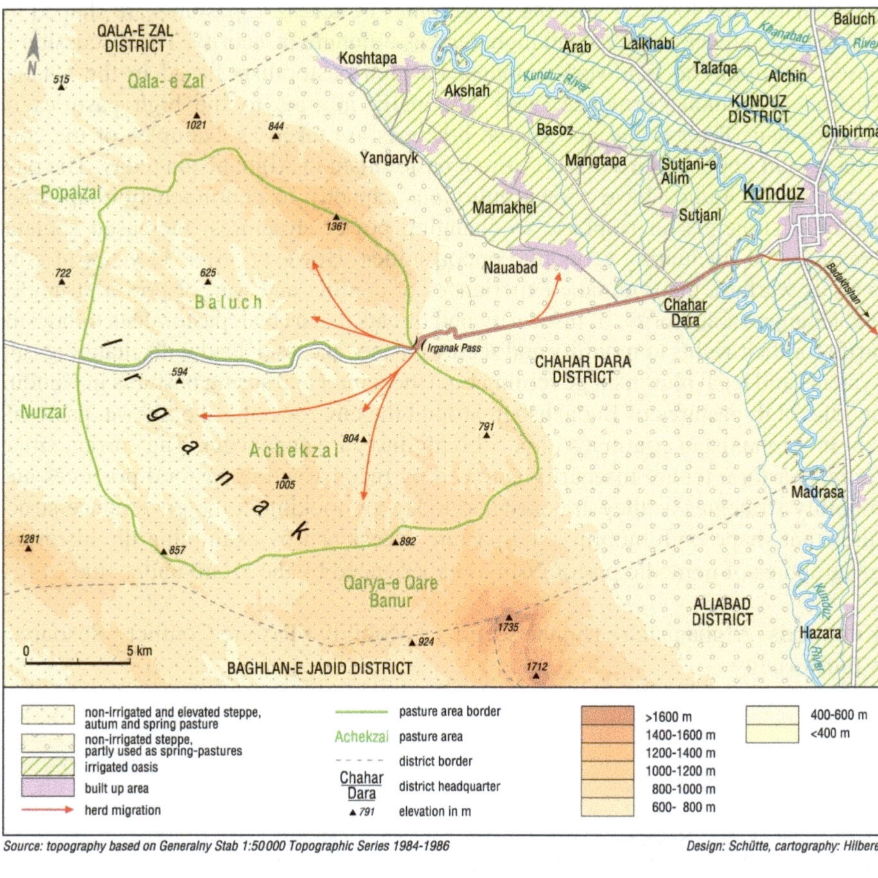

Source: topography based on Generalny Stab 1:50000 Topographic Series 1984-1986 Design: Schütte, cartography: Hilberer

Fig. 3.1 The village of Nauabad in the Kunduz oasis and the nearby autumn and spring pastures of Irganak

Fig. 3.2 Location of Achekzai summer pastures on the contested Shewa Plateau

Badakhshan, the largest and arguably most important mountain grazing area in all of Northern Afghanistan (Fig. 3.2). The case study is indicative for a variety of reasons. First, it shows how mobility and territoriality are organised and negotiated in a war-torn country with limited statehood and hybrid political orders, where power relationships and oppressive legal geographies surround and determine mobile practices. Second, it illustrates the 'power geometries of everyday life' (Massey 1993) that determine access to pastures and the mobility spaces of Afghan pastoralists. Third, the question of representation and the identity-alterity dichotomy characterising relations between nomadic pastoralists and sedentary groups highlight the ways identity is used to legitimate access to resources. Lastly, it provides examples of how people cope with disruptions to their spatial mobility and the partial breakdown of pastoral production and reproduction systems, shedding light on the resilience of Afghan pastoral communities.

This fourfold analytical framework is applied to the situation of Pashtun nomadic pastoralists, examining (1) the production of pastoralist mobility spaces and territoriality, (2) the power geometries of pasture access in spring and summer areas and the effects of legal pluralism, (3) issues of pastoralist identity formation and representation and (4) pastoralist resilience and coping strategies in the wake of protracted conflict. The analysis is based on data gathered during four field visits to the residential village of Pashtun herders between 2007 and 2009 and a further extended field visit to the Shewa Plateau in the summer of 2009. During these visits, open-ended interviews were conducted, both with migrating pastoralists from Kunduz Province and sedentary populations on the plateau.

3.2 Contexts: Pastoralism, Pastures and Herders in Afghanistan

In-depth studies on various Afghan pastoralist communities were numerous in the 1970s and provide important insights into the historical conditions of mobile livestock herding in the country (e.g. Jentsch 1973; Glatzer 1977; Barfield 1981).

As renowned specialists in sheep breeding, and as major suppliers of meat for the domestic economy in Afghanistan (Barfield 2008), pastoralists are required to traverse huge distances with their flocks to address seasonal shortages of pasture resources. This combination of animal husbandry and mobility produces challenging complexities that result in many hardships and insecurities: herd management has to be delegated, a mobility route has to be chosen, land tenure security has to be established, interaction with other social groups both en route and in the summer pastures has to be facilitated, and dependence on agricultural production for animal fodder purposes requires negotiation (Dyson-Hudson and Dyson-Hudson 1980). Specific groups use the same routes every year, often engaging in exchange relationships with settled populations that can be both characterised by cooperation and antagonism (Ferdinand 1969). However, pastoralists' experience and mobility spaces in Afghanistan have undergone momentous changes over the last two generations, and pastoralists today are faced with many critical challenges.

Pastures are an increasingly threatened resource in Afghanistan. Up to 70% of the country is used for grazing or for the harvesting of bushes for animal fodder or fuel, but access to certain pastures is, today, heavily contested and often the source of volatile conflict (Alden Wily 2004). The truth of this has been exemplified recently by the violent disputes about access to pastures between Pashtun herders and sedentary Hazara people in Wardak and Bamiyan Provinces, resulting in killings and displacements (IRIN 2007; Rassul 2010). In spite of an actually existing nomadic-sedentary continuum with movements in either direction triggered by changing economic and climatic conditions in Afghanistan (Glatzer 1981; Glatzer and Casimir 1983; de Weijer 2007), there exists an exclusionary identity politics that maintains the distinction between pastoralist herder and sedentary farmer. This, in turn, is often used to legitimate access to land resources.

At the same time, and adding to the problem, increasing pasture degradation facilitated by pasture conversion, overgrazing and overcutting poses ecological challenges.

Contested access to and conflicts over pastures along with ecological degradation has resulted in heightened land tenure insecurity (McEwen and Whitty 2006). Tenure insecurity refers to the degree to which land users lack confidence that neither the State nor other people will interfere with their access rights to the land for an extended period of time (Bruce 1998). For Afghan pastures, tenure insecurity has four broad dimensions: first, conflict over rights to pastures amongst groups of village residents and mobile groups; second, differences of opinion about the preservation of pastures between farmers with access to farmland and those without access to farmland, but with a dependence on livestock; third, contradictions between governmental agencies empowered by formal law establishing State ownership of pastureland and local communities which, by custom and necessity, use the pastures; and fourth, land-grabbing by powerful armed commanders who establish control over pastures. These tenure insecurities have distinctive forms and shape mobility spaces of pastoralists, as will be elaborated upon when examining the case of Pashtun herders in Northern Afghanistan.

Who are these mobile herders? This is not an easy question to answer. In Afghanistan, pastoralists are referred to by the generic Persian term 'Kuchi' (i.e. those who migrate), although this very broad label disguises differences between distinct social groups (Tapper 2008). There are, for instance, Persian- and Pashto-speaking pastoralists in Afghanistan, different ethnic groups, and there exist differences in practising pastoralism (i.e. long distance vs. short distance). Tapper answers the question by saying, 'It depends who is asking …, in what circumstances – time, place, audience – and for what perceived purposes. Bilingualism, and multiple and disputed identities, mean that boundaries between ethnic groups are neither precise nor territorial, but contextual and shifting' (Tapper 2008, 101). These shifts and contexts will be made apparent over the following, when looking at pastoral identity constructions and relations to sedentary populations in the frame of pastoralist mobility.

The case study focuses on pastoralists who partially reside in one specific village in the District of Chahar Dara, Kunduz Province (Fig. 3.1). Today, this district represents a stronghold of anti-government insurgency in Northern Afghanistan. It has become the venue of military clashes, insurgency attacks and NATO aerial bombings. Chahar Dara consists of various village clusters in the Kunduz River oasis, inhabited by people claiming different ethnicities, with a Pashtun majority. Many of these Pashtuns are engaged in pastoralism and irrigated agriculture, signifying an agro-pastoral continuum. The study settlement is inhabited by two distinct tribes stressing descent from the Durrani lineage: the Achekzai, originating from Spin Boldak in Kandahar Province and the Baluch, originating from Helmand Province (Balland 1996; Glatzer 1977, 114–118; Grötzbach 1972, 94).

Senior informants reported that outmigration from Southern Afghanistan was stimulated by severe drought conditions at the time, although it seemed to have happened at the same time as the Afghan ruler Amir Abdur Rahman Khan (reigning 1880–1901) implemented his resettlement policies. During his rule, Pashtuns were given land in Northern Afghanistan in preference to other groups with the aim of changing the ethnic composition of the North in an attempt at state-regulated 'Pashtunisation' (Barfield 1978).

In the year 1933, both tribes jointly built an irrigation canal and established village residency. Those who participated in the construction were allocated 24 *jerib* of land (about 5 ha), which was later expanded upon by individual households through land purchases from Uzbek neighbours.

The immigrant settlers profited from what has been termed the 'most successful development project in recent Afghan history' (Barfield 1978, 29) when the malaria-infested swamp lands of Kunduz were drained and transformed into fertile and productive agricultural lands. The base of agro-pastoralism was significantly enhanced by a higher productivity in crop and fodder farming whilst fertile pastures were accessible at the same time. In winter and spring, pastures in proximity to the residential village were used (cf. Fig. 3.1). These areas are still used today for livestock keeping and animal breeding, with clear and agreed upon demarcations between the pasture parcels used by Achekzai and Baluch. In summer, large distances of about 300 km had to be covered to migrate to the mountain pastures of Badakhshan, crossing the provinces of Kunduz and Takhar. This link between two highly fertile regions made the keeping of larger numbers of sheep viable.

With their newly established bases around Kunduz and access to both spring and summer pastures, both the Baluch and Achekzai were able to further invest in the husbandry of large fat-tailed sheep and matured to a thriving community. This outcome, however, became only possible through supportive policies from the central authority in Kabul, where the Pashtun King Zahir Shah (reigning 1933–1973) issued land titles (qawala) that guaranteed grazing rights for summer pastures to both the Baluch (in the mountain pastures above Kishim) and the Achekzai (in Shewa) in 1951. The documents acted as passes and safeguarded access to this critical resource from that time onwards and today still determine the destinations of pastoral mobilities.

Today, about every second of the 1,201 households in the study village still engage in far-reaching mobile livestock herding, and a majority of these households do so exclusively, with no access to agricultural lands. This is in spite of the changes stemming from the different phases of Afghan warfare that have significantly altered the mobility strategies of the study communities since 1978 and that continue to threaten the security of their pastoral livelihood systems today.

3.3 Mobility and Territoriality in the Pastoralist Experience and the Power Geometries of Pasture Access

Territoriality can be understood '... as a social (and political, economic, cultural) process that unfolds not only in place but through time. It thereby allows us to more easily see territories as social products' (Delaney 2005, 2). The social process of territoriality, however, is diverse, contested and fought over and therefore subject to unequal and changing power relations that determine who controls access to pasture resources and in what ways this control is exercised. In this sense, the power geometries of pasture access refer to the uneven positions of different actors in the agro-pastoral system of Northern Afghanistan in which the changing positionality of pastoralist groups shapes and constrains their mobility and access strategies. This fluid entanglement of mobility, territoriality and power and how it changes over time will be illustrated through the example of Pashtun herdsmen.

In a general sense, territoriality in the experience of the Baluch and Achekzai combines the need to establish secure access to spring and summer pastures and to traverse the spaces in between. In doing so, people classify certain pastures as 'ours'. The basis of this classification of pasture territory is diverse and shifting – contested and conflicting with claims of different groups to the same pastures, oppressive through the exercise of violent control over pastures through strongmen, or resting on shared community agreements establishing clearly demarcated pasture parcels. Whatever the case for the different pastures involved, territoriality in the pastoralist experience is always based on spatial mobility to connect the residence with spring and summer areas. The long March between these moorings on a 'road of insecurity' (Kreutzmann and Schütte 2011, 113) and their specific makeup is another illustration for the ways in which 'places and landscapes are continually practised and performed through the movement and enfolding of … people and things' (Cresswell and Merriman 2010, 7).

3.3.1 *Winter and Spring Pastures*

Winter and spring pastures used by the Baluch and Achekzai are located in an area called Irganak in proximity of the residential village, where tent camps are erected and selected members of a pastoralist household tend to the animals (cf. Fig. 3.1). The area was the scene of heavy fighting during the resistance against Soviet occupation and heavily mined and to date has not been entirely cleared. It is also reputed to be an area of retreat for insurgent groups today.

Access to Irganak pastures is based on shared community agreements and a clear definition of space and territory based on tribal affiliation. Certain pockets of pastureland are subdivided through a customary concept referred to in Pashto as mena. Literally meaning tent or locale, the term refers to a clearly defined geographic area of pastureland for which the exclusive user rights rest with a specific clan of herdsmen. The shape of mena and the location of their boundaries are orally transferred from generation to generation, with the size of each area originally determined by the size of an individual household or clan's herd. However, in the view of Pashtun pastoralists, a mena represents not only a specified area but also a system of rights. People do not claim ownership of the land in question, although the long duration of usage, stretching over many generations, does resemble something like it.

The shapes and boundaries of these areas have been collectively agreed upon by all pasture user groups residing in the extensive Kunduz oasis. A big meeting of elders from all pastoral groups was summoned in order to reach unanimous agreement on the location of boundaries demarcating the respective areas used by each tribal group. In doing so, the user communities established clear pasture territories in Irganak. It is, in fact, the tribal group (*khel*) that claims the right of use to a specific parcel to be subdivided into individual mena, illustrating the modes of social organisation prevalent amongst pastoral groups in Afghanistan.

The example shows how customary agreements have created distinct territories and shared rights to pasture areas that are clearly bounded. The practice adheres to the Islamic principle of shared consensus, but the distribution of user rights does not follow the provision of the written Pasture Law (GoA 2000a; Official Gazetteer 795) and is thus not acknowledged by state legislation. In fact, the representation of pasture territories in the official Pasture and Land Management Laws (GoA 2000a, b; Official Gazetteer 795) refers to grazing lands as national property under state control. Accordingly, the perception of government staff that 'all pastures belong to the state' may give rise to conflict. In terms of tenure security, however, the spring pastures and their mena resemble comparatively safe and mostly undisputed grazing resources. In times of abundance, after sufficient rainfall, people trade access rights to their pastures to herdsmen passing through the area. However, the pasture resources can be used for only about 5 months in a year (November–December and March–May). Over cold and snowy winters, the animals are kept near the house, and in the hot and dry summers of Kunduz, Irganak turns into a desert, forcing pastoralists to move with their animals to distant mountain pastures.

3.3.2 Summer Pastures

The summer pastures in Shewa (Fig. 3.2) are located between 2,600 m and 3,200 m a.s.l. and are used by the Achekzai, who utilised pastures on the Shewa Plateau for many generations. The more than 80-year-old Haji Samat remembers going to Shewa with his father during the reign of Amanullah Khan (1919–1929), at a time when access to pastures was undisputed and there was no need for official title. This corresponds with evidence derived from the literature, reporting that at the turn of the twentieth century, as many as 500,000 sheep belonging to mobile groups reached the Shewa Plateau from the plains around Kunduz, a number rising to 1.2 million in 1922 (Adamec 1972, 67; Koshkaki 1979, I, 220). The land title issued in 1951 by the Afghan king formalised this practice and specified the area of grazing land belonging to a certain clan by reference to major landmarks. These were the times when 'Kuchi were free', as one respondent remarked.

Today, things have changed. In spite all pastoralists carrying their land title with them at all times, access to the Shewa grazing lands has become insecure. It is now governed by different rule systems, in which the king's land title still defines the grazing area, but not the costs necessary to gain access. This process of change in territoriality as a contested social product was a gradual one. It commenced in 1978 with the Soviet occupation of Afghanistan and continued through the various shifts and drifts occurring in the wake of the civil war and Taliban rule and is still ongoing after the US-led invasion and installation of the current Karzai Government.

When Achekzai pastoralists first came to the area generations ago, the Shewa Plateau was already partially inhabited. From the beginning, there was some competition over the use of natural resources both amongst different pastoral groups and with the high plateau's permanent population. The latter is completely made up of people whose forefathers migrated to Shewa from Shugnan, today a district of Badakhshan province (Holzwarth 1990). The competition between pastoralists and Shughni grew in scope as the size of the settled population increased. It today consists of 14 villages made up of multiple hamlets dispersed all over the plateau.[1]

Shughni people engage in combined mountain agriculture at the upper limit of cultivation in which mobile pastoralism forms a significant component. Thus, both Achekzai and Shughni compete for the same resources. As the residential population grew, they increasingly started converting areas used as pastures into rainfed cultivation, thereby seeking support of official titles from different administrations and contributing to significant changes in land tenure relations on the plateau (Patterson 2004).

The Shughni as combined mountain farmers depend on fields and eventually succeeded in establishing a much needed greater agricultural basis by obtaining title documents from the Government of Rabbani (1992–2001)[2] and in so doing established ownership of land that formerly belonged to pastoralists. This enabled them to cope better with the harsh climatic conditions and make the most of the very short agricultural season on the plateau. However, competing legal documents over the same land issued at different times by different rulers caused some violent conflicts in the past between Shughni and various pastoralist groups, as documented by Patterson (2004, 17–31). Today, however, such violent clashes seem to have ceased, as both Pashtun and Shughni informants confirmed. Instead, disputes concerning trespassing of herds over agricultural lands and the location of pasture boundaries are usually settled by consensus between the different groups.

Still, in an area where one group resides permanently and others appear for seasonal utilisation, the situation is characterised by competition. It is facilitated through competing written titles for identical patches of land and manifests itself physically, for instance, through the common occurrence of rainfed agricultural fields on slopes and former pastures, by encroachment of villages on pasture lands or by pastoralist encampments overseeing a grazing area.

However, this competition is not the main determinant of pasture access today, as local strongmen were able to take control of the Shewa Plateau in recent times. This started happening in 2001, with the advent of the Karzai administration in Afghanistan. These strongmen, locally referred to as commanders, benefit from a 'culture of impunity' that prevails in Afghanistan (Mani 2003; Giustozzi 2009), and they have been able to newly determine pasture access rights by intimidation and physical power that has put the entire area under the law of the gun. What has occurred here, as was the case in many other areas of Afghanistan after 2001, was a particularly extensive incident of land-grabbing, where armed power holders assumed control over land and defined the rules of access (Gebremedhin 2007;

Giustozzi 2007, 2009). This process was supported through the redrawing of district borders and the formation of new districts in Badakhshan by the central government, in order to accommodate the interests of local strongmen and military leaders allying with Karzai and to (re)establish a self-serving patrimonial system in Badakhshan, where official positions such as district governors became an attractive resource to be exploited (Giustozzi and Orsini 2009; Goodhand 2009). The case of land-grabbing in Shewa is an important example of how Afghan institutions have been supplanted through power sharing deals by abusive stakeholders, who exert control through violence, patronage and corruption, often enjoying external backing (Giustozzi 2009; Grono and Rondeaux 2010).

This shift towards the 'law of the gun' in Shewa is the most recent trend of changing power geometries in Badakhshan, illustrating the ways pasture access is shaped in the context of limited statehood and corrupted institutions. The process has had severe consequences for both sedentary Shughni and migrating pastoralists. Dues for land access have to be submitted to commanders, now holding official positions in a new district named Arghanj Khaw that covers the Shewa area. For the Shughni mountain farmers, these dues comprise of taxes for pasture rights and payments for agricultural land title deeds, usually delivered in-kind. Pastoralists, however, are forced to pay large amounts in cash for pasture access in addition to donating livestock for strongmen in power. Paying in cash is a new phenomenon for pastoralists that commenced under the Karzai administration. This exposure to arbitrary rent-seeking of local strongmen in their ancestral summer grazing areas puts the already risky practice of animal husbandry under economic strain. As a respondent put it: 'From the time of the Saur Revolution [i.e. 1978] everybody raised taxes from us for using the pastures for which we already have a title. Baraki, Rabbani, Najibullah, all did so, but under Karzai it is the worst' (Achekzai pastoralist, September 2008). The revenues are forcefully extracted, often at gunpoint, and are used to maintain the power bases of commanders, and not to improve the welfare of local communities: 'Throughout 2001–2008, there was little or no sign of the revenue collected by local actors in Badakhshan being reinvested in the local economy and even less so in the provision of services to the population' (Giustozzi and Orsini 2009, 14).

Whilst such economic strain creates hardship for pastoral groups, the risk of eviction from pastures poses another threat that endangers pastoral livelihoods. In many cases, pastoralists have already been banned by the new rulers of the Badakhshan pastures, and whilst the Achekzai thus far have only been threatened, the Baluch faction of the study community was already forcefully evicted. For the first time, in 2009, they were not granted access to their ancestral pastures around the town of Keshem. Access rights to new pastures had to be sought elsewhere and paid for, and this lack of tenure security poses new uncertainties for their future.

Changing power geometries translate into changing systems of territorial control over pastures and restrict the mobility strategies of pastoralist. These trends hint quite literally at William Connolly's elaborations about the etymological roots of the word territory: 'To occupy territory … is both to receive sustenance and to exercise violence … Territory is sustaining land occupied and bounded by violence' (Connolly 1995, xxii). This interpretation is undoubtedly true for the Shewa pastures.

3.4 Road of Insecurity: Mobility Between Spring and Summer Pastures

Pastoral mobility is characterised by both conflict and cooperation with settled populations. Whilst contested access to land and water en route and damage to agriculture are obvious points of conflict, especially in drought seasons, mutually beneficial cooperation takes place too.

A flock of sheep raised by the Achekzai and Baluch is often maintained by various herd owners and easily number between 500 and 1,000 fat-tailed sheep. When migrating to Shewa, this large flock is accompanied by goats, camels, donkeys and horses to transport household materials and food. The sheer organisational effort required to engage in such long-distance migration is remarkable, and economic success as the ultimate goal of the entire venture depends on reciprocal relations with settled populations en route.

Entire households and their extended families migrate, carrying supplies for both outwards and return journeys and the entire 3 months stay in Shewa. A single migration unit may easily consist of 50 people. The more than 30 days of travel on the mobility route is punctuated by 26 overnight stops at defined locations (Fig. 3.3). At each of these places, fodder and water for the animals have to be guaranteed, the spacious tent dwellings for the migrating parties have to be erected and special care has to be taken to prevent animals from feeding on the agricultural fields of sedentary hosts. Arrangements are made with farming populations along the way to

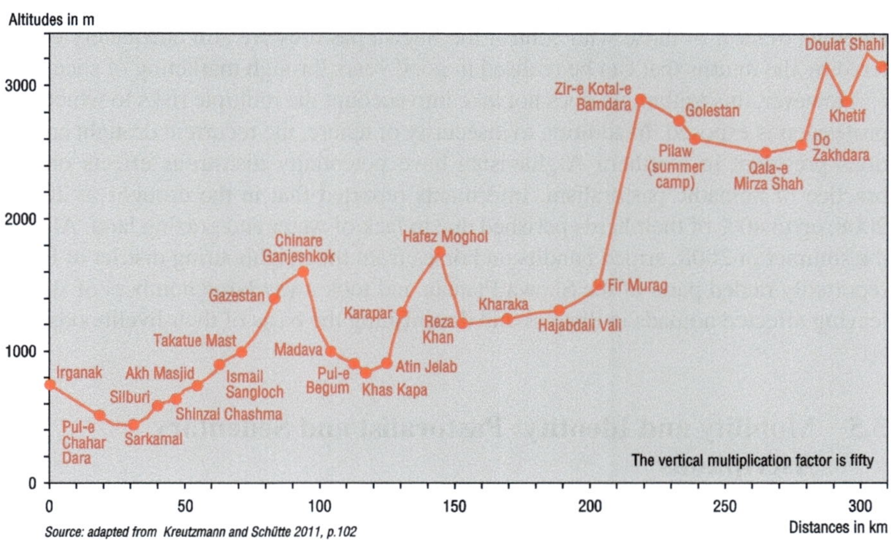

Fig. 3.3 Altitude-distance profile of the migration route of Pashtun pastoralists and the location of overnight stops between Chahar Dara and the Shewa Plateau

ensure that these necessities are available, and collaboration is organised differently on the onwards and return journeys.

At each stopping place, local farmers grow fodder crops for sale to pastoralists passing through in the spring season. This diminishes pastoralists' dependence on fertile and freely accessible grasslands whilst at the same time providing a secure income source for farmers. Animals are kept each night in designated spaces by permission of the sentinel of village grasslands, who is paid a grazing fee by herd owners. This way, lasting socio-economic relations between mobile and settled people have developed over time and can be counted upon when commencing the migration season.

Outbound, the stays at certain points en route usually last one or two nights, depending on the agro-ecological conditions in a given year and the availability of fodder and water. On the return trip from Badakhshan in late summer or early autumn, the stays easily extend to a week or longer so that the sheep maintain the weight gained during their months on the copious summer pastures. Encampments are erected on harvested fields, no payments are due and animals are allowed to roam freely to feed on crop residues. This practice also profits farmers when animal droppings fertilise their lands. Marketing of animals also takes place on return, and both Achekzai and Baluch sell about one-quarter of their fattened animals to farmers, butchers and on livestock markets to the larger urban areas of Faizabad, Kishim, Taloqan and Kunduz.

The continued prevalence of this type of pastoralism covering long distances in Northern Afghanistan indicates that it is still an economically viable undertaking, in spite of the detrimental conditions of heightened tenure insecurity and the oppressive rule systems that disadvantage pastoralists in the summer areas. Despite all this, pastoralists keep coming back to Shewa. Apparently, investments for fodder, for shifting entire households, for bribing officials, the arbitrary theft of animals and paying large amounts of cash to those who control the Shewa pastures are still adequately countered by the returns that can be realised in good years through marketing of sheep.

However, this arithmetic does not take into account the multiple risks to which the profession is exposed. In addition to insecurity of tenure, the recurrent drought conditions prevalent in Northern Afghanistan have potentially disastrous effects on the practice of nomadic pastoralism. Informants reported that in the drought of 2007–2008, up to 40% of their herds perished due to lack of water and grazing land. Also in the summer of 2008, armed bandits on horses from the neighbouring district of Ragh reportedly raided parts of the Shewa Plateau and took away large numbers of sheep, leaving affected nomads in distress and diminishing the basis of their livelihoods.

3.5 Mobility and Identity: Pastoralist and Sedentary Ascriptions

The fact of physical movement to gain and maintain pasture access, the embodied practice of movement, and the meanings that movement is given in their interplay contribute to form pastoral identities. Built, for instance, on tribal affinities and affiliations, on notions of belonging to a group and location, on commonality of social practice, spatial connectedness and social cohesion, the realities generated out of shared identities and the representation of difference and 'the other' – as in the dichotomy of mobile pastoralist and settler – have significance.

The Achekzai refer to themselves as Kuchi in their summer areas and whilst migrating, and differentiate themselves from others by stressing their practice of mobility. In this sense, mobility quite literally works as a lived relation, 'an orientation to oneself, to others and to the world' (Adey 2010, xvii). In the residential village, however, it is the tribal affiliation that serves as a distinguishing marker of identity and belonging. Being Achekzai, Baluch or Popalzai is literally connected to territoriality, as access to the spring pastures in Irganak is regulated according to such tribal memberships. Additionally, ethnic identities as Pashtuns are cultivated following a regained political salience of ethnicity that appeared to have been overcome during resistance in the 1980s (Tapper 2008). This was obvious during the Taliban reign when the Tajik and Shughni population of Badakhshan suspected the Achekzai and others to be Taliban associates, and it is also evident today when the Pashtun-dominated District of Chahar Dara is uniformly perceived as a Taliban stronghold. Ethnic identities are also employed when Achekzai claim that Pashtuns are universally oppressed by Tajik powerholders that today control the Shewa plateau and demand revenues. The ways different groups are represented are important, as social identities constructed in these ways are instrumental for claiming and establishing access to resources.

Consider the competition in Shewa between pastoralists and residential Shughni; the Achekzai relate to their distant summer areas as integral parts of their lives, livelihoods, their profession as animal breeders and as a group whose distinguishing marker is mobility. The spaces they use for grazing on the Shewa Plateau form an important element of their spatial identity as Kuchi that is employed to claim access to those pastures. Shughni stay in Shewa all year round and depend on mountain agriculture combined with short-distance pastoralism. They have to deal with extremely harsh winters, a very short and restricted agricultural season and distant market access, and are in many respects a disadvantaged group. The opportunity to obtain written title to land on the plateau that formerly belonged to pastoralist groups and has since been converted to rainfed agriculture was thus readily taken, although people were well aware that user rights had previously resided with pastoralists. This happened at a time when pastoral mobility to Shewa was made impossible because of the war that reigned in Kunduz and Takhar during the Taliban expansion in the mid-1990s. The Achekzai inferred from this development their 'obligation to mobility', or otherwise risk losing their pastures. Identity has become politicised in this way, as it has served factional interests resulting in land disputes, with all parties having competing title documents from different times and different rulers. Pastoral self-attributions as a social group that engages in meaningful mobility and social ascriptions from settled people that perceive Kuchi as powerful competitors over identical resources continually reproduce an essentialist dichotomy, which in the case of the study group always incorporated a spatial perspective and conflicts over space and resource access.

3.6 Resilience of Pastoral Livelihood Systems

Today, the mobility strategies of pastoralists are shaped by external restrictions. Seasonal migration to Shewa is perceived as mandatory based upon fears of loosing access to ancestral pastures to opposing groups should the space not be occupied each season. Still, the threat of eviction is always lingering. This was not the case during the time of King Zahir Shah when Achekzai nomads were represented as a group that enjoyed protection from the centre and substantially profited over other communities, ensuring unconditional mobility and access to pastures. With shifting power relations and war conditions holding sway over the country, a gradual decline in this protection altered the mobility spaces of pastoralists and has posed new and unprecedented threats to the community. Four distinct phases can be identified that are marked by alternating phases of sedentarisation and re-nomadisation, depending on political, socio-economic and environmental conditions:

1. A phase of comparatively less burdened mobility refers to the time prior to the Saur Revolution of 1978. The 'times of freedom' are remembered in the collective memory of Achekzai as periods with abundant pastures on the Shewa plateau. Mobility strategies were much more extended than today, and the annual migration period lasted up to 8 months. Following the summer period in Shewa, a delegate group of the Achekzai pastoralists used to go directly from the Shewa Plateau to the Afghan capital Kabul to market their fat animals and obtain the best prices for rams at Charikar, the capital city of Parwan. This entailed another 1-month journey across the Anjuman Pass (cf. Kreutzmann and Schütte 2011, Fig. 5, 114).
2. A phase of disrupted mobility occurred between 1978 and 1996 – the times of Soviet occupation and the subsequent disastrous civil war – when landmines on pastures, aerial bombardments as well as forced revenue extraction in-kind or blunt livestock theft by different factions of the resistance severely endangered the practice of mobile pastoralism.
3. A phase of sedentarisation and complete collapse of mobility characterised the reign of the Taliban between 1996 and 2001 when the route to Shewa was not passable due to heavy fighting. During this time, tenure relations on the plateau changed significantly. Tajik commanders established their power bases in Badakhshan, putting them in a position to grab vast pasturelands and forcefully extract revenues.
4. The current and prevailing phase of re-nomadisation and restricted mobility started in 2001 when the route was open but pastoralists discovered that their ancestral pastures were now under armed control and access possible only via cash payments.

Since 1978, pastoralism to Shewa has become an increasingly insecure venture exposed to many risks. It is characterised by a continuum of changing levels of livelihood insecurity and restrictions. However, pastoralism, as practised by the Achekzai and Baluch, has also proven its dynamism and flexibility. Depending on conditions, phases of sedentarisation and pastoralism may alternate as necessary. This flexibility and their access to irrigation land in the Kunduz oasis has allowed pastoralists to cope with the disruptions of spatial mobility and the temporal breakdown of pastoral production and reproduction systems that occurred during the previous 30 years. In the context of pastoralism in Afghanistan, the 'politics of mobility' (Cresswell 2010) are central to an understanding of how territorial practices unfold. This is to be understood in terms of changing local and regional power structures that have a bearing on how pastoralism is carried out. Access to those in power, as was the case during the reign of the Afghan king with his supportive policies, ensured that the Achekzai enjoyed undisputed access to fertile pastures. As Badakhshan gradually turned into an important stronghold of the Northern Alliances during the resistance, civil war and Taliban regime, Pashtun pastoralists had to endure growing restrictions in carrying out their pastoral practices. Access to pastures became contested when abounding title documents resulted in a form of legal pluralism facilitated by the multitude of stakeholders that exercise power in Badakhshan in order to extract dues from pasture users. The deeds issued by the king thus compete with various other title documents provided to a multitude of users by different rulers. The changing power relations that govern the pasture territories in Shewa benefitted different and often competing groups and resulted in legal geographies of pasture access that became increasingly determined by the exercise of power and not by state or customary law.

Rules of access are solely determined by strongmen and their militias, and the Afghan state is not included at all in any transfers of rights and resources. Lack of access to the new powerholders emanating from the Northern Alliances exposes pastoralism as practised by the Achekzai to new risks and vagaries that they cannot control. However, the example of the Achekzai also demonstrates the resilience of pastoral communities and their capacity to continuously adapt to changing political, economic and social conditions. Viewed through the lens of their changing mobility practices when striving to access the Shewa pastures, the intricate entanglements of power to which pastoral practices are exposed become visible.

Notes

1. The Shughni population in Shewa comprises about 700 houses distributed amongst those 14 villages, thereby accommodating nearly 1,900 households and 9,000 individuals. These households together cultivate about 12,000 *jerib* (approx. 2,400 ha) of rainfed and irrigated land, of which 73% is rainfed (data collected by Aga Khan Foundation in 2006).
2. Rabbani formally stayed Afghan president also during the reign of the Taliban. From 1996 on, he established his headquarters in Faizabad, the provincial capital of Badakhshan.

References

Adamec LW (1972) Historical and Political Gazetteer of Afghanistan, vol 1: Badakhshan Province and Northeastern Badakhshan. Akademische Druck- und Verlagsanstalt, Graz

Adey P (2010) Mobility. Routledge, London/New York

Alden Wily L (2004) Looking for peace on the pastures. Rural land relations in Afghanistan. Afghanistan Research and Evaluation Unit, Kabul

Balland D (1996) Dorrani. Encyclopaedia Iranica Online. Available at www.iranica.com. Accessed 12 Apr 2010
Barfield TJ (1978) The impact of Pashtun immigration on nomadic pastoralism in northeastern Afghanistan. In: Anderson J, Strand RF (eds) Ethnic processes and intergroup relations in contemporary Afghanistan, Occasional paper no. 15. The Afghanistan Council of the Asia Society, New York, pp 26–34
Barfield TJ (1981) The Central Asian Arabs of Afghanistan. Pastoral nomadism in transition. University of Texas Press, Austin
Barfield TJ (2008) Nomadic pastoralists in Afghanistan. Reconstruction of the pastoral economy. In: Janzen J, Enkhtuvhin B (eds) Present state and perspectives of nomadism in a globalizing world. Admon, Ulaanbaatar, pp 189–204
Bruce JW (1998) Review of tenure terminology. Tenure brief 1 – Land Tenure Center, University of Wisconsin, Madison, pp 1–8
Connolly WE (1995) The ethos of pluralization. University of Minnesota Press, Minneapolis
Cresswell T (2010) Towards a politics of mobility. Environ Plan D Soc Space 28(1):17–31
Cresswell T, Merriman P (2010) Introduction: geographies of mobilities – practices, spaces, subjects. In: Cresswell T, Merriman P (eds) Geographies of mobilities: practices, spaces, subjects. Ashgate, Aldershot, pp 1–15
de Weijer F (2007) Afghanistan's Kuchi pastoralists: change and adaptation. Nomadic Peoples 11(1):9–37
Delaney D (2005) Territory: a short introduction. Blackwell, Oxford
Dyson-Hudson R, Dyson-Hudson N (1980) Nomadic pastoralism. Annu Rev Anthropol 9:15–61
Ferdinand K (1969) Nomadism in Afghanistan. With an appendix on milk products. In: Földes L (ed) Viehwirtschaft und Hirtenkultur. Ethnographische Studien. Akadémiai Kiadó, Budapest, pp 127–160
Gebremedhin Y (2007) Land tenure and administration in rural Afghanistan: legal aspects: Asian Development Bank, Department for International Development: project report: Afghanistan Capacity Building for Land Policy and Administration Reform
Giustozzi A (2007) War and peace economies of Afghanistan's strongmen. Int Peacekeeping 14(1):75–89
Giustozzi A (2009) Empires of Mud. Wars and warlords in Afghanistan. Hurst, London
Giustozzi A, Orsini D (2009) Centre-periphery relations in Afghanistan: Badakhshan between patrimonialism and institution-building. Central Asian Surv 28(1):1–16
Glatzer B (1977) Nomaden von Gharjistan. Aspekte der wirtschaftlichen, sozialen und politischen Organisation nomadischer Durrani-Paschtunen in Nordwestafghanistan. Steiner, Wiesbaden
Glatzer B (1981) Processes of nomadization in West Afghanistan. In: Salzman PC (ed) Contemporary nomadic and pastoral people: Asia and the North. College of Willam and Mary, Williamsburg, pp 61–87
Glatzer B, Casimir MJ (1983) Herds and households among Pashtun pastoral nomads. Limits of growth. Ethnology 22(4):307–325
Goodhand J (2009) Bandits, borderlands and opium wars: Afghan state-building viewed from the margins, DIIS working paper 2009: 26. DIIS, Copenhagen
Government of Afghanistan (2000a) Law on Pastures of the Islamic Emirat of Afghanistan. Official Gazette 795, Ministry of Justice, Kabul
Government of Afghanistan (2000b) Land Management Law of the Islamic Emirat of Afghanistan. Official Gazette 795, Ministry of Justice, Kabul
Grono N, Rondeaux C (2010) Dealing with brutal Afghan warlords is a mistake. Boston Globe, 17 Jan 2010
Grötzbach E (1972) Kulturgeographischer Wandel in Nordost-Afghanistan seit dem 19. Jahrhundert, Afghanische Studien 4. Anton Hain, Meisenheim am Glan
Holzwarth W (1990) Vom Fürstentum zur afghanischen Provinz. Badakhshan 1880–1935. Support Edition. Gabler & Lutz, Berlin
IRIN (Integrated Regional Information Networks) (2007) Afghanistan: clashes over pastures threaten to ignite further conflict. www.irinnews.org/Report.aspx?ReportId=73373. Accessed 14 Sept 2009

Jentsch C (1973) Das Nomadentum in Afghanistan. Eine geographische Untersuchung zu Lebens- und Wirtschaftsformen im asiatischen Trockengebiet. Anton Hain, Meisenheim a. Glan

Koshkaki Mawlawi Borhân al-din Khân (1979) Qataghan et Badakhshân. Description du pays d'apres l'inspection d'un ministre afghan en 1922. Publications de l'URA n° 10. Mémoire n° 3. Travaux de l'Institut d'Etudes Iraniennes de l'Université de la Sorbonne Nouvelle 10. Tome I: Traduction, avec 35 cartes; Tome II: Texte persan et 35 cartes en fac-similé (translated by Marguerite Reut). CNRS. Centre de Recherches Archéologiques, Paris

Kreutzmann H, Schütte S (2011) Contested commons – multiple insecurities of pastoralists in North-Eastern Afghanistan. Erdkunde 65(2):99–119

Mani R (2003) Ending impunity and building justice in Afghanistan. Afghanistan Research and Evaluation Unit, Kabul

Massey D (1993) Power-geometry and a progressive sense of place. In: Bird J, Curtis B, Putnam T, Robertson G, Tickner L (eds) Mapping the futures. Local cultures, global change. Routledge, London, pp 60–70

McEwen A, Whitty B (2006) Water management, livestock and the opium economy: land tenure. Afghanistan Research and Evaluation Unit, Kabul

Patterson M (2004) The Shiwa pastures, 1978–2003: land tenure changes and conflict in Northeastern Afghanistan. Afghanistan Research and Evaluation Unit, Kabul

Rassul K (2010) Understanding conflict between nomadic and settled communities in Wardak's pastureland. Cooperation for Peace and Unity, Kabul

Schlee G (2005) Forms of pastoralism. In: Leder S, Streck B (eds) Shifts and drifts in nomad-sedentary relations. Reichert, Wiesbaden, pp 17–53

Tapper R (2008) Who are the Kuchi? Nomad self-identities in Afghanistan. J R Anthropol Inst 14(1):97–116

Chapter 4
Pastoral Production Strategies and Market Orientation of the Afghan Kirghiz

Ted Callahan

Abstract 'Closed frontiers nomadism' as an adaptation to specific political-geographic conditions represents a fusion of political, social and economic strategies. For the Kirghiz of the Afghan Pamir, surrounded by three international borders, the two main economic aspects of closed frontiers nomadism involve maximal utilization of pasture resources in an arbitrarily delimited geographic space and production of both primary (livestock) and secondary (wool, dairy) commodities for subsistence and trade. Market access remains one of the key issues facing the Kirghiz who throughout their recent history have been forced to adapt to geopolitical events far beyond their control but which directly impact their lives and livelihoods. In every instance, the Kirghiz have demonstrated great sophistication and resourcefulness in maintaining access to the twin pillars of their economy: productive pastoral resources (pasture and livestock) and market access. Kirghiz dynamism is reflected in their leasing of winter pastures in Tajikistan, cross-border trade with Pakistan, use of National Solidarity Programme funds to improve fodder production and manipulation of international aid agencies to receive annual food assistance. Such activities challenge the notion of pastoralism as a 'timeless' or 'traditional' activity, despite outward appearances based on a relatively limited material culture.

Keywords Kirghiz • Pastoralism • *Amanat* • Closed frontiers nomadism • Pamir • Afghanistan

T. Callahan (✉)
Boston University Arts & Sciences Anthropology, 232 Bay State Road, Boston, MA 02215, USA
e-mail: callahanted@gmail.com

H. Kreutzmann (ed.), *Pastoral practices in High Asia*, Advances in Asian Human-Environmental Research, DOI 10.1007/978-94-007-3846-1_4,

4.1 Introduction

In the popular imagination, pastoral societies are often wrongly believed to be autarkic and thus economically independent from the 'outside world', a fallacy critiqued at length by Khazanov (1994: 3). Such assumptions often derive from external perceptions of pastoralism as a 'timeless' or 'traditional' activity, based on the existence of a materially simple culture and often reinforced by the geographical isolation of many pastoral peoples. Isolation in marginal environments, however, generally serves to undermine autarky whilst at the same time obscuring the economic interconnectedness so essential for subsistence in remote areas.

The 1,149 Kirghiz living on the *Bam-i dunya*, the 'Roof of the World' in Afghanistan's Pamir Mountains, are an example *par excellence* of such a bucolic trope. For example, a recent article in *The New York Times* described the Pamir as 'a no-man's land', 'cut off the from Afghan heartland', its population 'sealed off from the currents of history' by virtue of dwelling in a place 'so remote that the people still live on a barter economy', and fated to 'probably remain cloistered in their world of wind and ice, as they have for centuries' (Wong 2010).

Isolation might be the most defining characteristic of life on the *Bam-i dunya* but in no way does it mean that the Afghan Kirghiz are either autarkic or unaffected by the outside world, so removed from the surrounding milieu that external events have little reverberation. The frequency of their emigrations over the past century, in an effort to maintain their independence and pastoral way of life in the face of political encroachment, vividly illustrates the truth behind Kreutzmann's statement that 'The lesson to be learnt from geopolitical interventions in peripheral mountain areas could be that decisions made in the core of empires always affect the livelihoods of people who have not been involved in the decision-making process' (2005: 23).

At the same time, whilst indisputably affected by economic and political events, often far beyond their control, the Kirghiz have demonstrated agency in their conscious decision to maintain a pastoral lifestyle, despite having been offered numerous opportunities to settle (Shahrani 2002: 230). Such choices are as much political and social as economic and are not merely guided by subsistence concerns alone (e.g. finding adequate pasture). In their decision making, the Kirghiz greatly benefit from having a subsistence economy and a semi-autonomous system of political authority. They are thus somewhat insulated from events and dynamics that would have a more immediate impact upon communities entirely dependent on markets and under direct government rule.

4.2 'Closed Frontiers' Nomadism

The isolation experienced by the Afghan Kirghiz and its effect upon their economic strategies is largely an artefact of political geography and as such must be understood in the broader historical context. Wakhan district, in Badakhshan province, encompasses the Wakhan and Panj River valleys (divided into upper and lower

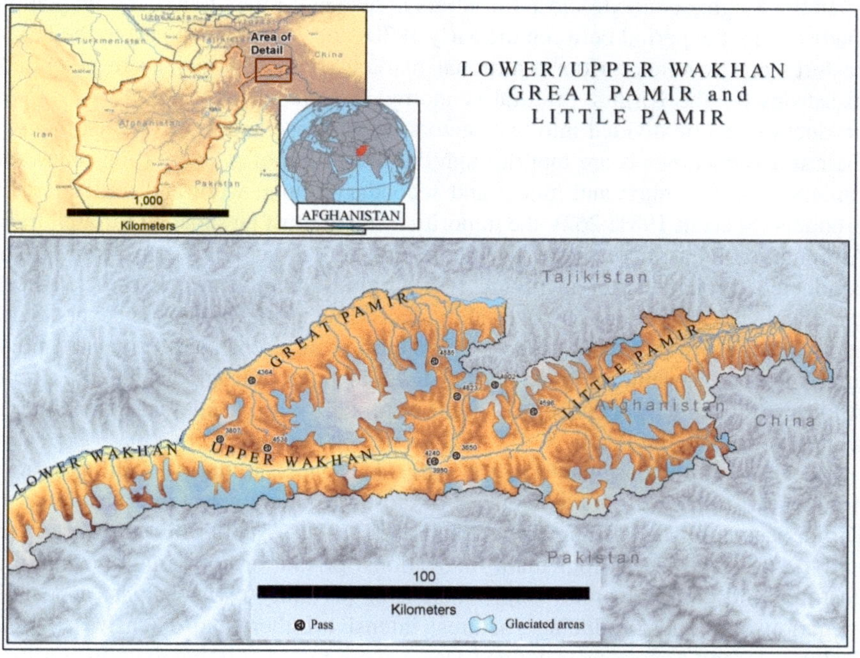

Fig. 4.1 The topography of Wakhan district (Source: Design by Jonathan Frishtick, data from author)

Wakhan, respectively), the Great Pamir and the Little Pamir (Fig. 4.1). These areas, often collectively called the 'Wakhan Corridor', were merged into a political unit in 1895, to serve as a buffer between British India and Tsarist Russia. Kirghiz movement across this new border was generally unimpeded until the 1930s when collectivization policies in the Tajik Soviet Socialist Republic (SSR) resulted in the closure of the border, forcing the Kirghiz to reorient their economy eastwards, towards China. Under pressure from Soviet cross-border raids, many Kirghiz fled Afghanistan to China in the 1940s, but they returned after the victory of the Chinese Communists in 1949.

By the midpoint of the twentieth century, the Wakhan-Pamir area was more cul-de-sac than corridor, surrounded by highly sensitive and very sealed international borders: to north, the Soviet Union (the Tajik SSR); to the east, the People's Republic of China; and to the south, the newly independent Dominion of Pakistan (which became the Islamic Republic of Pakistan in 1956). Fenced in by these closed international borders, the Kirghiz responded to the challenges of 'closed frontiers' (Shahrani 2002: 170) through a number of economic adaptations, which eventually produced significant political and social changes.

In the Kirghiz context, 'closed frontiers nomadism' refers to a set of adaptations made during the period between the early 1930s and 1949 in response to the loss of pastures and severance from traditional markets, both of which were potentially disastrous for the Kirghiz pastoral economy. Functionally, the Kirghiz economic production can be divided into two sectors: subsistence and trade. Whilst many of their subsistence needs are met through the utilization of primary (non-renewable carcass products: meat and hides) and secondary (dung, wool and milk) pastoral products (Sherratt 1981: 262), the majority of them must be satisfied through trade. Although cash has increasingly become a part of their economy, most trade is still conducted through barter involving two Kirghiz commodities: livestock (sheep, goats and yak) and secondary pastoral products (raw wool, wool horse covers, butter and *kurut*, dehydrated whey). These commodities are exchanged for any manufactured item or agricultural product the Kirghiz require from outside the Pamir. Additionally, the Kirghiz depend on trade to procure horses and donkeys, neither of which are bred in the Pamir.

Because livestock are the basis for both subsistence and trade, adapting pastoral production strategies to the exigencies of closed frontiers was a vital task, one the Kirghiz accomplished through what Shahrani (2002: 173) after his fieldwork in the mid-1970s identified as four main innovations: private or corporate ownership of pastures and camps; capital improvement of privately owned areas, especially irrigation of fodder-producing pastures; intensive rather than extensive seasonal pasture use; and a short, well-defined seasonal migration utilizing the same pastures and camps on a recurring basis. A fifth major innovation was the dispersal of livestock through a loaning system called *amanat*.

The net economic result of these innovations was the production of sufficient livestock for subsistence and a surplus for trade. The latter, however, also required access to markets where the Kirghiz could acquire staple goods. Accordingly, closed frontiers nomadism for the Kirghiz also involved political accommodation with the Afghan state and a reorientation of their economic production towards the markets of Afghanistan.

4.3 The 1978 Exodus and Its Levelling Effects

In the years following their return to Afghanistan from China in 1949, the Kirghiz successfully adapted to the new socio-economic conditions caused by closed frontiers (Shahrani 2002; Dor 1975; Dor and Naumann 1978). During the 1970s, the Kirghiz, symbolized by their famous khan, Haji Rahman Kul, were renowned for their pastoral wealth and enjoyed an amicable relationship with the Afghan government, which had exempted them from conscription and – after 1966 – taxes, and named Rahman Kul *pasbaan-e Pamir* – 'protector of the Pamir' (Shahrani 2002: 209). However, once again, they proved powerless against geopolitical events emanating far from the Afghan Pamir, which reverberated through this small community. In April 1978, the Saur Revolution successfully installed a Marxist government in Kabul. The khan's history of enmity with the Soviets across the border convinced him that he was in danger. Three months later, Haji Rahman Kul, together with 1,300 Kirghiz from the Little Pamir and nearly all of their livestock, fled across the border to Pakistan (Shahrani 1980: 11).

Finding their situation in Pakistan untenable, due to unfavourable conditions such as aridity and the non-availability of pastures where they could graze their livestock, and suspecting that Rahman Kul had exaggerated the threat posed by the Soviets, a number of Kirghiz decided to return to Afghanistan. Between 1978 and 1979, about 300 Kirghiz returned to the Little Pamir, joining the roughly 40 Kirghiz who had never left. Because the Soviet media had denounced the khan as a reactionary, fearful of the 'progressive forces' which had come to power in Afghanistan, the return of these Kirghiz and their supposed rejection of 'the feudal lord and landowner' Haji Rahman Kul (Akhmedzyanov 1978) offered an obvious propaganda coup. The Kirghiz returnees were provided with various forms of assistance (food, medicine and livestock) and supposedly even given an award by the Afghan government.

Initially, believing that more Kirghiz would soon be returning from Pakistan, they simply returned to their previous settlements. After Soviet forces occupied the Little Pamir in May 1980 and closed the Pakistani border (Amstutz 1986: 294), the Kirghiz realized that the others would not be coming back from Pakistan any time soon. They redistributed the pastures, fodder-producing areas (*koruk*) and seasonal camps (*aĭyl*) amongst themselves. Because the population was so small (about 340 individuals, divided into roughly 70 households) and livestock were comparatively few, most families were able to claim better lands than they had occupied previously. This redistribution was especially significant since, as Shahrani noted, 'During the initial period of settlement of the Pamirs [the early 20th century], the more powerful herders used their influence and/or wealth to secure the better pasturage and camping grounds' (2002: 176). Differential access to productive resources such as land had been one of the main drivers of inequality and the resulting economic stratification that resulted.

The other dimension of inequality was livestock ownership. Under Haji Rahman Kul, 33,550 sheep and goats (out of 38,600 in total) and 1,754 yak (out of 3,544 in total) were owned by just 18 out of 333 households, who parcelled out 22,346 of their livestock holdings through a system of loaning called *amanat* ('safekeeping') to over 200 households (Shahrani 2002: 181). Put differently, 5.4% of the population owned 84% of all the sheep, goats and yak, and 63% of their herds were loaned out as *amanat* to over 60% of the population, many of whom had few if any livestock of their own.

The *amanat* system serves two purposes in Kirghiz society: it allows wealthy livestock owners to disperse their livestock (and thus their risk), and it functions as a sort of welfare for households with insufficient livestock to meet their subsistence needs. However, in contrast to the wealthy herder, who by spreading out his assets has diversified his risk, the smallholder is continually exposed to having all or a greater proportion of his herds die. Distributing risk is crucial in the Pamir because many potentially disastrous events – snowstorms, epizootic outbreaks, theft and predation – are highly localized. *Amanat* is the pastoral equivalent of an investment strategy based on buying stock in several companies versus investing entirely in a single one.

As a pragmatic response to the problem of closed frontiers, the *amanat* system undoubtedly was one of the key innovations allowing the Kirghiz to remain in the Pamirs, especially for poor households. However, *amanat* was also one of the structural bases for inequality in Kirghiz society, stratified into two classes: a small, wealthy elite and the rest of the Kirghiz, extremely poor and dependent upon the rich for their livelihoods. Whilst not inherently exploitative, the *amanat* system often had that effect. Furthermore, for the *amanat* system to function to the degree that it did in the 1970s required a large segment of the population to own few, if any, animals of their own.

Most of the wealthy livestock owners in the Little Pamir brought their livestock with them to Pakistan. Amongst those Kirghiz who returned, only a few arrived back in Afghanistan with any animals, most having died en route to Pakistan or subsequently been sold there due to lack of pasturage. The wealthiest of the returnees was Astanabek, a former khan, who managed to return with 500 sheep and supposedly a large number of gold coins (his descendants are still the richest Kirghiz in either Pamir). The rest of the Kirghiz, however, were forced to rebuild their herds essentially from scratch. Within a year or two of returning, the average household herd consisted of five sheep, one yak and one donkey, thanks largely to the restocking assistance the Kirghiz received from the Afghan government and, later, the Soviet troops garrisoned at Bozoĭ Gümböz.

4.4 Patterns of Livestock Ownership

In the Little Pamir, both of these post-1978 levelling effects – the redistribution of land and the flattening of livestock ownership – temporarily eliminated much of the inequality that had previously been so characteristic of Afghan Kirghiz society. Since there was no comparable exodus amongst Great Pamir Kirghiz,[1] their society remained somewhat more stratified, with the majority of livestock still concentrated amongst a small, wealthy elite. Today, in both Pamirs, the *amanat* system is extant, but compared to the 1970s, it is much diminished in both scale and importance. The proximate cause is that livestock are now more evenly distributed throughout the society and most households own at least some animals. As a result, there are fewer individuals with sufficiently large holdings to loan out significant quantities of *amanat* livestock. The ultimate cause was the departure of most of the extremely wealthy herd owners, together with their livestock, in 1978.

One way to illustrate the divergent patterns of livestock ownership between the 1970s and today is by means of the Gini coefficient, a statistical measure of inequality based on the distribution of wealth. A numerical value of zero indicates perfect

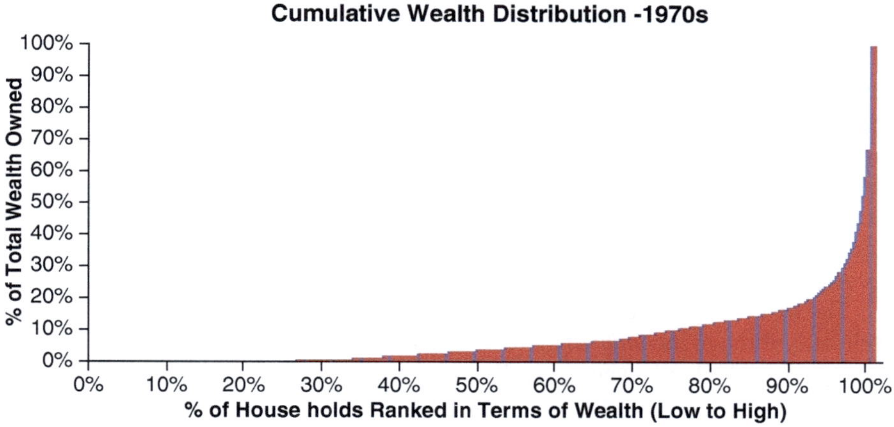

Fig. 4.2 This chart shows the percentage of total wealth (*the vertical axis*) held by the poorest X% of households (*the horizontal axis*) for the 1970s. The poorest 50% of households held about 4% of the total wealth. The Gini coefficient (inversely related to the *shaded area*) for these data is 0.857 (Source: Shahrani 2002 (the data were collected for the first edition of the book (1979) and refer to the time period 1972–1973))

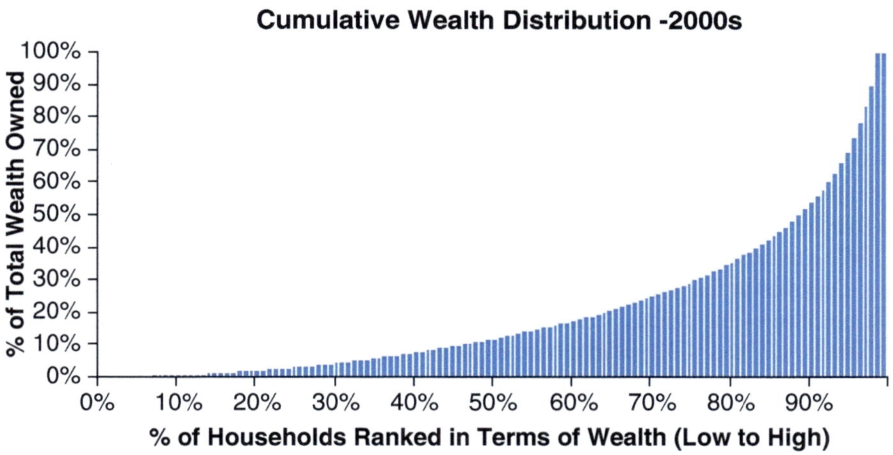

Fig. 4.3 This chart shows the percentage of total wealth (*the vertical axis*) held by the poorest X% of households (*the horizontal axis*) for the 2000s data. The poorest 50% of households held about 12% of the total wealth. The Gini coefficient (inversely related to the *shaded area*) for these data is 0.619 (Source: Data collected by author)

equality (i.e. everyone's wealth is the same), and a numerical value of one shows total inequality (one person has all the wealth). In the Little Pamir, the Gini coefficient for all livestock is 0.619 for the period between 2007 and 2008. Between 1972 and 1973, for the Pamir as a whole, the Gini coefficient is 0.857, demonstrating a much greater degree of inequality, as shown in Figs. 4.2 and 4.3.[2]

4.5 Recent Innovations

Despite the changes in land and livestock ownership that have occurred over the past 30 years, the pastoral production innovations initiated by Haji Rahman Kul are still practised. However, a number of important changes have happened, most of which followed the dissolution of the Soviet Union in 1991 and, with it, the decreased sensitivity of the Afghan Pamir as a restricted border region. Collectively, these recent adaptations illustrate the continuing dynamism and resourcefulness characteristic of Afghan Kirghiz pastoral strategies.

The first change involves outside investment in the Kirghiz's pastoral economy, primarily through the *amanat* system. Within Kirghiz society, *amanat* is still practised by wealthy herd owners, usually those with more than 250 sheep/goats and/or 70 yak. Because a herd this size could easily be managed by a single *aĭyl* or even just a couple of households, the distribution of a portion of their livestock as *amanat* suggests that the Kirghiz are still consciously diversifying their risk rather than simply outsourcing labour. In the Great Pamir, where economic stratification and the *amanat* system remain more prominent, average *amanat* holdings as a percentage of herd size are 25% for yak and 50% for sheep and goats.

More often, however, it is the case that *amanat* animals are loaned out by investors from outside the Pamir, as the *amanat* system offers an excellent vehicle for absentee ownership. The biggest investors at present are the very Kirghiz (and their descendants) who fled in 1978, remained in Pakistan for 4 years, and were eventually resettled in Turkey. Foremost amongst these Turkish Kirghiz investors is one of Haji Rahman Kul's sons, Muhammad Arif Kutlu (Rahman Kul died in Turkey in 1990). Besides the Turkish Kirghiz, other outside investors include the Wakhi (a neighbouring ethnic group, living in the Wakhan and Panj valleys), Afghan (Badakhshi) traders, Kirghiz from the Great Pamir (who have *amanat* animals in the Little Pamir) and a handful of government officials (Table 4.1).

One reason that outside investment occurs more frequently today than in the past is because of the greatly reduced sheep and goat population owned by the Kirghiz (Tables 4.2 and 4.3), nearly half of what it had been 40 years ago. Felmy and

Table 4.1 Outside investment in the livestock economy, Afghan Pamir

	Little Pamir		Great Pamir		Total	
Investor	Sheep and goats	Yak	Sheep and goats	Yak	Sheep and goats	Yak
Turkish Kirghiz	375	89	250–300	0	625–675	89
Badakhshi traders	300	20	200–300	6	500–600	26
Wakhi	100	38	100–200	8	200–300	46
Great Pamir Kirghiz	150	60	n/a	n/a	150	60
Govt. officials	UNK	UNK	150–200	40	150–200	40
Total	925	207	700–1,000	54	1,625–1,925	261

Source: Data collection by author, 2008–2010

Table 4.2 Livestock population (owned by Kirghiz), Little and Great Pamir (2008–2010)

Livestock	Little Pamir	Great Pamir	Total
Sheep/goats	10,608	9,000	19,608
Yak	3,120	1,275	4,395
Camels	53	82	135

Source: Data collection by author

Table 4.3 Comparison of Kyrgyz-owned livestock population, 1972–1973 and 2008–2010

Livestock	Shahrani (1972–1973)	Callahan (2008–2010)	% Change
Sheep/goats	38,600	19,608	–49.2%
Yak	3,544	4,395	+24%
Camels	83	135	+62.7%

Source: Shahrani (2002); author's data collection

Kreutzmann (2004: 115) suggest that reduced sheep and goat numbers were a result of the constrained market access the Kirghiz faced in the late 1990s when livestock were 'predominately bred for home consumption and utilization'. More recently, the continuing phenomenon of a smaller-than-expected sheep and goat population, relative to other commercial livestock such as yak, and despite comparatively robust market access, is likely attributable to greater mortality amongst sheep and goats during winter. For example, Wald (2005) reports that during an especially severe April storm (*jut* in Kirghiz), sheep and goat losses were typically 50–60%, compared to 10–25% for yak.

The smaller sheep and goat population at present, combined with a human population 63% of what it was 40 years ago, also explains why the Kirghiz have never seriously attempted to resettle a number of remote pastures in the Little Pamir that they abandoned in 1978. Most were occupied by Wakhi from the upper Wakhan, in some cases soon after they were vacated by the Kirghiz (Fig. 4.4). An additional reason, though one never mentioned by the Kirghiz, is that many of these areas belonged to the Wakhi prior to permanent Kirghiz settlement of the Afghan Pamir in the early twentieth century.

The loss of these pastures to the Wakhi has had little effect upon Kirghiz pastoralism primarily because fodder, rather than the availability of pasture, is the main factor limiting Kirghiz pastoral production. Even given ideal forage conditions, it is usually necessary to supplement grazing with fodder throughout the early spring and sometimes in winter as well. When natural grazing is not possible, in whole or in part, because either the winter pastures have been exhausted or owing to deep snow, typically the only option the Kirghiz have is to provide fodder to their livestock.

Provisioning adequate fodder has been a problem for the Kirghiz since they were cut off from their traditional winter pastures in Tajikistan in the 1930s; as Kreutzmann notes, 'Basically, winter pastures are missing everywhere today' (2009: 113). The importance of fodder explains the considerable effort the Kirghiz expend in trying

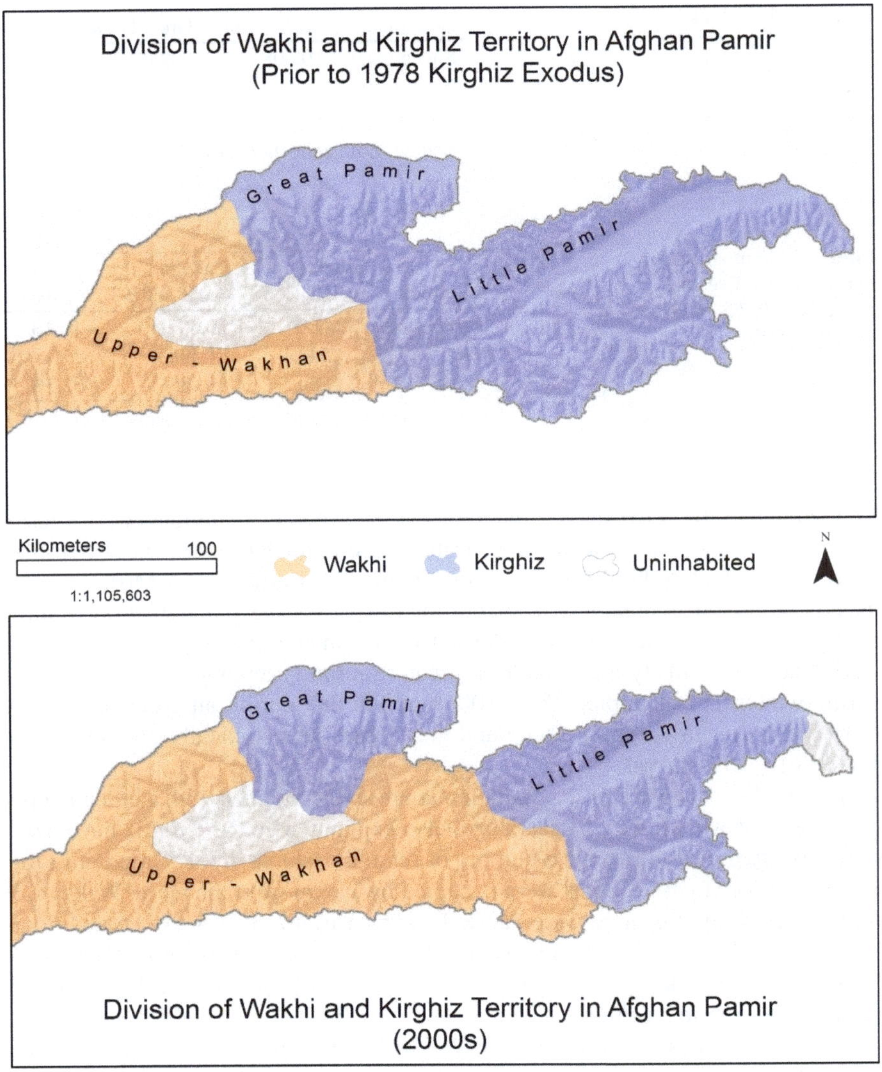

Fig. 4.4 Kirghiz and Wakhi rangelands, comparison of 1970s with present state of affairs (Source: Design by Jonathan Frishtick, data from author)

to boost its production through irrigation. Wherever possible, the Kirghiz try to irrigate fodder-producing areas (*koruk*) by digging basic channels (*aryk*) from natural streams and directing water so that it seeps down onto lower-lying *koruk*. With technical assistance provided as part of the National Solidarity Programme, larger and more extensive irrigation channels have been built, capable of transporting greater volumes of water over near-level ground.

Despite the considerable effort which goes into harvesting fodder every year, however, the total amount is typically modest. For example, one winter camp had just a single, large fodder bin measuring 7 m×4.3 m×2 m, which was capable of holding, when full, 60 m^3 of cut grass. This amount of fodder, during a bad winter, might have to support 250 sheep and goats, 6 camels, 16 horses and 6 donkeys, in addition to newborn livestock.

The Great Pamir Kirghiz, living across the narrow Pamir River from Tajikistan, have recently devised a simple solution to the limitations of fodder: when faced with deep snow, they take their livestock (sheep, goats and yak) across to Tajikistan, where the south-facing pastures tend to have less snow. Because of the 'tax' levied by the Tajikistani border commander on Afghan livestock grazing in Tajikistan (approximately one yak for every 100–150 livestock, or one percent), this option is only exercised *in extremis*, usually deep snow combined with dwindling fodder supplies.

4.6 Economic History

Closed frontiers not only necessitated pastoral production innovations, they also forced the Kirghiz to find new markets where they could trade pastoral products for staple goods. Prior to the 1978 exodus, Kirghiz commercial activity was limited to the territory of Afghanistan, with the exception of a few Wakhi traders from Gojal, Pakistan. During the summer, both Pamirs were visited by Afghan traders who would exchange market goods for Kirghiz livestock and secondary pastoral products, either directly or on credit. A few wealthy, entrepreneurial Kirghiz herd owners also participated in an annual livestock caravan to Kabul, where rates of exchange for Kirghiz commodities and prices of market goods were better than those offered by the traders servicing the Pamirs (Shahrani 2002: 207–8).

The Soviet occupation (1979–1989) disrupted this system but provided an unexpected economic boon. Wakhan district was one of the few areas in Afghanistan where the Soviets could reasonably hope to win the 'hearts and minds' of the population, which they did by helping those Kirghiz who returned from Pakistan rebuild their herds and then, from their base in the Little Pamir, bartering staple goods for Kirghiz livestock. Badakhshan, traditionally a food-deficit area with correspondingly low food security, also received subsidized cereals during the Soviet period (Goodhand 2000: 268), which kept the price of wheat low. The nearly 9 years that the Soviets occupied the Pamir are today remembered as a golden age, a time of peace and relative prosperity.

The civil war (1989–2001) that followed the Soviet withdrawal was the antithesis of the previous period. Besides suffering under the predatory rule of various warlords after they took over all of Badakhshan in 1992, the Kirghiz had to contend with the progressive ruination of the regional economy. As Pain (2010: 11) notes, 'The cost of living began to rise with the emergence of a siege economy in Badakhshan due to the blockades imposed by the surrounding Taliban forces'. Economic difficulties were further magnified by severe drought in Badakhshan in 2000 and 2001.

4.7 Differing Economic Trajectories

Two positive economic developments that occurred during the civil war were the aid the Kirghiz began receiving from Focus Humanitarian Assistance (FOCUS), an affiliate of the Aga Khan Development Network, and the opening of the Pakistani border in 1992 (Kreutzmann 2001: 55) as well as occasional bazaars held along the Tajik-Afghan border. Access to these markets, combined with food aid provided by FOCUS, was critical in enabling the Kirghiz to weather the worst economic period of the Afghan civil war, between 1999 and 2001.

Neither of these developments has benefitted the Kirghiz in the two Afghan Pamirs equally, though. The Little Pamir Kirghiz receive a disproportionate benefit from the food aid they are provided. And whilst the opening of the Tajik border was important for the Great Pamir Kirghiz, in the long run, the Little Pamir Kirghiz's access to the markets of Pakistan proved far more significant.

FOCUS, which already had considerable experience working in war-torn Tajikistan, began an annual aid delivery of food to the Kirghiz in 1998. This humanitarian aid – 'necessary due to the failure of Tajikistan to respond to requests from the Kirghiz nomads of the [Great] and Little Pamir regions to help them cushion the impact of severe supply shortages' (Kreutzmann 2000: 9) – was brought overland, across the Tajik border, where it was distributed to the Kirghiz in both Pamirs.

In 2006, the UN World Food Programme (WFP) began distributing food to the Kirghiz. WFP intended to provide 3 months' worth of food aid to the Kirghiz, consisting of (per household) 150 kg wheat flour, 18 kg legumes, 1.5 kg iodized salt and three 4-l cans of vegetable oil. The Little Pamir Kirghiz, however, inflated their population figures 85%, claiming 237 households (compared to an actual figure of 130). As a result, they receive an annual allotment sufficient for nearly 6 months. Unfortunately for the Great Pamir Kirghiz, WFP only gives them food aid sufficient for 101 households when there are actually 107. Because all of the legumes and much of the cooking oil are exchanged for other goods, including opium, and since the livestock savings the Kirghiz realize as a result of receiving food aid are often redirected towards luxury goods (electronics, tobacco and opium), food aid to the Kirghiz has been described as akin to an opium subsidy (Schaller 2004: 15–16). WFP is aware of this situation, but their attempts to switch to a food-for-work system have been resisted by the Kirghiz's patrons in government. According to one WFP regional director, 'For the Kirghiz, food aid is not an emergency. It's more like an obligation'.

4.8 Not-So-Closed Frontiers

Just as important for the Kirghiz as the windfall of several months' worth of food, provided for free and with no reciprocal obligations, has been the progressive opening of formerly 'closed frontiers'. Whilst the Chinese border remains sealed and the

Table 4.4 Commodity prices, October 2010 (all prices in Afghani – AFN; $1 = 45 AFN)

	Pamir	Faizabad	Kabul	Pakistan
(Turki) sheep, medium size	3,000–5,000	8,000	10,000–15,000	3,874
Black tea, 30 kg	1.5 sheep	4,800	5,700	4,110
Wheat flour (imported), 50 kg	0.5 sheep	1,200	1,150	800
Rice (domestic), 70 kg	0.75–1 sheep	4,000	2,500	1,520
Sugar, 50 kg	1 sheep	2,500	1,575–2,000	2,100
Opium, 1 *pau* (453 g)	1 sheep	n/a	n/a	n/a
Chinese blanket, medium	1 sheep	1,700	1,000–1,500	645
cloth (*tikka*), 40-m bolt	1 sheep	3,600–6,000	2,000–4,000	1,440

Source: Data collection by author

Kirghiz are still officially prohibited from entering Tajikistan, they are permitted to cross the 4,979 m Irshad Uween pass into Pakistan's Chapursan Valley, where every fall (late September to early October) a number of Pakistani traders set up a temporary bazaar where they barter manufactured goods (most from China) and Pakistani agricultural products for Kirghiz commodities.

In Tajikistan, following the end of the Tajik civil war in 1997, the Russian 'Moskva' border guards, backed by the 201st Motorized Rifle Division, assumed responsibility for securing Tajikistan's borders. Besides facilitating the movement of relief supplies to the Kirghiz across the border (Kreutzmann 2000: 9), they also permitted occasional bazaars to occur along the Tajik-Afghan border (Felmy and Kreutzmann 2004: 115). The latter have not occurred since 2005, when the Russian forces were withdrawn from Tajikistan, but the Soviet-era road into the Little Pamir (and across the Pamir River into the Great Pamir) is still used on occasion by visiting government officials and to deliver aid to the Kirghiz (Mortenson 2009).

Access to Pakistan, though, has been the dominant factor influencing the different economic trajectories of the Little and Great Pamirs. Because of high transport costs involved in bringing their merchandise up to the Pamir, the Afghan traders charge a substantial markup for their goods, typically 100–150% of the retail cost (Table 4.4). They also tend to deal in items with a high value-to-weight ratio, such as cloth and opium. Their real profit, though, comes from the livestock (sheep and goats) they acquire from the Kirghiz, which, when sold in urban markets, might fetch double or even triple their exchange value in the Pamir (Table 4.4). As livestock prices in Afghanistan have risen since 2001, due to an increasing demand for meat, more traders have been coming to the Pamir. A consortium of Pashtuns from Paghman (a town in Kabul province), seemingly well resourced, spend the summer amongst the Kirghiz and then together with their livestock walk to Kabul, a trip requiring about 50 days. The other traders generally sell their livestock in Faizabad, the provincial capital of Badakhshan.

For various reasons, the Kirghiz have not marketed their own livestock in Kabul (or any other urban centre) since the time of Haji Rahman Kul. In general, they would rather pay a premium for the convenience of not having to leave the Pamir. The Kirghiz are very much aware that they are offered a generally poor rate of

Table 4.5 Kirghiz commodities, 2009 rates of exchange in Afghanistan and Pakistan

Commodities, 2009 rates of exchange (for wheat flour)	Afghanistan (Wakhan) (kg)	Pakistan (Hunza) (kg)
1 sheep, medium size	100	152
1 goat (female or castrated male)	50	100
1 yak calf	100	152
1 3-year-old yak (uncastrated)	300	300
1 *seer* (7.61 kg) *kurut*	15–23	23–38
2–3 *seer* (15.22–22.83 g) butter	61	76
1 horse cover, fair (sheep wool)	61–114	61–76
1 horse cover, best (camel wool)	76–152	76–114
1 *booji* (23 kg) raw wool, yak	38	50
1 *booji* (23 kg) raw wool, camel	61–76	100

Source: Data collection by author

exchange by the traders, and as a result, they try to defer making substantial purchases until they go to Pakistan or down to the Wakhan valley in the fall to collect the WFP food aid. The latter event has become something of a major trade fair. Besides the WFP convoy, which distributes aid in two locations (Gaz Khan for the Great Pamir and Sarhad-e Broghil for the Little Pamir), a number of merchants (including opium dealers) as well as Wakhi trading partners of the Kirghiz arrive. The food aid is collected, and then the Kirghiz make the rest of their purchases for the winter, usually on credit. A second, lesser trade fair (minus the WFP aid) often occurs in May, as the Kirghiz are usually low on supplies after the winter.

Amongst the Great Pamir Kirghiz, both of their trade options – with the traders in the Pamir or down at Gaz Khan in Wakhan – are limited to Afghanistan, where most goods, including agricultural ones, are imported. Accordingly, prices within Afghanistan tend to be higher than in neighbouring countries. For example, Afghanistan requires annually 5.2 million metric tons (MT) of wheat to meet demand but rarely produces that much.[3] Most of the domestically produced wheat 'is consumed by the farming households and less than 10% of output reaches national food markets' (IRIN 2011). As a result, urban markets are dominated by imported wheat flour, mostly from Kazakhstan and Pakistan (from which Afghanistan imports, on average, 500,000–700,000 MT of wheat annually). Imported wheat flour, though of higher quality, tends to be significantly more expensive and is highly susceptible to fluctuations in price and availability. In 2010, Pakistan banned wheat exports after widespread flooding destroyed 500,000 MT of stored wheat (Anis and Pearson 2010).

The Little Pamir Kirghiz are able to avoid much of the cost and volatility associated with imported goods by travelling to Pakistan. Besides cheaper agricultural products, they are also able to buy manufactured items, many from China, for less than they would cost in Afghanistan. The better availability and lower cost of these goods in Pakistan mean that the Little Pamir Kirghiz have to exchange proportionally fewer pastoral commodities to satisfy their subsistence needs compared to what they would pay for the same market goods in Afghanistan (Table 4.5).

Access to the Pakistani market has also allowed the Kirghiz to trade a more diversified basket of commodities. In contrast to the Afghan market, for which the Kirghiz produce mainly sheep and secondarily goats and *kurut*, in Pakistan, there is demand for additional primary (yak) and secondary (*sarymaĭ*, clarified butter; horse blankets made from yak or camel wool; raw wool) pastoral products (Table 4.5). The market for secondary pastoral products is especially important since it allows the Little Pamir Kirghiz to trade renewable livestock products (wool and dairy) without incurring the gross loss of an animal.

The Pakistani market in the Chapursan Valley of upper Hunza, and access to it, is a fairly recent development. The Karakoram Highway (KKH) was finished in 1978 as part of a spurt of road building that resulted in 97% of the villages in the Hunza Valley becoming accessible by motorized transport (Kreutzmann 1993: 36). These road linkages, together with price controls on key staples and subsidized transport from producing regions in the south to deficit areas in the north, facilitated the flow of goods to formerly inaccessible, food-insecure regions (Kreutzmann 1991: 724).

For the Kirghiz, market opportunities in Pakistan resulted from the confluence of several factors: the construction of a road almost to the Afghan border, the greater availability of (subsidized) market goods, and a demand for Kirghiz pastoral commodities. The last factor occurred because the rise of more remunerative, non-agrarian economic opportunities for the youth of Hunza created a shortage of herding labour and led to a steep decline in domestic pastoral production. This decline, in turn, caused ‘a severe deficit in meat supply’ (Kreutzmann 1993: 30) and a lack of artisanal dairy products (*kurut*, *sarymaĭ* and others) (Kreutzmann 2004: 66). Since 2001, overall economic decline in Pakistan has led to a contraction of non-agrarian jobs, such as those linked to tourism. Poor economic prospects combined with a high cost of living in the cities have forced many younger men to return to their villages and resume animal husbandry as their primary occupation. Part of this process has involved the rebuilding of herds, especially yak, which had dwindled once the ‘task of shepherding’ devolved into ‘the hands of elderly men and women’ (Kreutzmann 2004: 65).

Pakistan, however, may offer limited advantage to the Kirghiz in the near future. Since 2007, prices have risen steadily, especially during the 2007–2008 global food crisis, when Pakistan banned wheat flour exports to Afghanistan in an effort to stabilize domestic prices. Between 2007–2008 and 2008–2009, Pakistan’s overall consumer price index (CPI) rose almost 21%, and the food CPI increased nearly 24% during the same period (State Bank of Pakistan 2010).

The inundation of a 19 km section of the KKH in January 2010 – the result of a landslide near Attabad, in Gilgit-Baltistan, which blocked the Hunza River and created a 21-km-long lake, 116 m deep in places – further contributed to rising prices and led to temporary supply shortages due to the closure of the KKH in the Hunza Valley. Although the Kirghiz were still able to trade in Pakistan in 2010, prices were reportedly higher than in the past, and for some goods, such as wheat, they were more expensive than in Afghanistan. The Little Pamir Kirghiz were still able to acquire all the market goods they needed for the winter, but at a greater overall cost than in the past. The long-term implications of any disruptions in the Pakistani market for the Little Pamir Kirghiz are uncertain, but already some Kirghiz in the Great Pamir are predicting that, given a continued decrease of purchasing power in Pakistan, the Little Pamir Kirghiz stand to lose their competitive advantage.

4.9 Conclusion

The continuing evolution of Afghan Kirghiz strategies to maximize pastoral production and maintain market access, both with the overarching goal of meeting their subsistence needs, illustrates how pastoralism is anything but an archaic, 'traditional' economic activity. Even on the isolated *Bam-i dunya*, Kirghiz pastoralists demonstrate resourcefulness and dynamism, though both are easy to miss as neither is readily apparent in terms of Kirghiz material culture, with the possible exception of the occasional satellite dish, incongruously situated next to a felt yurt.

Were the Kirghiz agriculturalists producing cash crops for export, it would be a simple matter to overlook their geographical isolation and see them in a broader regional context, one involving transnational urban markets, government policies and sophisticated trade networks. The reorientation of pastoral production, however, is more subtle; as Barfield (1981: xviii) observed, 'among the nomads, the sheep remained the same; it was changes in the regional economy that transformed the sheep from a "subsistence crop" to a "cash crop" without any conscious decision by the nomads themselves'.

Barfield's comment gets at the flip side of Kirghiz economic instrumentality: despite their resourcefulness in meeting the many challenges they have faced, Kirghiz agency is fundamentally reactive, a response to forces and events outside of their control and often beyond their comprehension. The Kirghiz are active agents in adapting their livelihoods to largely exogenous forces and constraints – in their boundaries, their markets, their political situation – which they were (and remain) powerless to negotiate. Accordingly, they have adjusted their behaviour opportunistically to changing exigencies, making strategic decisions based on desired political and economic outcomes whilst maintaining a deceptively traditional appearance of their nomadic pastoral society.

Notes

1. Unlike Haji Rahman Kul, the Great Pamir Kirghiz felt they had little to fear from the Soviets and in any case they were too distant from Pakistan to have made it across the border undetected.
2. All Gini calculations were performed by Prof. Michael J. Smith, Boston University School of Management. Prof. Smith also created the two wealth distribution charts (Figs. 4.2 and 4.3).
3. In 2010, a bumper wheat harvest throughout Afghanistan provided only 4.5 million MT (87% of demand).

References

Akhmedzyanov A (1978) The bloody path of the 'Independent Khan'. Izvestiya, translated in FBIS-SOV 3, no. 189:J1, J2
Amstutz JB (1986) Afghanistan: the first five years of Soviet occupation. NDU Press, Washington, DC
Anis K, Pearson M (2010) Sugar, wheat, rice worth $2.9 billion ruined by Pakistan floods. Bloomberg. 12 Aug. http://www.bloomberg.com/news/2010–08–12/wheat-stocks-sugar-cane-rice-crops-ruined-by-pakistan-s-worst-flooding.html
Barfield TJ (1981) The Central Asian Arabs of Afghanistan: pastoral nomadism in transition. The University of Texas Press, Austin
Dor R (1975) Contribution a l'etude des Kirghiz du Pamir afghan. Publication Orientalistes de France, Paris
Dor R, Naumann CM (1978) Die Kirghisen des afghanischen Pamir. Akademische Druck- und Verlagsanstalt, Graz
Felmy S, Kreutzmann H (2004) Wakhan Woluswali in Badakhshan: observations and reflections from Afghanistan's periphery. Erdkunde 58(2):97–117
Goodhand J (2000) From holy war to opium war? A case study of the opium economy in North Eastern Afghanistan. Central Asian Surv 19(2):265–280
Integrated Regional Information Networks (IRIN) (2011) Afghanistan: imported wheat prices raise concerns. IRIN News. 13 May. http://irinnews.org/Report.aspx?ReportID=92704
Khazanov AM (1994 [1983]) Nomads and the outside world, 2nd edn. The University of Wisconsin Press, Madison
Kreutzmann H (1991) The Karakorum highway: the impact of road construction on mountain societies. Mod Asian Stud 25(4):711–736
Kreutzmann H (1993) Challenge and response in the Karakoram: socioeconomic transformation in Hunza, northern areas, Pakistan. Mt Res Dev 13(1):19–39
Kreutzmann H (2000) Hard times on the roof of the world. Frankfurter Rundschau 258:9
Kreutzmann H (2001) Überlebensstrategien der Kirgisen Afghanistans. Nomaden auf dem Dach der Welt. Geographische Rundschau 53(9):52–56
Kreutzmann H (2004) Pastoral practices and their transformation in the north-western Karakoram. Nomadic Peoples 8(2):54–88
Kreutzmann H (2005) The significance of geopolitical issues for development of mountainous areas of Central Asia. Presented at the "strategies for development and food security in the mountainous areas of Central Asia." Dushanbe, Tajikistan, 6–10 June. (n.p.)
Kreutzmann H (2009) Transformations of high mountain pastoral strategies in the Pamirian Knot. Nomadic Peoples 13(2):102–123
Mortenson G (2009) Stones into schools: promoting peace with books, not bombs, in Afghanistan and Pakistan. Viking, New York
Pain A (2010) Afghanistan livelihood trajectories: evidence from Badakhshan. AREU, Kabul
Schaller GB (2004) The status of Marco Polo sheep in the Pamir Mountains of Afghanistan. The Wildlife Conservation Society (n.p.), 36 pp
Shahrani MN (1980) The Kirghiz, now of Pakistan. Cult Surviv Q 4(4):11–13
Shahrani MN (2002 [1979]) The Kirghiz and Wakhi of Afghanistan, 2nd edn. University of Washington Press, Seattle
Sherratt Andrew (1981) Plough and pastoralism: aspects of the secondary products revolution. In: Hodder I, Issac G, Hammond N (eds) Pattern of the past: studies in honour of David Clark. Cambridge University Press, Cambridge, pp 261–305
State Bank of Pakistan (2010). Inflation Monitor, Oct 2010. (n.p.). www.sbp.org.pk/publications/Inflation_Monitor/2010/Oct/IM_Oct_2010.pdf
Wald B (2005) Notes on Marco Polo sheep and local people in the Southern Pamir Mountains of Tajikistan and the Wakhan corridor of Afghanistan, late winter/early Spring 2005. (n.p.), 7 pp
Wong E (2010) In icy tip of Afghanistan, war seems remote. New York Times 28 Oct, Section A1. http://www.nytimes.com/2010/10/28/world/asia/28wakhan.html

Chapter 5
Livelihoods of the 'New Livestock Breeders' in the Eastern Pamirs of Tajikistan

Tobias Kraudzun

Abstract This chapter focuses on the adaption of livelihood strategies based on livestock breeding in the Eastern Pamirs to the new politico-economic conditions in post-Soviet Tajikistan. Furthermore, its effects on the changing socio-economic structures are analysed. Peculiarities like the introduction of a household responsibility system and the disorder during the years of the Tajik civil war eased private livestock appropriation and delayed all further steps to reorganise the societal set-up. By recording and interpreting household biographies, it could be revealed how individuals perceived opportunities and chances during the transitional period and how their economic decisions altered their livelihoods. Given the growing gap of socio-economic disparities within the community, the significance of the livestock-based economy could be shown in terms of the division of labour, pasture user rights, herd management, entrepreneurial and trade opportunities.

Keywords Post-socialist transformation • Livelihood trajectories • Livestock economy • Eastern Pamirs • Tajikistan

5.1 Introduction

Around the turn of the millennium, the collective farms (*kolkhozy*) were dissolved in the Eastern Pamirs, and the distribution of their livestock assets was implemented formally in an egalitarian way. The rationale allocated a certain number of livestock

---

T. Kraudzun (✉)
Department of Earth Sciences, Centre for Development Studies (ZELF), Geographic Sciences, Freie Universität Berlin, Malteserstrasse 74-100, D-12249 Berlin, Germany
e-mail: tobias.kraudzun@fu-berlin.de

H. Kreutzmann (ed.), *Pastoral practices in High Asia*, Advances in Asian Human-Environmental Research, DOI 10.1007/978-94-007-3846-1_5,

units for each member of a household, being a resident of the sub-district of the respective *kolkhoz*. However, a decade later, most households were not successful in starting a business in animal breeding, and substantial inequalities in livestock ownership between the households exist today. On first sight, this observation seems surprising, as the vast majority of the Eastern Pamirs' population identifies livestock breeding as their most important income source. Only a few economic alternatives exist in this peripheral high mountain region. The livestock economy is affected by the meagre local purchasing power. Other opportunities for enterprises are still less developed. This leads us to the question: What hindered the majority of the households to participate successfully in the livestock economy after the implementation of an egalitarian distribution mechanism for livestock assets?

The transitional setup in the Eastern Pamirs differed from that of adjacent regions in post-Soviet republics. Just before the dissolution of the Soviet Union, a reform was introduced in the district, when responsibility for economic success of the then-state enterprises was transferred back to individuals: animals could be leased out to people who were formerly employed as herders, bureaucrats and teachers. Hence, many of them had no livestock herding and breeding knowledge. The post-independence civil war in Tajikistan delayed all further steps to reorganise the societal setup. The absence of law enforcement by the state caused a kind of power vacuum that encouraged the herders to act on their own behalf. They took opportunities to appropriate collective livestock and to control the pastures. Consequently, during the 1990s, livestock assets decreased substantially until the late dissolving of collective enterprises.

Since the livestock distribution, independent decision-making has been a challenge to the 'new livestock breeders'. Their previous experience was guided by a strong dominance of Soviet geopolitical interests that ensured high levels of supply for livestock production and their well-being.

In the following, the present-day livelihoods of the Kirghiz and Pamiri communities in the Eastern Pamirs will be described and interpreted in respect to changes in the socio-economic setup that significantly affects the applied livelihood strategies. It will be shown how individuals have perceived opportunities and chances during the transitional period and how this perception influenced their actions as stakeholders. The presentation of the evidence is guided by the following specific questions: How did the households relate to changing politico-economic frame conditions, and what shifts in strategies were undertaken in order to sustain their livelihoods? How is the socio-economic structure of the local population affected, and how has pasture use been impacted in the Eastern Pamirs?

5.2 Fieldwork-Based Approaches

Today, pastoralism is practised by small-scale, household-based enterprises. This overlapping of economic and social life qualifies the household as the elementary unit of analysis and interpretation. The findings presented here are based on extensive fieldwork conducted starting 2003, though mainly between 2007 and 2010 over a span of 14 months in the Eastern Pamirs.

The utilisation patterns of pastures as the crucial natural resource were recorded in order to gain insight into the spectrum of pastoral practices. Interviews with 280 pastoralists took place in two sub-districts selected for my case study over a period of 3 years. I visited and interviewed my counterparts repeatedly in their seasonal pasture camps. Data on previous and present-day usage rights, pastoral practices, provision of winter fodder, along with composition, size and ownership of the herds were gathered. Another sample of 70 households was selected to record their life histories, where information was collected about their perceptions, experiences, livelihood decisions, their rationales and the outcomes during the different phases of change in the Eastern Pamirs.

The informational base was supplemented by data on pasture use, pastoral production and economy in the collective and state farms during the Soviet era as well as the subsequent transition period. Documents were consulted that have been archived by farms, district and province administrations. Enquiries about historical contexts are based on work in regional archives and provided helpful insights into the current situation. Key informants and eyewitnesses were interviewed for their authentic views and personal experiences with and during the transformation processes.

5.3 District Murghab in the Eastern Pamirs

The Eastern Pamirs, a high mountain desert of Central Asia, are located in the easternmost part of Tajikistan. Two characteristics mark this region. First, after the colonisation by the Russian Empire, the region experienced significant geopolitical changes. Since the 1930s under Soviet power, the area became a borderland that was closed, detrimentally affecting regional mobility and exchange patterns (Kraudzun 2011; Kreutzmann 2009a, b). Second, the combination of a sparse vegetation cover (Vanselow 2011; Walter and Breckle 1986) and its peripheral location contributes to limited choices for livelihoods based on local resources. Pastoralism seems to be a prime adapted strategy.

Nowadays, Murghab district is bound in the east by the autonomous province of Xinjiang in China, in the south by Afghan Badakhshan, and the northern boundary was established after the dissolution of the Soviet Union between Tajikistan and Kyrgyzstan. The Eastern Pamirs comprise the major part of the Pamir Plateau and are widely congruent with the Murghab district which is part of the autonomous province of Gorno-Badakhshan in Tajikistan (cf. Fig. 5.1).

The high mountain plateaux have been used by members of two different communities for centuries. Combined mountain farmers originating from the deeply incised Western Pamir valleys cultivated food crops on irrigated terraces there and drove their cattle herds to the Eastern Pamir pastures during summers. The second group consisted of Kirghiz pastoralists. Some evacuated from conflict-prone regions

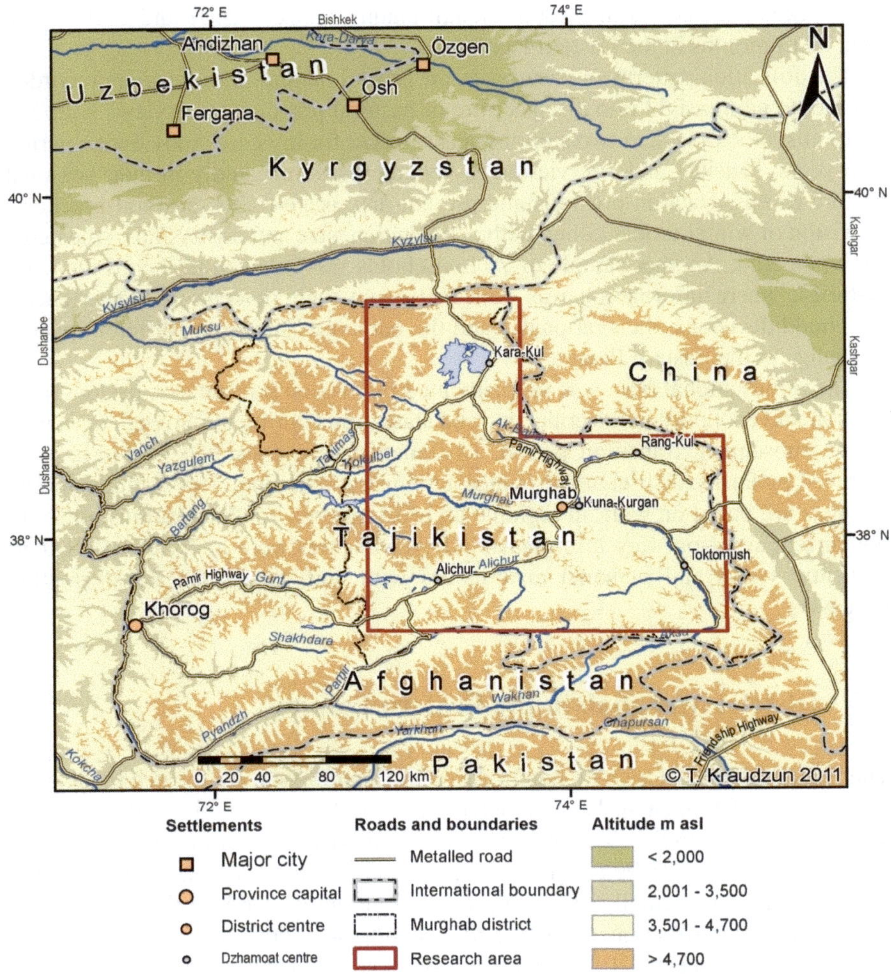

Fig. 5.1 The location of the research area Murghab district within Central Asia

to the remote Pamirs, initially on a seasonal basis and later on permanently (Dor 1975; Maanaev and Ploskikh 1983; Stankevich 1904).

In former times, the pastoralists depended on the exchange of their livestock surplus with basic foodstuffs and other goods in agricultural markets such as Kashgar and Osh. Furthermore, their pastures were spread over a wide range of the Pamirs. Beginning with the establishment of the Russian administration, the trans-boundary mobility of the pastoralists was increasingly constrained. A prominent example of 'closed frontier nomadism' is that of Kirghiz pastoralists in the neighbouring Afghan Pamirs.[1]

The valuable pastures of these high plateaux are situated on an altitude that ranges from around 3,500 m asl in the lowest valleys up to 4,700 m asl. The mountain peaks reach to more than 7,000 m in the northern and Western mountain ranges. The orographic environment and its latitudinal location cause low temperatures throughout the whole year, with an annual average of only −1°C to −3°C in the valleys. Moreover, the climatic situation is marked by extreme aridity with precipitation values below 100 mm per year (cf. Agakhanianc 1965; Walter and Breckle 1986).

Based on these natural preconditions, an extensive and seasonal livestock herding of yaks, sheep and goats seemed to be a self-evident option for agricultural activities. The population density today is still low in this district, with an area of 38,300 km² but only 14,000 inhabitants, consisting of 77% Kirghiz and 23% Pamiri people. Half of the population is living in the district capital and economic centre Murghab, the other half in far-flung villages and hamlets (Statotdel Murgab 2008b).

5.4 The Historic Legacy

5.4.1 The Soviet Modernisation Project: Secure Livelihoods

Prior to the Soviet collectivisation, the pastoral production was organised by kinship-based groups who moved jointly to seasonal pastures. These communities were marked by huge disparities in the socio-economic status between households. The whole group operated as a political and economic unit and was governed by the head (*bay*) of a wealthy household, who usually owned most of the group's livestock (Gotfrid and Gafiz 1930:30; Shibaeva 1973:104).

After the establishment of the Soviet power in the region, the major goals were to persuade the local population of the advantages of state-led, integrated rural development and to convince them of the utility of a completely different organisation of economic activities, sometimes with force – to produce in collective farms. Every endeavour was made to improve the supply with foods and goods[2] (cf. Fig. 5.2). Eleven *kolkhozy* were established quickly in 1940–1941 and finally reorganised into four *kolkhozy* and one *sovkhoz* specialised in yak breeding (Sovnarkom TadSSR and KPTad 1943; Taipov 2002:71). The district's assigned role within the Soviet economy was to breed livestock to be delivered to processing plants in lowland regions (TK[3] 05.04.09, VV 05.07.08, Antonenko 1985; Kleandrov 1974). However, the economic performance was dissatisfying; only during a short period of the 1960s did the *kolkhozy* produce a surplus (Statotdel Murgab n.d.).

The main rationale for the exceptional endeavours introduced to the Soviet Pamirs was based in the geopolitical significance of the long external boundary shared with China and Afghanistan. The control of this vast border region is dependent on appropriate infrastructure, effective administration and, last but not least, the pastoralists' loyalty and knowledge. Consequently, the Pamir highway, the first

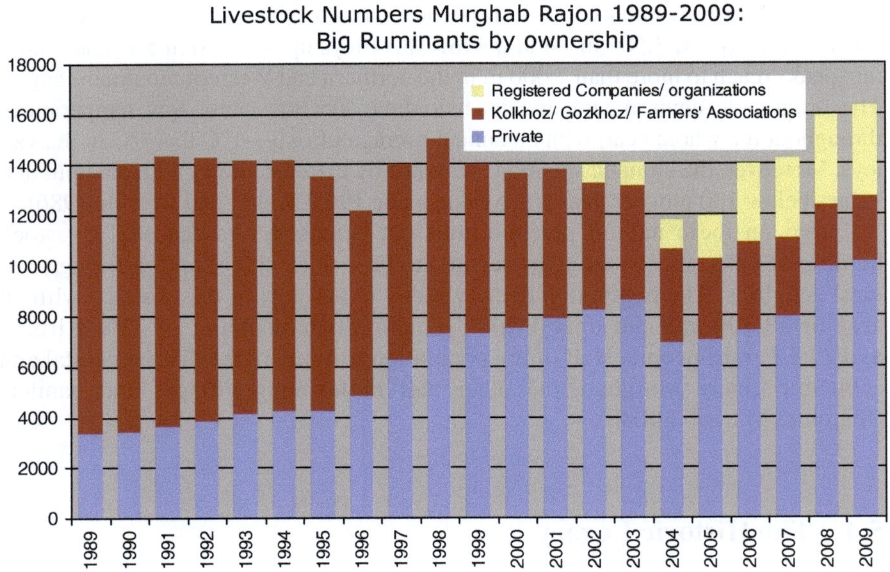

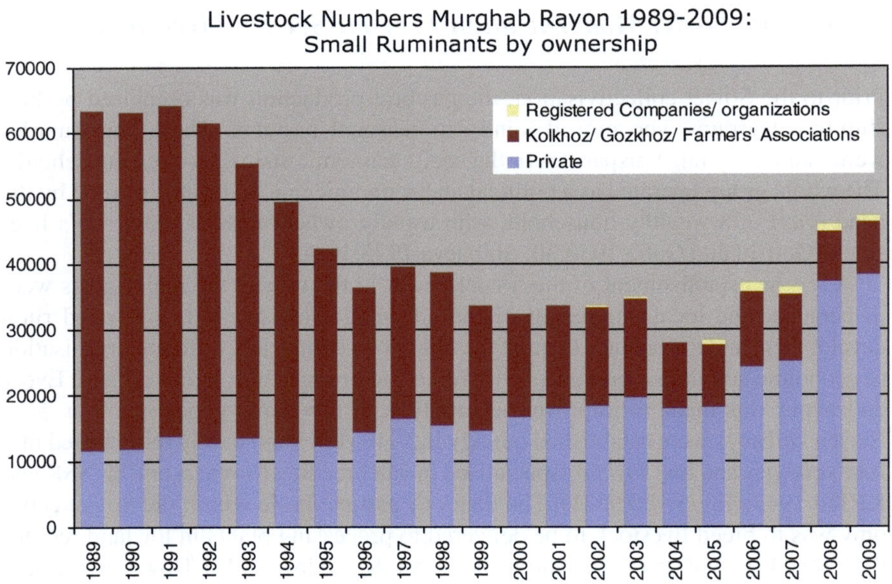

Fig. 5.2 Livestock numbers in Murghab Rajon (Source: Statotdel Murgab time series n.d.)

metalled road in this high mountain region, was advanced over a distance of more than 700 km from Osh to Khorog (cf. Fig. 5.1). Furthermore, modern buildings with administration, education and social infrastructure were erected starting with the establishment of Murghab as a district in 1932 (Taipov 2002). With the progress of

Photo 5.1 'Moscow supply': Countless trucks of the transport trust 'Pamirskoe avtotransportnoe upravlenie' (PATU) in Murghab that was financed and controlled by the supranational administrative level of the Soviet Union (Source: El Registan 1936) (Special thanks go to Markus Hauser for providing the author with the scanned photograph from his Pamir Archive Collection)

the development efforts, many people were employed in non-agrarian sectors. To establish a modern administration and high living standards, the region was in need of a variety of professions. Many young specialists, mainly Pamiris originating from the Western Pamir, came to work in the administrational centre of Murghab (AA 28.05.08, LH52 13.05.09). Although they were directed by the state after their professional education, most of them were pleased by the privileges that stemmed from this. The allowance paid for living under high mountain conditions attributed an additional 50% income, combined with a comparably rich provisioning with consumer goods – called vernacularly 'Moscow supply' (cf. Photo 5.1) – resulted in satisfying livelihoods (BT 18.05.09).

The livelihoods and living conditions of the pastoralists were changed by the reorganisation of the agrarian sector. After the collectivisation, most of the pastoralists were settled in villages, where the production infrastructure of *kolkhozy* was set up. Mechanisation and the centralised supply with production inputs like fuel, combustible fuel, forage as well as a supply infrastructure for the population with foods and goods required many specialised workers and the establishment of a sophisticated administration and accounting system. Living and working in the *kolkhoz* centres, most of the households left the arduous herders' life on the pastures behind them. The renowned feature writer of the *Izvestia* newspaper El Registan pictured it as follows:

> Moscow has climbed the 'Roof of the World'! It has come armed with knowledge, technology, culture, to change the backward lifestyle of this peripheral area. (1936:Ch. 4; translated by the author)

However, the core of the pastoral production in the Eastern Pamirs remained unchanged. Herders still moved to seasonal, sometimes very remote, pastures, where they lived permanently and took care of the herds. A great amount of effort was made to ease their work and improve their living conditions. A farm manager of the Soviet period remembers:

> [In Soviet times] each herder had to engage only in his core mission: herding the animals. Unlike today, every herder even had a mount and an assistant to control the herd. For each additional task he was assigned additional workers. … lambing brigades, … shearing brigades, … hay making brigades. (TU 11.04.09)

The herders received backstopping for most labour-intensive phases like lambing and shearing. A supply with emergency forage in times of snow-covered winter pastures and with combustibles during the gestation period was limiting climate-related losses in the herds. Sales vans of the state shops delivered foods and goods to the pasture camps, even regular movie screenings with mobile cinemas were organised. The herders, who lived with their families in the pasture camps, were regularly checked by mobile doctors, and their children were educated in boarding schools.

To conclude, livelihoods were secure, and socio-economic living conditions were good, given the region's peripheral location, and especially in comparison with neighbouring countries. On the other hand, private economy and trade were quite uncommon. For example, a '*kolkhoznyj bazar*' like in most Soviet towns, where the population usually sold the private agricultural surpluses, did not exist. The then-mayor of Murghab remembers: 'The shops were full, there was no need for [private] trading, and any privately organised distant trade was outlawed as "*spekulacija*"' (KK 06.05.08).

5.4.2 The Transitional Decade: Uncertainties and Opportunities

The framework for the transition period was partly set in the late Soviet period, when the Soviet power was increasingly failed to compensate for the inefficiency of the *kolkhozy*. A reform – similar to the household responsibility system in China (cf. Kreutzmann 2009b, 2011; Lin 1996) – was introduced locally in order to transfer responsibility for the economic success of state/collective enterprises to individual employees/*kolkhoz* members. From 1988 on, livestock was leased out to the 'new livestock breeders' formerly employed as herders, bookkeepers and teachers. A herd and a set of seasonal pastures were allocated for each livestock tenant, and he was assigned an account in the *sovkhoz* bookkeeping. He was made responsible for the economic efficiency of 'his' small virtual enterprise. All production inputs, like transportation services or fodder, were booked on the tenant's virtual account as expenditures and had to be balanced with the delivered animals as production output (TK 05.04.09).

This tenancy system worked well in the last years of the Soviet period, but it collapsed a few years after independence. The hyperinflation of the Russian Rouble, in combination with growing uncertainties of the management of the livestock-purchasing enterprises, contributed to the discontinuation of the hitherto existing exchange relations. The supplies to the Pamirs ceased, and the state farms were in need of everything to make a living except their own production based on livestock. Then, with dramatically inflating prices and without livestock sales, most of the tenants ran into debts to the *sovkhoz*. The vast majority of goods needed to be imported and became very expensive, whilst cash was scarce and losing rapidly in value.

The livelihoods suffered from this economic crisis, as salaries were missing and household savings were devaluated by the inflation. Households were in a desperate situation because after witnessing five decades of social security in the Soviet Union, they had no experience of how to cope with this severe economic crisis. At first sight, subsistence economy and migration were two appropriate options to sustain a livelihood.

Taking the first option, livestock was the most obvious resource. Private ownership was strictly limited in the Soviet time, so private herds were still insignificant and most livestock was owned by the *sovkhoz*. These animals were accessible particularly for livestock tenants. They took opportunities to appropriate state livestock, either for immediate consumption, to barter for daily needs, or to integrate into their private herds.

> The *sovkhoz* sheep were very cheap. To tell the truth, for 4–5 bottles of vodka you got a big sheep … or 20 packs of cigarettes.… 1991, 92 – in those days there was still much *sovkhoz* livestock, but the law enforcement was not active anymore, it was complete anarchy, you just gave little to the tenants and got the best animals, they utilised the livestock as if it was theirs. (LH37, 24.05.09)

During the absence of cash, livestock was the main currency for local exchange. Usually, the appropriated animals of the *sovkhoz* herds were declared by the herders as loss. The oblast' level criticised the sharp decrease in livestock and tried to intervene with repeated inspection teams.[4] After all the subsequent audits and livestock censuses, most of the tenants had a negative balance on their *sovkhoz* accounts and were missing considerable amounts of the entrusted animals. Though, even if the damages had to be compensated, the missing assets calculated in prices of the initial tenancy system were easily repayable after the hyperinflation. In retrospect, the tenancy system can be seen as a 'simulation' of privatisation with the difference that individuals who failed were bearing only limited consequences.

Emigration was a second option: In 1993, the records mention 798 people (not including undocumented migration) who left the district, four times more than in the subsequent year (Statotdel Murgab n.d.). This development is in stark contrast with previous immigration. The post-independence civil war in Tajikistan forced many migrants from the lowlands into the Pamirs and further isolated the region (Herbers 2001). In order to alleviate the disastrous supply situation, the Aga Khan Foundation organised humanitarian food supplies within the framework of the 'Pamir Relief and Development Project' (Bliss 2006:300). This meant the import of a yearly supply of 300–1,800 tons of foodstuff for the district Murghab (MSDSP 2005). With the advent of these provisions, the exchange relations based on the Soviet system ceased entirely.

After the failure of the tenancy systems, the *kolkhozy* returned to the earlier Soviet-style management. Meanwhile, the number of the small livestock had decreased by 45% since independence. On the other hand, the animals in private herds accounted already for about one third, growing to more than a half of the district's total livestock before the official privatisation took place (cf. Fig. 5.2). De jure, very few pastures were allocated for the strongly limited private livestock during Soviet time. However, most of the former tenants continued to work as herders for the collective livestock – very often on the plots that had been allocated for them before. They felt already like proprietors on the pastures which they had worked as tenants.

5.4.3 Final Reorganisation: Distribution of the Collective Assets

At the turn of the millennium, the four still existing collective farms were dissolved. Subsequently, their former members were gathered to form new collective structures called *associations of dekhan farms*.[5] They inherited the land titles for the pastures and the remaining livestock. However, the animals were distributed equally amongst their members. Depending on livestock assets and village population, a household[6] – consisting of 5.8 members on average – could receive between 1.2 and 5 head of big livestock and 6.2 and 56 head of small livestock.[7] Yet the impact of the redistribution remained limited for most of the households, given the remaining share of only 62% small livestock and 37% big livestock of the 1991 livestock assets to be distributed (cf. Fig. 5.2).

Since the procedure was initiated by the district administration and communicated to the population as a distribution of the livestock, the population perceived it as straight privatisation of the *kolkhoz* assets. The distribution mechanism led to a legal ambivalence, in which the administration counts the distributed livestock de jure as owned by the *associations of dekhan farms*, whereas the people have integrated the animals de facto into their herds and perceive them as their own.

The second most important asset of the *kolkhozy* was the means of transport. All vehicles were privatised straightforwardly. Usually, the *kolkhoz* drivers had the first possibility to buy the trucks. Although the prices were favourable, cash was hardly disposable within most households in that time.[8] Only a few individuals, who had already started a successful business in the 1990s, were able to equip themselves with the needed transportation.

The same applies for the third most important asset of the *kolkhozy*. Buildings and stables were offered for sale right after the dissolution of *kolkhozy*. Initially, the users would have preferred to rent them. Only recently, these assets have become contested because their ownership can serve to strengthen and underline claims for certain pastures.

The livestock distribution had some immediate effects. Despite the livestock allocations for their private herds, most households were uncertain of how to start a business that would ensure their livelihoods. This insecurity and inexperience triggered many decisions to sell their animals immediately. The then-head of the district administration's agriculture department assesses the outcome pessimistically:

> Most of the people did not know how to develop their own livestock business. ... They regarded the [distributed] livestock as a gift from above. For the first time after many years they were in a position to make their purchases – they just had to sell one sheep on the bazaar. ... After a year, these people had spent everything, and no livestock was left. (MD 17.05.09)

The main livestock buyers were households that were economically successful already in the 1990s. By this formula, the livestock distribution had only limited levelling effects.

5.5 Present-Day Livelihoods: Failure and Success

After the peace agreement of 1997, Tajikistan's economy began to recover after 5 years of civil conflict. Following this, the state was increasingly able to raise wages and pensions[9] – a development that improved the livelihood situation in the district of Murghab, where most people with formal jobs are employed by state-funded organisations.

In addition, an external actor stopped playing a leading role for the local economy: until 2003, the Russian federation was financing and managing the control of Tajikistan's borders with Afghanistan and China. Many local people were hired as contracted border guards. With an extraordinary high salary for local standards, every employed soldier and labourer was usually able to entirely support his household. A former contract soldier assessed that 200–300 households benefitted in this way (JI 19.04.09). Furthermore, many products were purchased by the affluent Russian soldiers on the local market. The considerable inputs to the local economy ceased when the task was handed over to the poorly financed Tajik border guards. This created a significant income gap in many local livelihoods.

5.5.1 Livelihood Options in the Local Economy

Today, only one-third of the population of working age people are employed formally, of which 68% are engaged in the state sector. Official statistics show a share of employment of only 15% in the villages, whereas in the district centre, 45% of the working-age population are declared to be employed (Statotdel Murgab 2008b).[10] About 17% of the households supplement their livelihoods with informal jobs, and 21% receive remittances from household members who out-migrated in order to seek employment. In sum, these resources combine to an average of 450 TJS per household, but the income is distributed very unequally. More than two-thirds of the Murghab households spend less than the 400 TJS necessary for minimal needs,[11] 10% with at least 1,000 TJS can afford sufficient amounts of consumer goods and

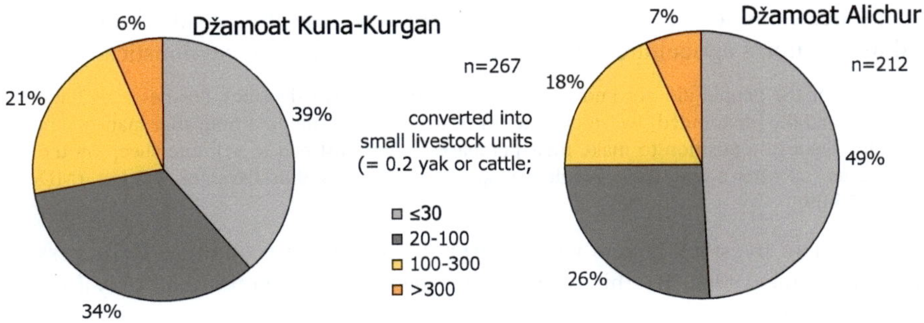

Fig. 5.3 Herd sizes of pasture users in two sub-districts of the Rajon Murghab (Source: Selkhozotdel 2007b)

only 2% of the population, with more than 2,000 TJS, can purchase everything they desire.[12] Interpreting these figures in comparison with the previously mentioned minimum needs level, we might conclude that some additional livelihood sources are still missing in our discussion.

Informal incomes are often irregular and remain unaccounted for. For example, bartering complements the livelihoods to a large extent. For many households, livestock is the main productive asset that generates a surplus beyond their limited regular income. Furthermore, animals are the main household savings. To maintain their herd productivity, every household needs a set of yielding seasonal pastures, good relations to a knowledgeable herder and/or a sufficient disposable workforce. In the district, an average household owns 18 small livestock and five yaks for subsistence and barter (Statotdel Murgab 2008a). Significant and big disparities are reflected in the variegation: between a third and a half of pasture users are in a position to sustain their livelihoods out of their small herds (Fig. 5.3). Furthermore, *teresken*, a dwarf shrub, is extracted for energy purposes, supplementing livelihoods as a barter or cash commodity (Samimi et al. 2011). Altogether, the majority of the households have difficulties earning livelihoods sufficient to meet their basic needs, not to mention its robustness against economic shocks.

5.5.2 Household Biographies and Livelihoods

The population of the Soviet Pamirs witnessed an egalitarian societal organisation with widely equal living conditions and opportunities. The sharp decline after the disintegration of the Soviet exchange and supply system was felt by all citizens. The 1990s were marked by negative trends of the regional economy and external emergency aid. At the same time, private livestock numbers developed steadily, whereas collective ones lowered (cf. Fig. 5.2). The enforcement power of the district and *kolkhoz* administrations was weak. Private and collective animals were mixed in the herds, providing many opportunities for herders to appropriate livestock. Since the official statistics were usually extrapolated from reported numbers, significant corrections became necessary when inspections figured out the real numbers.[13]

Anyway, most of the herd owners were not able to make their private livestock economy a viable livelihood. The reasons stated in interviews were almost identical. Most named the lack of financial resources and insecurities about economic decisions, especially after the well-remembered failed experiment with earlier tenancy schemes. There were few incentives for establishing sustainable livelihoods when emergency aid covered much of the basic food requirements for a period of 11 years (MSDSP 2005). Developments in western Tajikistan were not encouraging as well: the civil war abated only slowly. The Pamir region remained more or less isolated from adjacent markets that could and would purchase livestock products.

But why were some households able to successfully start a livestock business – even before the dissolution of *kolkhozy*? In order to illustrate the triggering factors, some examples of household biographies will be presented that provide some hints as to how people perceived the events of the transitional period and what they based their decisions on for adapting and coping with the perceived realities of the time.[14]

Egembay (age of 50, LH08) is an extraordinarily wealthy livestock owner representing a five-person household. The foundation for his success was laid already 20 years ago. His father was amongst the few who exploited gaps in the tight Soviet control system by using his position to gain assets. As a truck driver delivering supplies to the pastures, he was in permanent contact with many herders. Amongst those were herders who did not keep the maximum number of permitted private livestock.[15] He convinced these individuals to declare his illegal livestock as their own by filling-up their quota. In this way, he managed to gather a herd of 10 small and 50 big livestock by the end of Soviet control. Egembay himself continued to manage the growing herd with his father. They sustained the high livestock numbers even during challenging times. Additionally, his brother's son, as well as the family of his daughter, got involved in the family enterprise. In order to exercise effective control over the hired herders, they rotate living in the pasture camp. Their example shows how acquired experience and good access to valuable information can support the quick start of a livestock business. Egembay continues to be successful because his management is in tune with present-day requirements, resulting in large private herds of 1,500 small and 500 big livestock.

Janybek (age of 56, LH03) was one of the *kolkhoz* specialists who has a profound knowledge about livestock breeding and management. In the final years of the Soviet period, he worked as a breeding expert (*zootekhnik*) and head of a farm department. During the difficult transformative periods, he occupied positions as a bookkeeper and even became the chief economist of the *kolkhoz*. At the same time, he was successful in his private livestock breeding, maintaining a herd of 60 small and 15 big livestock. However, he lost most of his animals in devastating weather of April 1998, when the pastures were covered with crusted snow (*zhut*), and commented: 'Knowledge about livestock helps nothing here. It is God who distributes … in the Soviet Union, the adherence to the rules of herd management guaranteed good results'.

Photo 5.2 Herding animals on the winter pasture on the Alichur Pamir near Chatyr-Tash. Given the hard job and the great responsibility for the valued livestock, most hired herders feel underpaid and unattended (Photograph Tobias Kraudzun, March 25, 2009)

Accepting this damage as god's will, he changed his focus to trade and opened a shop in the Murghab bazaar. However, he had to acknowledge that his business did not generate earnings. His only benefit was to purchase supplies at cheaper prices in Osh. He had to start breeding again, though out of the ten heads of small livestock allocated to his household during privatisation, half was missing when he went to the herder entrusted with them. After taking over the remainder, with difficulties he managed to develop a medium-sized herd that is alternating between 60 and 100 small livestock – depending on occurring and extraordinary expenses. His and his wife's pensions are the main cash income source for the five-person household. To economise, his sons regularly gather a small team and drive to pastures eligible for digging *teresken*. The dwarf shrubs are sold or used for heating the home and cooking. Recently, Janybek took out a loan to buy 15 small yaks. However, he is not sure that the additional returns will outweigh the high reimbursement rates and pay back his investment.

Ismail (LH36) was not involved in herding during the tenancy system, but was working as a driver for the Russian border troops for a few years. His strategy was to use his salary to set up a herd. The investments were quite low in the period of weak control, when many animals were extracted from the *kolkhoz* herds and offered at reasonable prices. Living in Murghab town, he contracted his herd for maintenance to a herder (cf. Photo 5.2). Whenever a large number of offspring and adult animals were lost, he would release the shepherd and hire another one. His judgement on hired herders is typical:

> You cannot please them, no matter how good you support them. The first season, they are working well. Then they get out how to embezzle animals. You have to kick them out. If they cannot defraud, they leave claiming they were not supported. (Ismail, April 19, 2009)

Although he replaced the herder four times, the herd size did not increase as he aspired. In 2007, he ventured to open a shop, where he expected to earn around 600 TJS monthly. But soon after the initial investment, he had to support his sister who had to undergo an expensive surgery. Ismail sold animals to cover the expenses, thus diminishing his herd from 150 small and 30 big livestock to 120 small and 20 big livestock. He seems to be quite optimistic that he might recover the herd loss, but largely it depends on the faithfulness of the hired herder.

These three examples illustrate a number of factors that have influenced the chances for success in private livestock business. Significant organisational and management skills were required to build a herd during the period of post-Soviet *kolkhozy*. Although the limitations for private livestock ownership were relaxed, there were officially only a few pastures allocated to graze private animals. Every owner had to make informal arrangements with herders – or to be one himself. On the one hand, the post-independence disorder and the lack of control eased opportunities to acquire livestock. There was a window of opportunity to increase one's own herd either free of cost or for very low investments. On the other hand, only a few people availed these chances and developed long-term visions in the livestock sector.

The often stressed livestock distribution of the former *kolkhozy* had no long-lasting and significant impact on households. First, less than half of the population benefitted from the distribution. Second, many animals merely existed on the distribution lists. Third, the small numbers of distributed livestock were of little importance for already successful herders. Poorer people, for whom the distributed livestock was a considerable input, were too inexperienced to maintain their herds. Most of them lost their animals in the first year by selling them in order to cover daily needs.

5.6 Conclusions

Only some households became successful livestock entrepreneurs. In contrast, the vast majority of the local people did not take the appropriate economic steps. This was mainly influenced by three factors. First, the realities of the current economic crisis in comparison to their previous experience were discouraging for them. Second, their perception of being first and foremost dependent on external support was somewhat endorsed by a steady supply of humanitarian aid during a period of 11 years, by the big economic impact of the Russian border troops' presence and by the continuing large-scale subsidies for the local administration from the national level. Third, many of the 'new livestock breeders' were discouraged by the collapse of the tenancy regime which can be seen as failed 'simulation of privatisation'.

Another positive factor is the availability of workforce committed and experienced in livestock breeding and management within the network of relatives. The lambing and calving season is crucial for the productivity of a herd. During this period, the product of several factors influences the survival of offspring. These include the motivation of, control over and support for the hired herder. Furthermore, protecting the animals against predators all year round, but especially on the winter pastures, keeps livestock losses low. These dependencies make the relationship between livestock owners and herders decisive (cf. Photo 5.2). The regular complaints about fraud of animals on the side of the owners, as well as about scant support and unfair treatment on the side of the hired herders, indicate the negotiation processes between both as pivotal.

Specialised professional knowledge and experienced professionals, as they were employed in the Soviet system that allocated and divided labour in the livestock sector, have been beneficial in some cases in increasing herd productivity. However, it has rarely helped to develop coping strategies to tackle and adaptation strategies to overcome the present economic crisis.

At present, when after a period of decline livestock numbers have recovered to levels known from the Soviet period, the competition between pastoralists has increased. The availability of accessible fertile and suitable pastures is shrinking. The chances of 'newcomers' to access seasonal pastures are rather limited. The growing competition for pastoral resources has effected that most pasture users do not relocate their livestock four times a year to seasonal pastures anymore. The rotational cycles have been adapted to the changed valuation of pasture resources. De jure, the associations of *dekhan* farms distribute pastures according to users' herd sizes. De facto, many individual claims are enforced by justifying them as based on customary law dating back to the tenancy period. The changed appreciation and the growing competition have led to evidence of under- and overgrazing. Pastures near the settlements and some without strong user claims are heavily frequented and seem to get degraded. Large areas that are claimed by powerful and influential livestock owners are partially under-utilised (Kraudzun 2009; Samimi et al. 2011). Conflict seems to be inevitable in such situations, constraining the further development of the livestock economy, as well as the sustainable use of natural resources.

However, the pastoral livestock economy is embedded in a wider range of livelihood activities. The Eastern Pamirs are linked to national and regional exchange networks and are part of the domestic and international labour market. Survival of pastoral households is augmented by migrants' remittances, trade and commerce. Thus, the human-environmental system is modified by substantial external agents, contributions and dependencies. The case of the Eastern Pamirs exemplifies that an understanding of local and regional resource utilisation needs to be embedded in a wider set of communication, cooperation and exchange relations. Nevertheless, the pastoral practices of Kirghiz and Pamirian mountain dwellers form a significant backbone for the survival of their households in a harsh environment.

Acknowledgements The author is grateful to the Volkswagen Foundation that made the empirical research in the Pamirs between 2007 and 2009 possible by funding the research project 'Transformation Processes in the Eastern Pamirs of Tajikistan. Changing Land Use Practices, Possible Ecological Degradation and Sustainable Development'. Many interview partners in Tajikistan have supported my enquiries. All support by herders, bureaucrats and development practitioners living and working in the Pamirs is gratefully acknowledged.

Interview Partners

AA: Adylbek Atabaev, 28.05.08
BT: Bekzhol Taipov, Murghab 31.03.09, 18.05.09
MD: Murzabay Dzhooshbaev, Murghab 22.10.07, 09.04.09, 17.05.09, 03.09.09
JI: Jurmat Ismailov, Murghab 19.04.09
KK: Kökönbek Kamchibekov, Murghab 06.05.08
TK: Teshebay Kolchokabev, Murghab 05.04.09, 16.05.09
TU: Tashbay Usenov, Kamar-Ötök 11.04.09
VV: Vakhid Vakhidov, Dushanbe 04.07.08, 05.07.08

Notes

1. This term was coined by Nazif Shahrani (2002 [1979]); cf. the Chap. 4 in this volume by Ted Callahan.
2. For example, more than 7,500 tons of goods were brought into the Pamirs in the planning year 1936–1937 to supply about 29,000 people who were living in the region in 1935, according to regional archival documents (GosArkhiv-GBAO 1936:1-3-27).
3. Full names of quoted interview partners are listed below.
4. In the state farms of Murghab district, animal loss increased in 1993 by 4,051 head of sheep and goats and 670 yaks in comparison to the precedent year, constituting 10% respectively, 7% of all stocks (USKH GBAO 1994).
5. In this case, each member household of the dissolved *kolkhoz* was regarded as a *dekhan* farm (literally peasant farm). De jure, a formally registered business was addressed by this denomination; de facto, it just referred to a household as an economic unit. For an overview of other types of *dekhan* farms, see Robinson et al. (2010).
6. Statistical tables based on the official population register show smaller household sizes, but de facto households sharing a budget and a table include often relatives and young families. The figure referred to here is derived from a survey conducted in Murghab in 2008 (Kreczi 2011).
7. By default, all inhabitants of a village were members of the *kolkhoz*. However, they amounted only to 47% of the district population (Statotdel Murgab 2000; Selkhozotdel Murgab 2007a). The population of the district centre Murghab and of the sub-district Alichur, with a persisting state farm, were not considered.
8. Trucks, as the most desired means of transport for seasonal herd relocations, were sold for the equivalent of 80–200 USD. This comparatively small amount equals the wages earned in 30–70 months (based on wages in 1999).
9. From 1998 to 2003, the officially fixed minimal wage has increased almost sixfold (Goskomstat RT 2004).
10. In official statistics, every occurrence of employment is considered, including part-time positions with salaries insignificant for earning a livelihood.
11. An average Murghab town household comprises 5.8 members and requires a minimum of about 400 TJS monthly for purchasing only the imported consumer goods. This calculation is based on consumption data (AgentStat 2010) refined with information from my own enquiry

of household consumption in Murghab. The vast majority of foods and dry goods have to be imported from outside. Only locally produced goods and raw materials extracted from nature can be substituted through workforce and support networks. The respective costs are omitted from this minimal value, including them would increase the margin of basic needs to 700 TJS or even to 946 TJS (Kreczi 2011).

12. Numbers in this paragraph are calculated from a livelihood survey conducted in Murghab in 2008 by Fanny Kreczi (2011). I gratefully acknowledge her sharing of these data.
13. After an inspection, both numbers for private and collective livestock had to be corrected for all livestock by 9% – the private upwards, the collective downwards. A similar correction was necessary after an inspection took place in June 2007: 6% for yaks and even 11% for small livestock – again upwards for private and downwards for farmers association's livestock (MD 17.05.09, cf. Fig. 5.2).
14. All biographical interviews were conducted with the household elder. Other household members often helped out with specifying dates and adding to memories. Nevertheless, the biographies reflect merely the assessments of the household head, disregarding the intra-household variations. All names of my interview partners have been anonymised.
15. In Murghab district of the late Soviet period, it was allowed to keep an amount of six small livestock and one yak with a calf. Livestock beyond that was 'contracted' by the *kolkhoz* for a (low) acquisition price set by the state.

References

Agakhanianc OE (1965) Osnovnie problemy fizicheskoj geografii Pamira. Izd. Akademija Nauk Tadzhikskoj SSR, Dushanbe

AgentStat (2010) Osnovnye pokazateli obsledovanija bjudzhetov domashnykh khozjajstv respubliki Tadzhikistana. Agenstvo po statistike pri presidente Respubliki Tadzhikistan, Dushanbe

Antonenko BA (1985) Socialisticheskie agrarnye preobrazovanija v Gorno-Badakhshanskoj avtnomnoj oblasti. In: Masov RM (ed) Ocherki po istorii Sovetskogo Badakhshana. Izdat. Donish, Dushanbe, pp 220–253

Bliss F (2006) Social and economic change in the Pamirs (Gorno-Badakhshan, Tajikistan). Routledge, London

Dor R (1975) Contribution à l'étude des Kirghiz du Pamir afghan, Cahiers turcica 1. Publications Orientalistes de France, Paris

El Registan (1936) Dem Autonomen Sowjetgebiet Gorno-Badachschan gewidmet. UdSSR im Bau 1936 (12)

GosArkhiv-GBAO (1936). fond 1, opis 3, delo 27 Khorog: Gosudarstvennyj arkhiv Gorno-Badakhshanskoj Avtonomnoj Oblasti

Goskomstat RT (2004) Regiony Tadzhikistana – statisticheskij sbornik. Gosudarstvennyj komitet statistiki respubliki Tadzhikistan, Khorog

Gotfrid L, Gafiz M (1930) Krasnyj flag na Kryshe mira. Pamirskie ocherki. Tadzhgiz, Tashkent/Stalinabad

Herbers H (2001) Transformation in the Tajik Pamirs: Gorniy-Badakhshan – an example for successful restructuring? Central Asia Surv 20(3):367–381

Kleandrov IM (1974) Èkonomika Sovetskogo Gornogo Badskhshana. Sovremennoe sostojanie i problemy razvitija. Donish, Dushanbe

Kraudzun T (2009) Mobile Viehwirtschaft im tadschikischen Pamir. In: Toktoraliev B, Kreutzmann H (eds) Herausforderungen für die Mensch-Umwelt-Beziehungen in Hochgebirgsregionen Tadschikistans und Kirgistans. Osh, pp 141–152

Kraudzun T (2011) From the Pamirian frontier to international borders: exchange relations of the borderland population. In: Bruns B, Miggelbrink J (eds) Subverting borders. Doing research on smuggling and small-scale trade. VS Verlag für Sozialwissenschaften, Wiesbaden, pp 166–186

Kreczi F-S (2011) Vulnerabilities in the Eastern Pamirs. Berlin Geographical Papers 39. Centre for Development Studies (ZELF), Berlin

Kreutzmann H (2009a) Geopolitical perspectives on cross-border exchange relations. In: Kreutzmann H, Beg GA, Richter J (eds) Experiences with and prospects for regional exchange and cooperation in mountain areas. InWEnt, Bonn, pp 20–57

Kreutzmann H (2009b) Transformations of high mountain pastoral strategies in the Pamirian Knot. Nomadic Peoples 13(2):102–123. doi:10.3167/082279405781826092

Kreutzmann H (2011) Pastoralism in Central Asian mountain regions. In: Kreutzmann H, Abdulalishoev K, Zhaohui L, Richter J (eds) Regional workshop in Khorog and Kashgar. Pastoralism and rangeland management in mountain areas in the context of climate and global change. GIZ, Bonn, pp 38–63

Lin JY (1996) Success in early reform: setting the stage. In: Garnaut R, Guo S, Ma G (eds) The third revolution in the Chinese countryside. Cambridge University Press, Canberra

Maanaev ED, Ploskikh VM (1983) Na 'kryshe mira' (Istoricheskie ocherki o pamiro-alayskikh kirgizakh). Mektep, Frunze

MSDSP (2005) Analiz o postupivshchij gumanitarnoj pomoshchi po linii fonda Aga-Khana po Murgabskomu rajonu s period sentjabr 1993 po dekabr' 2005 goda. Mountain Societies Development and Support Programme, Murgab

Robinson S, Whitton M, Biber-Klemm S, Muzofirshoev N (2010) The Impact of land reform legislation on pasture tenure in Gorno-Badakhshan – from common resource to private property? Mt Res Dev 30(1):4–13

Samimi C, Vanselow K, Kraudzun T, Aknazarov K, Kreutzmann H (2011) Der Ostpamir nach der Unabhängigkeit Tadschikistans – Landnutzungsverhältnisse und Ökologie. In: Anhuf D, Fickert T, Grüninger F (eds) Die Erde als Ökosystem im Wandel. Passauer Kontaktstudium Erdkunde 11. Passau

Selkhozotdel Murgab (2007a) Analiz sravnenie chislenosti razdannogo skota s kolichestvom skota na 1.1.2007 g. Upravlenie selskogo khozjajstva Murgabskogo rajona, Murgab

Selkhozotdel Murgab (2007b) Svedenie o perschjota skota na vsekh dzhamoatakh rajona. Upravlenie selskogo khozjajstva Murgabskogo rajona, Murgab

Shahrani N (2002) The Kirghiz and Wakhi of Afghanistan: adaptation to closed frontiers and War. University of Washington Press, Seattle

Shibaeva JA (1973) Zhivotnovodstvo u kirgizov Vostočnogo Pamira. Trudy Instituta Ethnografii (98):99–132

Sovnarkom TadSSR, CK KPTad (1943) postanovlenie ot 02.08.1943 g.: O sozdanie jakovochesk-ovo khozjajstva v GBAO. Sovet narodnykh komissarov Tadzhikskoj SSR i Centralnyj komitet KP/b Tadzhikistana, Dushanbe

Stankevich BV (1904) Po Pamiru. Putevye zapiski B.V. Stankevicha. Russkij vestnik 1904 (8; 9; 10; 11):624–657; 235–269; 456–499; 623–628

Statotdel Murgab (2000) Svodnoe o polovom i poluvozrostnom sostave naselenie v dzhamoatah Murgabskogo rajona na 1 janvarja 2000 goda. Otdel statistiki Murgabskogo rajona, Murgab

Statotdel Murgab (2008a) Bjulletin po zhivotnovodstvo. Otdel statistiki Murgabskogo rajona, Murgab

Statotdel Murgab (2008b) Svodnoe o polovom i poluvozrostnom sostave naselenie v dzhamoatah Murgabskogo rajona na 01 janvarja 2008 goda. Otdel statistiki Murgabskogo rajona, Murgab

Statotdel Murgab (n.d.) Dinamicheskie rjady po Murgabskomu rajonu. Otdel statistiki Murgabskogo rajona, Murgab

Taipov B (2002) Sary Kol tarykhyndyn kyskacha ocherkteri. Murgab

USKH GBAO (1994) Prikaz No. 2 ot 12.01.1994 g.: O rezultate vnedrenija arendnogo otnoshenie i sostojanie razvitie zhivotnovodstva v khozjajstvakh Murgabskogo rajona, Upravlenie sel'kogo khozjajstva Gorno-Badakhshanskoj Avtonomnoj Oblasti. Khorog

Vanselow KA (2011) The high-mountain pastures of the Eastern Pamirs (Tajikistan) – an evaluation of the ecological basis and the pasture potential. Dissertation thesis, Universitätsbibliothek der Universität Erlangen-Nürnberg, Erlangen

Walter H, Breckle S-W (1986) Ökologie der Erde: Geo-Biosphäre. Fischer, Stuttgart

Chapter 6
Kirghiz in Little Kara Köl: The Forces of Modernisation in Southern Xinjiang

Hermann Kreutzmann

Abstract The Chinese modernisation programme has affected even the remotest high mountain pastures of Xinjiang. The dynamics of such processes are exemplified in a case study from the Chinese Pamirs and are analysed in regard to their importance for adaptation and development. Based on a diachronic examination of transformation processes related to Kirghiz nomads, special emphasis is put on four stages of transformation that might lead from mobile pastoralism to a township settlement. Developments affecting the Kirghiz nomads of Little Kara Köl may be classified as transformations in space and time, resulting in the integration of this marginal region into the permanently settled areas of Xinjiang. Externalities in the field of social structure and political administration have supported this integration and growing dependence on the commercial and service centres of the foreland oases.

Keywords Modernisation • Resettlement • Kirghiz • Pamir • Kizil Su • Xinjiang

6.1 Introduction

How does the modernisation process affect pasture areas in peripheral mountain regions? Conventional thinking binds modernisation to the urban centres of economic activity, whilst the periphery is affected last or never. Two explanatory models prevail in the discussion of socio-economic change. From a viewpoint of

H. Kreutzmann (✉)
Human Geography, Department of Earth Sciences, Centre for Development Studies (ZELF) and Institute of Geographical Sciences, Freie Universität Berlin, Malteserstrasse 74-100, D-12249 Berlin, Germany
e-mail: h.kreutzmann@fu-berlin.de

H. Kreutzmann (ed.), *Pastoral practices in High Asia*, Advances in Asian Human-Environmental Research, DOI 10.1007/978-94-007-3846-1_6,

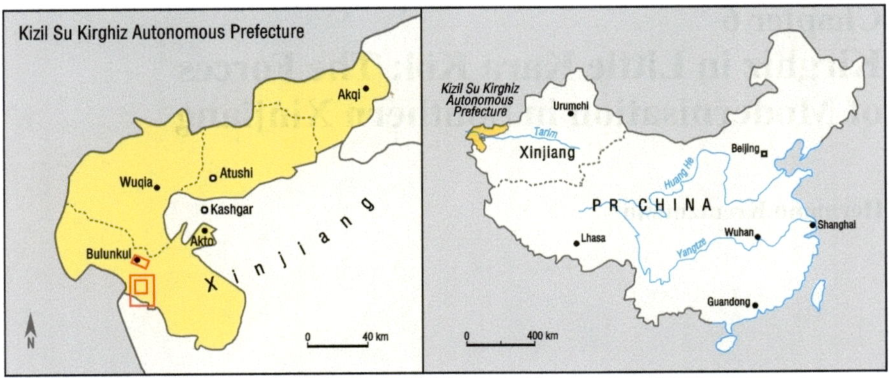

Fig. 6.1 Kizil Su Autonomous Prefecture within the Xinjiang Uyghur Autonomous Region, People's Republic of China

modernisation theory, classifications apply that highlight the backwardness of peripheral societies, their outdated economic strategies and their adherence to traditional behavioural patterns. By contrast, the paradigm of sustainability has changed the interpretation of adaptive strategies of mountain pastoralists who are seen as landscape managers of niche resources. The limited potential can be perceived as the base for an appropriate economy that conserves and protects such marginal regions. Nevertheless, the antagonists of such a view promote the notion of resource degradation and the need for evacuation, centralisation and modern pasture management. Both perspectives have influenced development models in the pastoral sector and resulted in practical approaches towards mobile animal husbandry. In this case study, the changes that have occurred to the Kirghiz pastoralists living in Little Kara Köl of present Akto County (Fig. 6.1) within Kizil Su Kirghiz Autonomous Prefecture (Xinjiang, People's Republic of China) will be examined in the light of external interventions leading to recent modernisation processes with significant impact on their livelihoods.[1]

6.2 Encounters

Early records of Kirghiz utilising the high pastures of the Pamirs link them to the fertile oases of the Fergana Valley and the Tarim Basin. Seasonal migration between the low-lying oases – where most of autumn, winter and spring time was spent – and the summer pastures in the various Pamirs (Curzon 1896; Dor and Naumann 1978) characterised mobility patterns and corridors of encounters and activities. Documents from the early nineteenth century reveal that Kirghiz pastoralists got involved in exchange and political relations with representatives of the Manchu dynasty in Kashgar. Kirghiz participated in a mutually valuable gift exchange by offering horses and sheep in return for precious Chinese goods. On the basis of those agreements, they enjoyed the right to freely offer their marketable goods such as horses, yaks and animal products in the bazaars of Kashgar (Di Cosmo 1993, 25–27, 1994; Millward 1998, 97).

Conflict and bellicose confrontations accompanied by plundering of caravans and competing territorial claims regularly caused the evacuation of pastures and abridgement of mobility patterns (Abramzon 1963; Bezkovic 1969; Centlivres and Centlivres-Demont 1983; Dor and Naumann 1978; Dubeux and Valmont 1848, 118; Millward 2007, 96, 109–118; Shahrani 1979). The Kirghiz – or Kara-Kirghiz as they were called then by travellers – gained the reputation of being untrustworthy and hostile to settled farmers whose property they frequently looted and destroyed according to Dubeux and Valmont (1848, 116):

> Les Kara-Kirguizes du plateau de Pamère sont en hostilité constante avec les provinces chinoises qui avoisinent leur territoire, et en particulier avec les pays d'Yarkende et le Tibet. Ils font des excursions dans ces deux provinces pour voler des hommes, des femmes et des enfants qu'ils réduisent en esclavage, et pour détrousser les caravanes: aussi les magistrats chinois condamnent ils à mort impitoyablement et sans aucune forme de procès tous les Kirguizes qui tombent entre leurs mains. Une pareille conduite, bien qu'elle paraisse souvent injuste, est cependant justifiée par des crimes antérieurs et par la nécessité d'inspirer de la crainte à ces bandits; car les Kara-Kirguizes ne vivent que du vol et des produits du brigandage, à tel point qu'ils ne respectent pas même les propriétés de leurs parents ou de leurs amis; et, lorsque quelqu'un d'entre eux a été victime d'un vol, il cherche à prendre sa revanche sur celui de ses voisins qu'il croit pouvoir dépouiller plus facilement. [The Kara-Kirghiz from the Pamir Plateau live in constant hostility with the Chinese provinces that border their territory, and especially with Yarkand and Tibet. They make journeys to these two provinces in order to capture men, women and children whom they enslave, and to rob caravans; just as well, the Chinese magistrates mercilessly condemn to death and without any trial all the Kirghiz who fall into their hands. This attitude, even though it often seems unjust, is however justified by previous crimes and by the necessity to inspire fear to these bandits; the Kara-Kirghiz live only on robbery and stolen products to the extent that they do not even respect the properties of their parents or friends. When one of them is victim of a robbery, he will try to take revenge on the neighbour whom he believes to be easier to strip of his possessions.]

Their audacity and daringness were linked to all kinds of perils and threat once they lost local battles and were called to account by Chinese authorities who often made an example of them and applied capital punishment for violation of property rules. The Danish Captain Ole Olufsen (1904, 91) comments on their behaviour:

> The Kirghiz of High Pamir, not without reason, are looked upon as wandering gipsy robbers whom it is best to drive away as soon as they show themselves in the neighbourhood.

These pejorative assessments could be counteracted by reverting to statements that estimate pastoralists as victims of exploitation, forced labour and arbitrariness. The remote pastures of Little Kara Köl and other locations in the Pamirs and the Western Kun Lun Shan functioned as a southern outlet for pressure that was exerted from the northern oases of Fergana and Alteshar – the six cities in the Tarim Basin. In general, pastoral migration oscillated between low-lying winter pastures (Kirg. *kštau*, aul; Taj. *kishlaq*) – in general, below 2,000 m – and high pasture settlements (Kirg. *dzailou*, Taj. *aylaq*) above 3,500 m. Although animal husbandry prevailed as the principal income resource, significant contributions were derived from transport services, arrangements for protecting caravans or exerting territorial control for authorities and furnishing supplies to military outposts. The ethnonym Kirghiz was synonymously used for 'nomad' by contemporary travellers (General Staff India 1907, 90–93, 1929, 217; Hartmann 1908, 30–31; Schultz 1910, 250–252). This observation especially holds true for the Chinese Pamirs where administrative terms were strongly linked to occupational status. The Kirghiz of Little Kara Köl belong to the southern subgroup of *Ichkilik*, in which mainly *Kesek*, *Nayman*, *Qipchaq* and *Teyit* patrilineages (*oruq*) prevail (Bregel 2003, 78–79; Shahrani 1979, 150–166). All of these groups are to be found in the Afghan and Tajik Pamirs and in Xinjiang (Sarikul and Little Kara Köl). The evidence presented here refers to the latter community.

6.3 The Winds of Change in Little Kara Köl: Kirghiz Pastoralists in Kizil Su

A contemporary reference at the end of the nineteenth century is provided by the observations of Sven Hedin, who twice visited the Kirghiz of Kara Köl at the foot of Mount Muztagh Ata during his Pamir expeditions. Two quotations from both expeditions:

> Der Bek von Su-baschi (südlich des Kleinen Kara-kul) hat mir mitgeteilt, die Gegend um den See sollte von 300 Teit-Kirgisen mit 60 Jurten bewohnt sein. Er sei Häuptling über 286 Jurten, von denen jedoch die größte Anzahl östlich der Mur-tag-Kette gelegen sei.... Die obigen Angaben zeigen, wie spärlich das Plateaugebiet von Pamir bewohnt ist, und es kann nicht anders sein für ein Land, wo Kälte und Stürme herrschen und wo die Grasvegetation eine große Seltenheit ist.[The Beg of Su-bashi (south of Little Kara-kul) told me that the region around the lake is inhabited by 300 Teit [Teyit] Kirghiz in 60 yurts. He is supposed to be the headman of 286 yurts of which the majority is located east of the Muztagh Range - meaning snow mountains... The above given figures show the low population density of the Pamirian Plateau. It could not be different for a region where cold spells and storms prevail and where grassy spots are rare.] (Hedin 1894, 303)

Sven Hedin observed that

> ... die Pflege der Heerden und die jährlichen Wanderungen, die damit im Zusammenhang stehen, das einzige Interesse der Kirgisen. Den Sommer bringen sie auf den Jeilaus, den Sommerweideplätzen, zu, die auf den höheren Abhängen des Mus-tag-ata und der Pamirgebirge liegen; ihre Winterweiden, Kischlaks, in den Tälern suchen sie auf, wenn der Schnee und die Kälte sie von den Bergen herunterjagen. In jedem Aul sind die meisten Bewohner miteinander verwandt. Sie ziehen stets auf dieselben Jeilaus und Kischlaks, und kein anderer Aul hat das Recht, ohne Uebereinkommen in das von einer Familie durch ihre Tradition gewonnene Gebiet einzudringen [... the care for their animal herds and the inter-related annual migrations are of sole importance to the Kirghiz. They spend the summer in the *dzailou*, the summer pastures, which are located at the higher slopes of Muztagh Ata and the Pamir Mountains; their winter pastures, *kishlaq*, in the valleys are visited when snow and cold drive them down the mountains. In nearly every *aul* (winter settlement) most of the inhabitants are relatives. They migrate always to the same *dzailou* and *kishlaq*, and no other *aul* possess a right to penetrate there without the consent of the traditional owners] (Hedin 1899, I, 269)

These contemporary observations confirm that the high pastures at the foot of Muztagh Ata were the summer abodes of Kirghiz communities – identified as yurt-based households (Kirg. *akoi*, Taj. *xirgó*) – who were migrating towards Kashgar through the Gez gorge, a tributary to the Kizil Su river (Fig. 6.2a). The winter pastures of these groups were located at the fringes of the Kashgar oasis in Akto and Yengisar. On Thursdays the weekly market in Kashgar Town attracted all kinds of traders and buyers; especially Kirghiz pastoralists became animal vendors here in autumn to get the means for bartering necessary goods for the upkeep of their households (Gillet 1937, 9; Hedin 1899, I, 271; Schultz 1921, 44–48). At the turn of the century, there are already remarks about single households that remain within the Pamirs for the whole year (Hedin 1899, I, 163, 180, 208, 277). Until the Chinese Revolution, Kashgar, Yengisar and Yarkand remained reference points for barter and supplies.

In the aftermath, significant structural changes took place that can be related to three subsequent phases: collectivisation and creation of people's communes (1958–1976), introduction of the pastoral household contract responsibility system (1978–2009) and the resettlement phase since 2009 (Fig. 6.2b–d).

6.3.1 Collectivisation and Communes

The initial far-reaching structural change commenced with the attempts to promote collectivisation. Following a phase of introducing mutual aid teams in the first half of the 1950s, a transitional structural setup consisted of the agricultural production cooperatives that again were superseded by people's communes after 1958. The full-fledged collectivisation process reached the Little Kara Köl Pamir when the 'nomadic' communities were reorganised in pastoral production brigades (Chin. *dadui*) and people's communes (*gongshe*). Besides forced sedentarisation in their former winter settlements, the Kirghiz pastoralists were exposed to new breeding techniques, veterinary services and scientific experiments to optimise resource exploitation including the expansion of crop farming in high altitudes (McMillen 1979, 158–159; Lo 1961, 101). The people's communes were oriented towards the headquarters of the Kizil Su *zizhizhou* (autonomous district) in Atushi (Fig. 6.1). The administrative centre itself was founded in 1952 as a model town to exemplify the modern design, infrastructure and facilities that were representing the advent of a new era. The tragic disaster connected with the 'Great Leap Forward Campaign' was felt in the Kara Köl Pamir as well and led to the re-introduction of limited private herding communities based on traditional structures (*aul, uru*). Only since the mid-1960s has the slow recovery process from the prior major losses taken off. The people's commune and the production brigades as centres of administration and social organisation remained in force until the end of the Cultural Revolution. For the Kirghiz of Little Kara Köl, the pastoral people's commune was established in Bulunkul[2] in 1959, whilst the production brigade was located in Subashi (Photo 6.1).

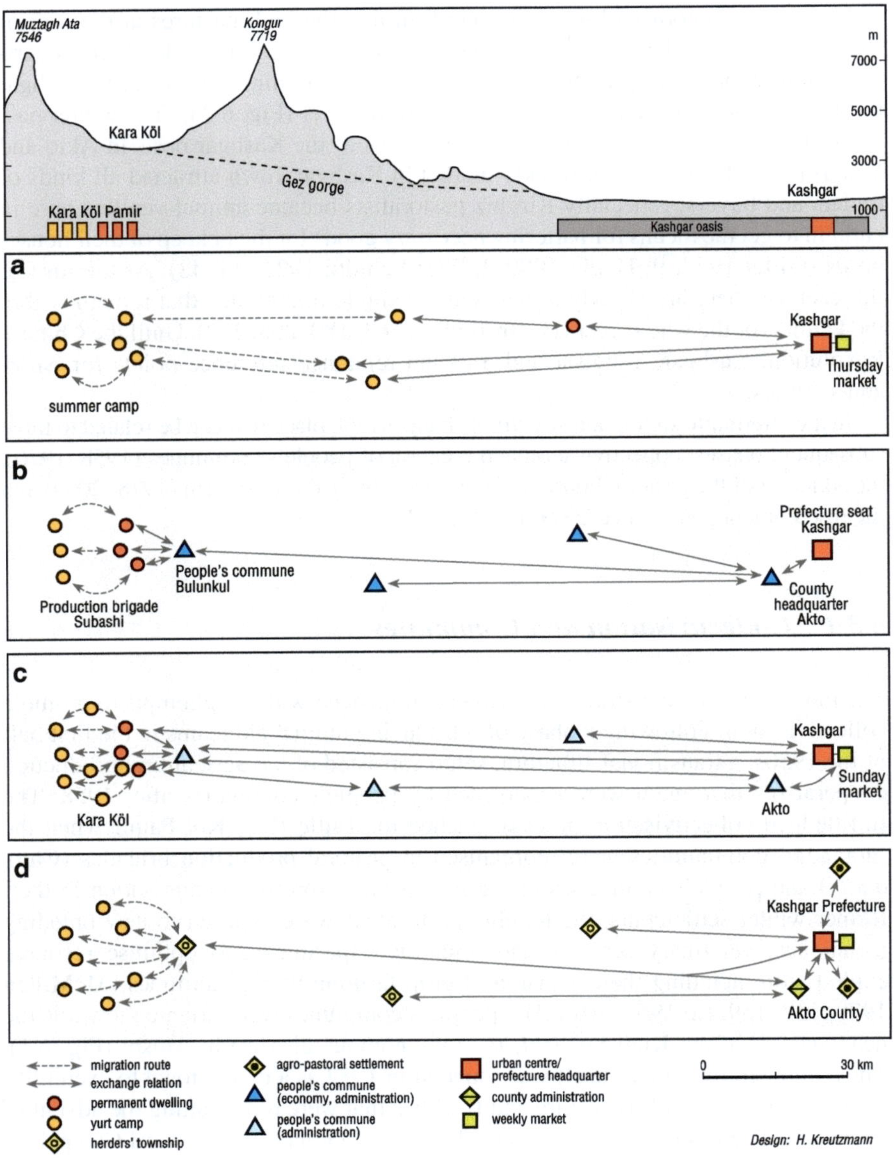

Fig. 6.2 Transformation in pastoral strategies in Little Kara Köl Pamir. (**a**) Pre-revolutionary mobility pattern (pre-1949), (**b**) people's commune phase (1958–1976), (**c**) production responsibility system (1978–2009) and (**d**) resettlement phase (post-2009)

The result of the sedentarisation process was a concentration of hamlets around the nuclear centres identified by the administration of the autonomous district. A more significant change was the confinement of all pastoralists within the Pamir all year round. The former seasonal migration between winter camps in the Kashgar

Photo 6.1 The Kirghiz winter camps Subashi and Aktash (3,650 m) form part of the production brigade of Subashi *dadui*, which offers infrastructural assets common to most rural settlements in Xinjiang (Photograph © Hermann Kreutzmann, July 24, 1991)

oasis and summer settlements in Little Kara Köl ceased to exist (Fig. 6.2a). Kirghiz pastoralists adapted their livestock keeping and migratory patterns to the high altitudes of Kara Köl (3,600–4,100 m). Compensatory measures included special loan schemes for stock purchases, the increase of biomass production by irrigating natural pastures and stockpiling of fodder (Abramzon 1963, 206).

The migration cycles have shrunk to distances of 3–15 km in comparison with the 280 km formerly covered between Kashgar and Little Kara Köl (Fig. 6.3). Three to four months of opulent fodder supplies are followed by lean months under harsh climatic conditions (Photo 6.2). The fattened animals and livestock products had to be delivered to the livestock department at the county headquarters in Akto which in return was expected to supply all necessary goods. The previous weekly Thursday market in Kashgar had ceased to exist and was replaced by governmental departments that took responsibility for food circulation. In return, all necessities for life in the pasture settlements were provided at nominal cost. Ma Yin et al. (1990, 246) was convinced that the county was self-sufficient in all necessary food items in the medium term. Though very restricted in their movements and decision-making, the Kirghiz secured their basic livelihoods and their living conditions improved. Despite harsh environmental conditions of survival, the animals raised in these productive pastures are high-quality products.

At the end of the people's commune phase, in 1976, the pastoralists of Subashi shared 0.5 horses, 0.3 camels, 3.5 yaks and 74.9 sheep and goats on average per household.

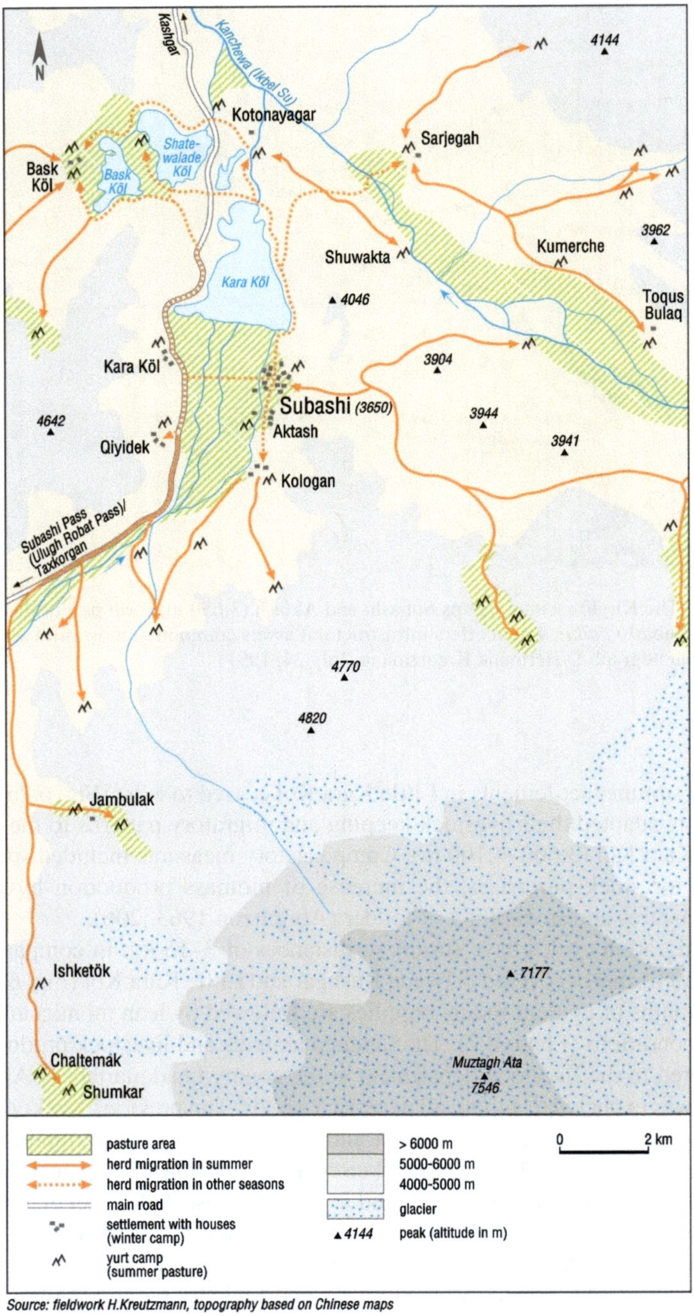

Fig. 6.3 Adaptation of annual migration patterns to structural interventions. After being confined to the Little Kara Köl Pamir as perennial grazing grounds, Kirghiz pastoralists as members of Subashi production brigade (*dadui*) developed a short-distance, energy-efficient shifting of pastoral camps in an altitudinal range between 3,600 and 4,100 m

Photo 6.2 Kirghiz pastoralists in Bulunkul (Photograph © Hermann Kreutzmann, October 15, 2008)

The total number of livestock amounted to around 10,300 animals, which was four times the number in 1958 (Myrdal 1979, 25). Besides state ownership of flocks, private property rights for a limited number of animals had been assured for the pastoralists. This phase was framed by two repressive periods: the 'Great Leap Forward' (1958–1960) and the 'Cultural Revolution' (1966–1976). Both campaigns aimed at 'modernising' the supposedly backward pastoralists and were inspired by policies of regional autonomy. The intermediate phases of greater tolerance took back some of the excesses, but generally the modernisation strategy epitomised by forced sedentarisation and incorporation into governmental exchange structures persisted.

6.3.2 *Pastoral Household Contract Responsibility System*

In the aftermath of the 'four modernisations' (*si hua*) announced by Premier Zhou Enlai in January 1975 and implemented for all layers of society – including the rural communities and the agricultural sector – by Deng Xiaoping, the people's commune was reduced to an administrative and logistical unit of secondary importance. Bulunkul was renamed as an autonomous village again in 1984, a legal term that is still in force today.

The liberalisation of the economy led to a tripling of livestock numbers from 10,300 in 1976 to above 30,000 in 1991. The upward trend continued and crossed the threshold previously identified by the local authorities who had estimated the carrying capacity of accessible pastures at 40,000 animals (Myrdal 1979, 25). By 2009, the livestock numbers had risen to 55,738 animals including 9,496 yaks that had received bio-food certification. The annual turnover has reached a sale of 35,623 mainly fat-tailed sheep and goats including their skins and about 51 tons of sheep wool and 1.5 tons of goat hair. The data[3] underline the growth pattern that was stimulated by the reforms and by the introduction of the pastoral household contract responsibility system (*jiating lianchan chengbao zerenzhi*). The reforms acknowledged the dual character of animal husbandry in the Little Kara Köl Pamir. Yaks, horses and camels were privately kept for the sustenance of the pastoralist households, whereas the majority of sheep and goats were part of the communal arrangements to be delivered to the county livestock department (Fig. 6.2c). In the early phases, government control was focused on the marketable fat-tailed sheep (*dumba*) and goats. The private access to livestock markets was stimulated by the 1981 (re-)introduction of Sunday bazaars in Kashgar and Yarkand that featured large-scale livestock sections. The permission of private entrepreneurship boosted the growing livestock numbers that were marketed in the bazaars, leading to a greater increase in animal herds in the remote Pamirian pastures than in the lower parts of the county close to Akto. Their livestock compete very well on the profitable markets in the urban oases along the southern silk route (Tarim Basin). The Kirghiz are respected as one of the most affluent livestock-breeding communities of the region. During fieldwork in 1991, we established household herd sizes ranging from 50 to 600 sheep and goats with an average of 140, which reflects a doubling of livestock since the end of the 'Cultural Revolution'. Nevertheless, within Xinjiang, Kizil Su *zizhizhou* still occupied the last position in the per capita income ranking (Giese and Zeng 1993, 192). The comparatively poor standard of living within a growing Chinese economy might have triggered modernisation programmes that are intended to bring significant change to the remote corners of the Pamirian pastures.

6.3.3 Resettlement: Modernisation Strategy and Coping with Pasture Degradation?

Development and modernisation have never stopped being central strategies for poverty alleviation and rural uplift within the People's Republic of China. Xinjiang *zizhiqu* and Kizil Su *zizhizhou* are no exception. The diagnosis as phrased by An et al. (2011) is:

> With the rapid population growth and people's increased demand for animal products, the prefecture is faced with problems of weak infrastructure, severe degradation of natural pastures and decreasing grass productivity, as they attempt to promote the development of pastoralism. Due to the constraints of traditional production and management methods, the prefecture is still plagued by some prominent issues such as low productivity of pastoralism,

> low economic returns and lack of opportunities for pastoralists in remote areas to increase their generally low incomes. To promote pastoralism and economic development in pastoral communities, in recent years, the national, regional, and prefectural governments have taken various measures to promote the transformation from traditional to modern pastoralism and to uproot poverty and backwardness in nomadic and pastoralist communities of the remote mountain areas. (An et al. 2011, 123)

In a major move to improve the living conditions in remote locations including mountain regions, the central planning institutions have recently developed so-called resettlement schemes (Fig. 6.2d) that are concentrating infrastructural assets, social amenities, veterinary services and marketing facilities. The rationale takes into account the growth in livestock numbers and their impact on the available pasture resources and is based on two assumptions:

1. Modernisation can be implemented only in urban contexts, and if modern lifestyles are to be achieved, then economies of scale need to be applied in the periphery. These considerations affect the whole rural setup and ask for a concentration that is reflected in townships, the focal points of the resettlement scheme.
2. Nature protection requires coping strategies that reduce grazing pressure and fight degradation by concentrating settlement space and facilities for animal husbandry and enhancing fodder production.

The strategy of resettlement aims to combine both goals by concentrating the whole resident population in the sole township of Bulunkul *xiang* (Fig. 6.4) and by reducing the number of persons engaged in pastoralism. In 2009, all 1,466 households generated 14.23 million yuan from animal husbandry, equalling 83% of Bulunkul's income compared to less than 4% from other agricultural activities.[4] The rest is shared amongst construction and industrial activities and income from transport and other services. The aim is to reduce the number of people engaged in agricultural activities in order to direct them to other occupations. The township aims at providing housing[5] (Photo 6.3) and social infrastructure for all herders' households that are presently scattered in the area of the former people's commune of Bulunkul (cf. Fig. 6.4).

During the first phase, 450 out of 1,600 households will be resettled in housing schemes that are supported by federal and regional funds as well as by rich cities such as Shanghai.[6]

Within the township, a central stable is designed for concentrating the limited livestock. Six veterinary doctors are supposed to offer their services to the first batch of households, each of which will enjoy a supply of 300 kg of grass and 30 kg of high-nutrition fodder for each animal during the cold season. From 1954 to 2009, the number of livestock in Bulunkul *xiang* increased by a factor of 21.6 to more than 55,000 animals, crossing the given threshold of a calculated 'carrying capacity'.[7] Concentration and reduction aim at a qualitative boost and at optimising pastoral practices. The modernisation strategy includes controlling the livestock, regulating internal and external fodder supply, providing professional veterinary services and marketing livestock. The envisaged side effect is the 'professional' approach to

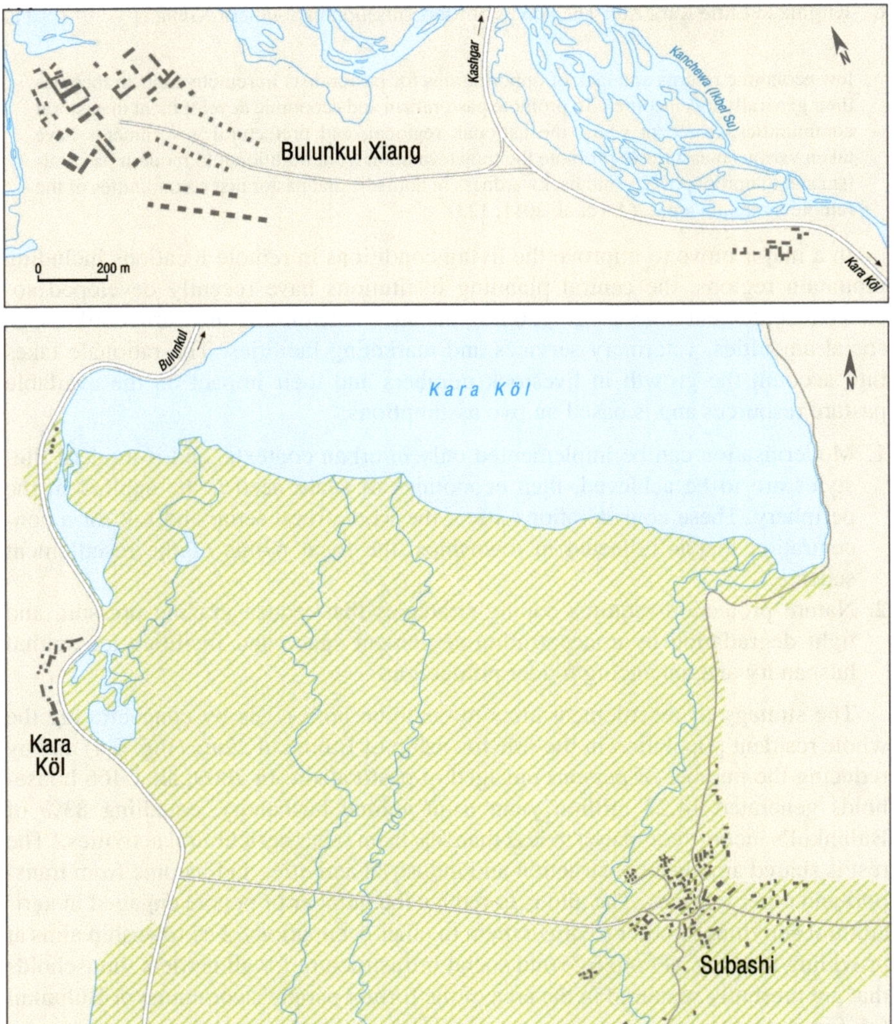

Source: representation based on Chinese maps and satellite images

Fig. 6.4 The scattered settlement pattern of Subashi (*bottom*) with corresponding hamlets is the result of adaptation processes, the establishment of a people's commune and the introduction of the pastoral household contract responsibility system. With the implementation of the resettlement scheme, Bulunkul *xiang* (*top*) is supposed to become the modern township where all households will eventually have their permanent abodes

Photo 6.3 The new resettlement scheme in Bulunkul (Photograph © Hermann Kreutzmann, July 19, 2010)

pastoral practices. In only 2 years (2006–2007), Akto County created '760 ha of man-made fodder pasture, 360 ha of fenced and improved pastures, and nine breeding stations were completed' (An et al. 2011, 125). The holistic approach is the latest and far-reaching strategy that is perceived as a comprehensive model for future development in pastoral communities.

A second strategy that has been implemented in Kizil Su is that of agro-pastoralism. An et al. (2011, 121) report 1.6 million heads of livestock in Kizil Su and attribute two-thirds of pastoral activities to high mountain pastoralism and one third to agro-pastoralism within farming areas. The latter strategy is a further step towards removing pastoralists further away from high mountain pastures and resettling them in the vicinity of highly productive agricultural areas such as the Kashgar oasis. Both approaches lead to a concentration of settlements at focal points and detach pastoralists from their former abodes. Whether the second goal of enhancing nature protection can be achieved in such a manner needs to be seen in future, especially when the diagnosis seems to be alarming:

> According to a recent survey, the prefecture's degraded pastures account for 94.66% of the total area of pastures, of which there's 45.83×10^4 ha lightly degraded pastures, about 12.97% of the total; 157.39×10^4 ha moderately degraded pastures, or 44.55% of the total and 131.17×10^4 ha heavy degraded pastures, or 37.13% of the total. Compared with the 1980s, vegetation coverage has decreased by 50% or more, the grass height has decreased by 8–15 cm, the proportion of fine grasses has decreased by 20–40%, the proportion of weeds has increased by an average 20–45%, and grass production has decreased by 30% or above. The degradation of pastures is serious …

> Pastoralism, in fact, was developed at the cost of heavy use of pasture resources. The predatory practice of harming the eco-environment for economic benefits will not only result in lower productivity of pastures, but will also weaken the ecological functions of the pastures. The country has been, therefore, compelled to invest heavily in ecological restoration and reconstruction (An et al. 2011, 125–126).

In consequence, the resettlement programme and different forms of agro-pastoralism have been implemented. In Kizil Su, the process is in its infancy, whilst in other regions – especially in Inner Mongolia and Tibet – the effects of the restructuring are visible already.

6.4 The Way Forward: Coming Down from the Mountain Pastures?

The Chinese government has put major emphasis on implementing a comprehensive strategy that finds regional and local expressions, as introduced above. The latest stage of transformations in the pastoral sector has affected livelihoods and practices to such a degree that – for some persons and households – pastoralism might soon be terminated as a profession and occupation. Traditional pastoralism cannot give herders the 'pleasures of modernisation' as Zhao Xichun phrased it in his presentation during the pastoralism conference in Kashgar 2010.[8]

Moving herders to the foot of the mountains – to agro-pastoral settlements – or to resettlement townships within the mountains – as in the case of Bulunkul – aims to provide pastoralists with amenities that allow them 'to live in a stable manner and to get rich' (ibid.). More specifically, the scope of programmes and packages becomes clear in the design presented by the chief scientist of Animal Husbandry in Xinjiang:

> To get rid of poverty and underdevelopment in pastoralist communities and transform their traditional production and living patterns, the Party Committee and Government of Xinjiang Autonomous Region have been promoting the strategic deployment of pastoralist settlements since the 1980s, with the purpose of improving living and working conditions for pastoralists and improving fodder and sheds for cattle in cold seasons. The Sedentarization program has taken into full consideration the environmental and lifestyle factors of the region.
>
> To improve the living and working conditions for the pastoralists, we have brought the scattered households to the flat areas from the mountain slopes, providing them with fully-equipped dormitories and necessary production facilities, and hence, providing a better quality and standard of life. To supply livestock with enough fodder and shelter in cold seasons requires building high-standard crop-based fodder pasture in the flat areas where conditions of water, soil, light and heat are appropriate. The irrigated crop-based pastures are expected to satisfy the needs of fodder and grass for cattle. This way the sedentarization of pastoralists can be easier, more stable and generate more income for them. Water is a precondition for cultivating a man-made forage base. Therefore, the pastoralists in Xinjiang follow a mode of sedentarization in which locations are determined based upon water source, grass variety by location, animal variety by grass variety and human settlement by animal variety.
>
> Under the leadership and support of the Party Committees and the People's Government at various levels, different districts in Xinjiang carry out coordinated planning of dormitories, animal sheds, forage bases and infrastructure for water and power supplies for the

settlements. A series of favorable policies have been put forward to promote the development of pastoralist settlements. Through efforts of many years, Xinjiang has established a number of workable models such as the village-based concentrated settlements, 'flower-arrangement' settlements and mobile settlements. Standards for sedentarization have been institutionalized, including 'three accesses' (access to water, road and electricity); 'four availabilities' (availability of living quarters, animal sheds, fodder pastures, and fodder/silage storage spaces) and 'five complementary facilities' (school, clinic, shopping, cultural center and technical support service). At the end of 2009, Xinjiang had 106,500 settled pastoralist households, accounting for 38.6% of the total number of pastoralist households. After settlement, the winter mortality rate for livestock has been reduced to 1.5% from the 10% experienced in the 1990s. The living standards of pastoralists have been constantly increasing, with the per capita income rising from 1,050 Yuan in 1996 to 2,480 Yuan in 2008 (Xinchun 2011, 183–184).

Since 2000, the pace of resettlement has been enhanced and a re-planning of the function of natural pastures has been undertaken. In this framework, Little Kara Köl Pamir has been identified as one of the resettlement schemes that are regarded as one of 'the biggest livelihood projects in Xinjiang'. The ultimate path leads to a transformation of livelihoods that incorporates the

... resettlement of herders as labourers in other areas. The remaining herders can engage in pastoralism and economies of scale, increase the market supply and thus increase the pastoralists' income leading to a modernisation of mountain pastures and the modernisation of pastoralism. Voluntary participation of herders in resettlement programmes is based on hearings at township level, annual sessions of party committees. Wishes can be articulated through proper channels (ibid.).[9]

It will be seen in the near future how the lifestyle, economic position and social organisation of Kirghiz herders may have been changed. The first decade of the twenty-first century is experiencing a transformation process in pastoral practices that is inspired by postulates of modernisation theory of the previous century. The process of modernisation once again has found its target in marginal regions of the periphery. The latest approaches could lead the Kirghiz down the mountain slopes back to the oasis. More than two generations after their mobility cycles were confined to the Pamirian pastures, the opportunity arises to leave the harsh mountain conditions. The people who are part of this new movement are different from those who once were collectivised in Little Kara Köl Pamir. Their destination will be different as well.

Acknowledgements Archival and empirical evidence presented in this chapter was collected during the last 20 years, with generous funding provided by the Deutsche Forschungsgemeinschaft (DFG) and the Volkswagen Foundation which is gratefully acknowledged.

Notes

1. The current state of affairs is described by An et al. (2011, 123): 'At present, the prefecture has contracted out approximately 80.5% of the pastures'.
2. The name of Bulunkul *dadui* (production brigade) was changed into Fanxiu *dadui* in 1967. In 1984, it was renamed as Bulunkul *xiang* (autonomous village). The settlement covers an area of 4,585 km^2 of which 31,290 ha are identified as pastures and 665 ha as forage grassland for fodder production.

3. Economic and livestock data for 2009 were kindly provided by Ms. Zhao Lixin, Xinjiang Academy of Agricultural Sciences, Urumchi.
4. Compared to the distribution in Akto County – 26.5% in the livestock sector and two-thirds in remaining agricultural activities – the importance of pastoralism in Bulunkul is obvious. Areawise, only 3.6% are utilised for agriculture, whilst 96.4% are classified as mountain pastures (China Statistics Press 2010).
5. The government's expenditure per house was estimated at 160,000 yuan (app. 24,000 USD).
6. Akto County's Poverty Alleviation Programme (*fu min an ju*), Reservoir Resettlement Programme (*ku qu yi min ban qian*), Nomads Settlement Programme (*you mu min ding ju*) and the Jiangxi Affordable Housing Dwelling Ceremony (*Jiang xi bao zhang xing zhu fang mu min ru zhu yi shi*) are applied in Bulunkul *xiang*. The Reservoir Resettlement Programme has facilitated the resettlement of 451 households (1,643 persons) from the Gez Gorge within Kizil Su due to the implementation of the Bulunkou-Gonggeer Hydropower Project.
7. Whilst the increase in Bulunkul was significant, during the same period, the livestock numbers in the prefecture grew only by a factor of 2.8 (An et al. 2011, 126).
8. Cf. for the proceedings of the conference Xinchun (2011).
9. The quotations are taken from the English simultaneous translation of the presentation by Zhao Xinchun, deputy director and chief scientist of Animal Husbandry, during the regional workshop in Kashgar on July 20, 2010; cf. Zhao Xinchun (2011).

References

Abramzon SM (1963) The Kirgiz of the Chinese people's republic. Central Asian Rev 11(2):196–207

An S-Z, Dai Jian L, Zhaohui MY, Xiao-Yuan L (2011) Pastoralism in China's Xinjiang Kizilsu Kirghiz autonomous prefecture. In: Kreutzmann H, Abdulalishoev K, Zhaohui L, Richter J (eds) Regional workshop in Khorog and Kashgar. Pastoralism and rangeland management in mountain areas in the context of climate and global change. GIZ, Bonn, pp 119–131

Bezkovic AS (1969) Nomadenwirtschaft und Lebensweise der Kirgisen – 19. bis Anfang des 20. Jahrhunderts. In: Földes L (ed) Viehwirtschaft und Hirtenkultur. Ethnographische Studien. Akadémiai Kiadó, Budapest, pp 94–111

Bregel Y (2003) An historical atlas of Central Asia. Brill, Leiden/Boston

Centlivres P, Centlivres-Demont M (1983) Frontières et phénomènes migratoires en Asie centrale: Le cas de l'Afghanistan de 1880 à nos jours. In: Centlivres-Demont M (ed) Migrations en Asie. Migrants, deplacées et réfugiés. Société suisse d'ethnologie, Bern, pp 83–114

China Statistics Press (2010) Xinjiang statistical yearbook 2010. CSP, Beijing

Curzon GN (1896) The Pamirs and the source of the Oxus. Geogr J 8: 15–54, 97–119, 239–264

Di Cosmo N (1993) Reports from the Northwest: a selection of Manchu memorials from Kashgar (1806–1807) (Papers on Inner Asia 25), Bloomington

Di Cosmo N (1994) Ancient inner Asian nomads: their economic basis and its significance in Chinese history. J Asian Stud 53(4):1092–1126

Dor R, Naumann CM (1978) Die Kirghisen des Afghanischen Pamir. Akademische Druck- und Verlagsanstalt, Graz

Dubeux L, Valmont V (1848) L'Univers, ou histoire et description de tous les peuples, de leurs religions, moeurs, coutumes, etc.: Tartarie. In: Tartarie, Beloutchistan, Boutan et Népal, Dubeux, Louis, V. Valmont, and Xavier Raymond and in Afghanistan, Xavier Raymond. Firmin Didot, Paris

General Staff India (1907) Military report on Kashgaria prepared in the division of the Chief of the Staff. Intelligence Branch, Simla (IOL/MIL/17/14/74)

General Staff India (1929) Military report on Sinkiang (Chinese Turkestan). Government of India Press, Calcutta

Giese E, Zeng G (1993) Regionale Aspekte der Öffnungspolitik der VR China. Geographische Zeitschrift 81(3):176–195
Gillett MC (1937) Report on Vice-Consul Gillett's tour to Keria via Yarkand, Karghalik and Khotan. Kashgar (IOL/P&S/12/2336)
Hartmann M (1908) Chinesisch-Turkestan. Geschichte, Verwaltung, Geistesleben und Wirtschaft (= Angewandte Geographie. Hefte zur Verbreitung geographischer Kenntnisse in ihrer Beziehung zum Kultur- und Wirtschaftsleben 3. Reihe, 4. Band). Gebauer-Schwetschke, Halle a. S.
Hedin SA (1894) Forschungen über die physische Geographie des Hochlandes von Pamir im Frühjahr 1894. Zeitschrift der Gesellschaft für Erdkunde zu Berlin 29:289–338
Hedin SA (1899) Durch Asiens Wüsten. Drei Jahre auf neuen Wegen in Pamir, Lop-Nor, Tibet und China, 2 vols. Brockhaus, Leipzig. English translation: (1899) Through Asia, 2 vols. Harper, New York/London
Lo JP (1961) Five years of the Sinkiang-Uigur autonomous region, 1955–1960. China Q 8:92–105
McMillen DH (1979) Chinese communist power and policy in Xinjiang, 1949–1977. Westview Press, Boulder/Folkestone
Millward JA (1998) Beyond the pass. Economy, ethnicity, and empire in Qing Central Asia, 1759–1864. Stanford University Press, Stanford
Millward JA (2007) Eurasian crossroads. A history of Xinjiang. Columbia University Press, New York
Mydral J (1979) The silk road: a journey from the High Pamirs and Ili through Sinkiang and Kansu. Pantheon, New York
Olufsen O (1904) Through the unknown Pamirs. The second Danish Pamir expedition 1898–1899. William Heinemann, London
Shahrani MN (1979) The Kirghiz and Wakhi of Afghanistan. Adaptation to closed frontiers. University of Washington Press, Seattle/London
von Schultz A (1910) Volks- und wirtschaftliche Studien im Pamir. Petermanns Geographische Mitteilungen 5:250–254
von Schultz A (1921) Kaschgar (Chinesisch-Turkestan), Stadt und Landschaft. Mitteilungen aus dem Seminar für Geographie der Hamburger Universität, Hamburg
Xinchun Z (2011) Pastoralism in Xinjiang: implementing the pastoralist settlement programme to promote sustainable development of pastoralism in Xinjiang. In: Kreutzmann H, Abdulalishoev K, Zhaohui L, Richter J (eds) Regional workshop in Khorog and Kashgar. Pastoralism and rangeland management in mountain areas in the context of climate and global change. GIZ, Bonn, pp 182–188
Yin M et al (eds) (1990) Die nationalen Minderheiten in China. Verlag für Fremdsprachige Literatur, Beijing

Chapter 7
Legal Arrangements and Pasture-Related Socio-ecological Challenges in Kyrgyzstan

Andrei Dörre

Abstract This chapter has two objectives. The first one is to emphasize the important role of legal institutions for the emergence of natural resource-related social and ecological problems in Kyrgyzstan's post-Soviet era. Second, it advocates for a participatory approach to the creation of institutional regulations regarding the management and utilization of resources. It is important to consider that grasslands have an economic importance from the macroeconomic national level down to the level of local households, as well as crucial ecological meanings. For these reasons, pastures are significant for the development processes of Kyrgyzstan's whole society. The hypothesis to be explored here is that formal institutions, especially top-down-initiated legal rules implemented since 1991, are decisively contributing to the formation of socio-ecological pasture-related challenges. I argue that it could be misleading and insufficient to explain these problems through neo-Malthusian arguments. The causes are much more complex. Utilization practices applied by the actors can be understood as results of the interplay of economic necessities, weak legal institutions, legal uncertainty and a related lack of reliable planning opportunities. In this way, inappropriate and unstable legal arrangements are stimulating the processes of socio-economic stratification and disintegration of the society as well as those of pasture degradation. Based on findings obtained during field studies in the walnut-fruit forest region, this article advocates for the principles of an integrated sustainable development of Kyrgyzstan's society in economic, social and ecological matters. Management responsibilities, access and utilization rights need to be matched to the specifics of local contexts and legitimized through participatory approaches. Including the local population in the institution-building process can make a decisive contribution to the development of Kyrgyzstan by balancing different interests.

A. Dörre (✉)
Human Geography, Department of Earth Sciences, Centre for Development Studies (ZELF), Institute of Geographic Sciences, Freie Universität, Malteserstrasse 74-100, D-12249 Berlin, Germany
e-mail: adoerre@zedat.fu-berlin.de

H. Kreutzmann (ed.), *Pastoral practices in High Asia*, Advances in Asian Human-Environmental Research, DOI 10.1007/978-94-007-3846-1_7,

Keywords Kyrgyzstan • Transition • Legislation • Pastures • Resource conflicts • Degradation

7.1 Kyrgyzstan's Fragile Post-Soviet Society

For hundreds of years, animal husbandry has played a crucial role in Central Asian economies, societies and cultures. Until the early socialist times, pastoralists exploited the spacious and agronomic uncultivable grasslands of the region predominantly as natural forage grounds by applying spatio-temporal mobility patterns between seasonal pastures. After a forced sedentarization campaign under Soviet rule, an intensified pasture utilization was introduced as part of the socialist-style, scientifically informed and large-scale agro-industrial sector, which was composed mainly of state and collective farms. Because grasslands cover a vast area of more than 90,000 km² mainly in mountain regions, the national economy of the KSSR became highly specialized as the 'wool factory' within the Soviet system and its spatial division of labour. The republic also achieved a pivotal position within the production of milk and meat. Simultaneously, resulting pasture-related ecological problems were combated by different, frequently technical measures as well as by applying modified seasonal mobility patterns as rotational grazing based on long-time pastoral experiences and practices of nomadic peoples. These efforts were partially successful (Isakov 1974, 3–14; Brylski et al. 2001; 1, Ludi 2003, 119, 123; Undeland 2005, 18–21; Baibagushev 2011, 103–106).

As a consequence of the disintegration of the Soviet Union in December 1991 as well as of the rapid change to capitalism, the first years of Kyrgyzstan's independence were marked by a sharp economic downfall attended by economic disentanglements and deindustrialization with severe impacts on the people. In consequence of the retrenchments and liquidations of the state-owned and collective enterprises, a lot of former employees, especially in rural areas, lost their workplaces and secure wages. Simultaneously, the state radically cut the provision of economic, social and legal services. The people had to cope with this situation by creating new income strategies (Ludi 2003, 119–120; Undeland 2005, 22; Schmidt 2006; Schoch et al. 2010, 212). Public discontent with the political and socio-economic conditions of Kyrgyzstan's post-socialist everyday life, especially the insufficient provision of public goods through the state, has ended twice in dismissals of the government so far (in 2005 and 2010). The increased socio-economic uncertainties have led to a growing dependency on nature-based resources for both the national economy and individual households, especially in rural areas (Brylski et al. 2001, 1; IBRD 2001; Ludi 2003, 122; Schmidt 2006; Schoch et al. 2010, 212–213). In this regard, a series of violent incidents, in part related to rivalries about access to and utilization of nature-based resources like cropland and others, caused much human suffering and material losses. The communal violence in the regions of Osh and Zhalalabad in June 2010 was only the most recent and brutal top of a whole chain of such incidents (Tishkov 1995; Beyer 2010; ICG 2010; Trilling 2010; Melvin 2011).

Today Kyrgyzstan's national economy is characterized by a low diversification level with a remarkable strong agriculture of nearly 21%, a relatively weak industry of 28%, and a considerable large service sector of more than 51% of the GDP [4.6 billion US$, 2010 est.] (CIA 2011). According to official government data – which has to be handled with caution because of imprecise collecting and opaque calculating methods – in 2007, almost 35% of the whole labour force [app. 2.15 million in 2007] worked within the primary sector, whilst in 2008, the animal husbandry contributed more than 41% of the sectoral value creation (NSKKR 2009, 89, 178–179). At the same time, the nationally and individually important livestock production is decisively based on the use of the relatively cheap natural fodder stock of seasonal pastures and hay meadows (Brylski et al. 2001, 1–2; Schmidt 2001, 109; Kazybekov 2007, 5; Shamsiev et al. 2007, 52). In this relation, pastures as the main production basis have to be considered as a crucial economic resource. Additionally, grasslands provide diverse services within ecological processes such as water and nutrient cycles as well as soil formation. Together with mountain forests, they are protecting existing watersheds and runoff regimes. By doing so, pastures protect the society from effects of uncontrolled water drain, mass transport and extreme spatio-temporal runoff variations of watercourses and therewith from high social costs induced by these processes. Besides, pastures are important as sites distinguished by high biodiversity (Brylski et al. 2001, 1–2, 11).

In general, the current state of Kyrgyzstan's society can be characterized as the result of crisis-laden processes. The society is endowed today with different threats to its integrity, notably a high economic dependency on directly usable nature-based resources such as pastures, a weak national economy, inefficient public structures, an emergent socio-economic stratification and – as I hypothesize – unsound formal legislation. This chapter will attempt to prove this assumption.

7.2 Kyrgyzstan's Pasture-Related Socio-ecological Challenges

Nowadays, despite the vast grasslands covering nearly 46% of the country's territory and around 90% of all agricultural lands, a temporary livestock reduction in the 1990s due to the economic crisis, and several attempts to reform the pasture-related legal framework and management system, a number of challenges arose in differentiated spatial patterns, scopes and intensities (Ludi 2003, 119–120; Undeland 2005, 22; SAEPFUGKR and UNDPKR 2007, 19–20; Baibagushev 2011). Some of them are new phenomena, notably those problems with a strong social dimension as their core, whilst others are known already from the Soviet times, especially ecological ones. These challenges are outlined below through selected documented incidents (cf. Fig. 7.1) and statistical data (cf. Table 7.1).

Even before the inter-ethnic violence in June 2010, news agencies reported at the end of May about violent incidents between inhabitants of Uzbekistan's exclave Sokh in the western Fergana Basin and citizens of Kyrgyzstan living in the surrounding areas. The matter of the dispute was the briefly noticed introduction of

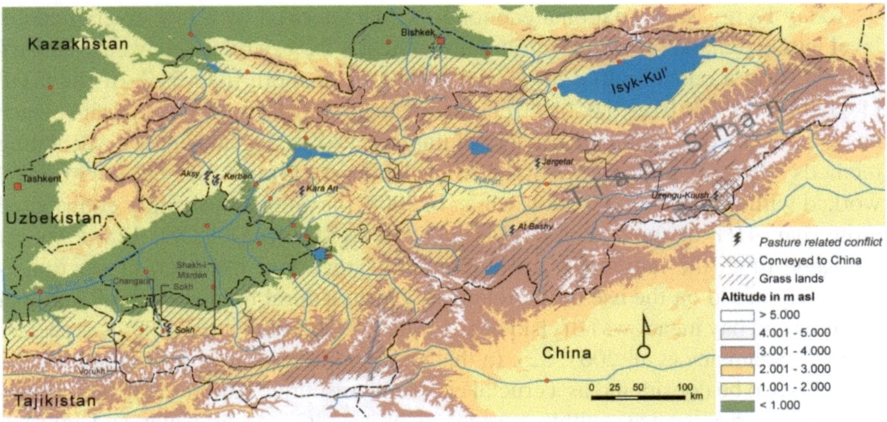

Draft by Andrei Doerre (2011) based on AN KSSR and GK KSSR IPK 1982, GUGK 1987: 18-19, Jarvis et al. 2008, ADB 2010: 122, sources mentioned in the text
Projection: Universal Transverse Mercator

Fig. 7.1 Kyrgyzstan's grasslands and selected pasture-related social conflicts (Design: Andrei Dörre)

new utilization rules of the administration of the Batken Region for the usage of pastures located in their territory. As a written agreement could not be presented, the new regulation forbids the Sokhis to use the grasslands that had been used by them for a long time and on which they rely economically. Over a few days, the conflict developed up to the point of an international military skirmish between the two countries, but fortunately this was stopped (Fergana.ru 2010; Mikhajlov 2010; TCA 2010).

This incident is only one event in a row of pasture-related social conflicts that have happened in different regions over the last years: In 1999, after secret intergovernmental negotiations, a nearly 1,000-km²-large area named *Uzengu-Kuush* was ceded to China. Despite of glaciers, the area consists of grasslands. Parliament members demanding a clarification of this case were criminalized. As happened in the Aksy Rural Municipality of the Zhalalabad Oblast in March 2002, popular protests were knocked down by state forces (ICG 2002, 17–18; Kreutzmann 2004, 8; Zhuk 2010). Another cross-border pasture rivalry repeatedly appeared during the first decade of the 2000s in the same region between citizens of Kyrgyzstan's settlements of Kerben on the one hand and inhabitants of Uzbekistan's Yangikurgan District of the Namangan Region on the other (Urumbaev 2004; UNEP et al. 2005, 19). For a couple of years, a conflict has been smouldering between the inhabitants of Zhai Terek and Arslanbob, two settlements of the rural municipality of Arslanbob within the region of the walnut-fruit forests, also located in the Zhalalabad Region. The subject there is the rivalry over an intensively used summer pasture called *Kara Art* located close to these settlements. Both sides are claiming the resource for themselves to execute inter alia conflicting exploitation forms of animal grazing and tillage, respectively (UNEP et al. 2005, 19; Mamaraimov 2007, 1, 4). In 2008, nearby the settlement of Jergetal in the Naryn Oblast Kyrgyz, shepherds faced

Table 7.1 Shifts in the average biomass production (dry matter) and degradation of Kyrgyzstan's seasonal pastures

Observation period	Average yield of all grasslands (about 90,900 km² in 2010), rounded in kg/ha	Seasonal pasture yields, rounded in kg/ha: Winter (22,850 km²)	Spring-autumn (26,970 km²)	Summer (38,890 km²)
1986–1994	220 ⇘	135 ⇘	210 ⇘	275 ⇒
1997–2004	210	85	170	275
	Average proportion of degraded areas of all pastures, rounded in percent	Proportion of seasonal pastures degraded areas, rounded in percent		
1985	24 ⇗	12 ⇗	16 ⇗	35 ⇘
2002	25	16	26	29

Sources: SAEPFUGKR and UNDPKR (2007, 23), Shamsiev et al. (2007, 52), and Baibagushev (2011, 106)

Chinese gold miners. Their exploration activities caused a massive lowering of the quality as well as a reduction in the area of the pastures on which the shepherds were dependent (Steimann 2008, 6 and see Chap. 7 in this volume). In the early summer of 2009, inhabitants of two settlements in the At Bashy Rural Municipality of the same region blocked the strategically important highway between Bishkek and the Torugart Pass on the border to China in protest against a putative long time lease of vast grassland territories to China. However, this proved to be a rumour (Kasybekov 2009).

This selection of non-transparent political decisions, local and regional rivalries as well as competing utilization practices is neither a complete list of pasture-related social conflicts nor of pasture problems, but it shows that this resource is an important political and economic issue in Kyrgyzstan and the Central Asian Region.

Furthermore, ecological damages, notably degradation, which is described as a 'substantial decrease in either or both of an area's biological productivity or usefulness due to human interference' (Johnson and Lewis 1995, 2), endanger the fulfilment of the ecological functions of pastures and contribute to the man-made shortage of this income-relevant resource. New qualities and spatial patterns of degradation have appeared as a result of fundamentally changed economic organizational forms, maladjusted uses and ineffective resource management after 1991: The amount of winter, spring and autumn pastures classified as degraded has grown. In contrast, the share of degraded summer pastures has slightly decreased (cf. Table 7.1; Brylski et al. 2001, 13; Penkina 2004; Undeland 2005, 8–9; Shamsiev et al. 2007, 53, 71; SAEPFUGKR and UNDPKR 2007, 18, 23–24).

The different trends in the post-Soviet time can be explained partly by the particular spatial location of the pastures, as well as the weak state of infrastructure as water supply facilities, roads and bridges. These issues result in relatively high levels of investment needed before the possible use of the affected pastures. Due to the small level of disposable capital, a lot of local households in rural areas are not able to visit distant pastures and to include them into their livelihood portfolio. Therefore, they keep huge livestock numbers all year round on winter- and spring-autumn pastures which are generally located near to settlements, water sources and along roads. These activities contribute to the widespread degradation of those areas. In contrast, because only a relatively small number of users are able to visit remote pastures, a large share of them is underused (Ludi 2003, 121; Penkina 2004). Both settings lead to a deterioration of the animal food supply, resulting in a lowering of the livestock's productiveness and reproduction.

Hence, for individual actors, degradation means an immediate culmination of resource scarcity and with it a prospective increase of resource-related rivalries. On the other hand, because of dominance of the agrarian sector in Kyrgyzstan's national economy, degradation leads to severe macroeconomic losses (Ludi 2003, 121; Undeland 2005, 8; SAEPFUGKR and UNDPKR 2007, 20, 24; Baibagushev 2011, 103, 107–110, 113). Consequently, it is closely interwoven with social challenges such as income generation, political stability, food security and others. Therefore, pasture degradation as a symptom, result and cause of undesirable processes and underdevelopment can be assessed as a threat for the country's social integrity.

I argue that it is myopic and insufficient to explain the emergence of the pasture-related socio-ecological problems through neo-Malthusian arguments about the responsibility of an increasing population and the excessive usage of local actors. The causation of the described challenges is much more complex and embraces historical preconditions, the current socio-economic and political frame as well as resource management practices. I also claim that the formal legislation in general as well as informal power relations between relevant actors and organizations decisively contribute to the formation of the described problems in concrete local contexts. In my opinion, legal uncertainty caused by repeated substantial changes of the pasture legislation and by ambiguities of several simultaneously valid regulations hereby plays an important role. Such situations are used by powerful actors to enforce their individual interests against the resistance of others. These practices are fostering the socio-economic stratification of the population. I will attempt to prove this hypothesis in a two-step approach. First and generally, I will outline the changing meanings and ambiguities of the legally sanctioned 'management competences' and 'allocations of pasture entitlements' as two features strongly effective in pasture-related socio-ecological problems. Second, I will assess the certain effects of the legal framework on the ground by examining a concrete situation in the walnut-fruit forest region.

7.3 Methodology: Review of Legal Norms, Interviews, Observation

The line of my arguments and evidence is based on insights gained during field studies between 2007 and 2010, in which I applied different research methods.

I regularly reviewed pasture-related legal sources as laws, regulations and decrees with simultaneous and critical consideration of changes of the most central contents and passages during the post-Soviet transformation. I looked at questions of property and ownership, aspects of management responsibilities as well as allocation practices.

I talked with experts for two reasons. First, I wanted to explore the research object, i.e. to learn about the basic aspects, relations and dimensions of pasture-related socio-ecological challenges. Second, I asked for assessments and explanations of the transformations, meanings and mutual relations of different legal regulations. Those people were chosen for interviews that could provide information resulting from their professional experience and knowledge. In this regard, it was especially interesting to talk to people who assumed an institutional function within the pasture-related context in order to gain insider information. Therefore, my interview partners represented governmental and non-governmental organizations such as the Pasture Department of MAWPRI, the GIPROZEM, the GOSREGISTR, the SAEPFUGKR, public administrations as well as the IBRD and the non-governmental legal advice project LARC.

Following these interests and armed with a basic understanding, I went to the research area and conducted non-participatory observations of pasture utilizations, as well as guided interviews with pasture users and other place-based and non-place-based actors. The guidelines encompassed questions about the individual entitlements, perceptions and assessments of the legislation in general and its effectiveness and implications in situ as well as individual usage practices. The aim of this approach was to compare the legal demands with the reality on the ground.

7.4 A Challenge on Its Own: Kyrgyzstan's Pasture Legislation

Kyrgyzstan's resource relations required a new legislation after the dissolution of the Soviet Union. For pastures that were and still are defined legally as 'agricultural usable lands, which show grass-like vegetation and that are used as a fodder base for livestock' (PPPAIP 2002 par. 1.1; ZOP 2009 art. 1), several formal attempts were made to define the general legal guidelines for the subjects of management, allocation, utilization and others. A central, and so far unchanged, regulation adopted from Soviet times is that in contrast to the privatized arable land of former collective farms, pastures as well as the lands of the so-called forestry and water funds with very few exceptions lasted exclusive state property. Generally, they cannot be transformed into private property (ZKRK 1991 art. 2; ZKKR 1999 art. 4 par. 2; ZKKR 2003 art. 4 par. 2; ZOP 2009 art. 3 par. 1). Despite this initial clear arrangement inconsistencies and discontinuities, ambiguities and intricacies were and still are characteristic for Kyrgyzstan's post-Soviet pasture legislation. First, this becomes apparent by comparing the diachronically changing regulations for management responsibilities as well as procedures of the entitlement allocation. Second, and according to the specific spatial locations of pastures, diverse regulations are valid simultaneously (cf. Table 7.2). Third, many of them proved to be impractical and were hard to apply.

Table 7.2 Legal regulations of pasture management and allocation since 1991

Legal norm	Main regulations on 'management competences'	Main regulations on 'pasture allocation'
ZKRK (1991)	Administrations at *rayon* level: pastures in their own territories Administrations at *oblast* level: pastures lying across *rayon* borders Council of Ministers: pastures extended over *oblast* borders	Permanent and long-term lease of up to 25 years to private, state-owned and collective farms and others
LKKR (1993)	Government and SAEPFUGKR: forest pastures	Through *rayon* and *oblast* administrations
UPKR-MNRGPZAR (1995)	Administrations at local level: pastures nearby settlements *Rayon* administrations: intensive used pastures Ministry of Agriculture and Food: remote pastures	No statement
ZKKR (1999)	Aiyl okmotu: pastures near to settlements Rayon administrations: intensive used pastures Oblast administrations: remote pastures	Use via temporal lease only, few exceptions
ZKKR-UZSN (2001)	No statement	Possible only via leasing contract
PPPAIP (2002)	Aiyl okmotu: pastures near to settlements Rayon administration: intensively used pastures Oblast administration: remote pastures SAEPFUGKR: forest pastures	Leasing contracts of up to 10 years via auction and via provision to local communities and vulnerable people
ZOP (2009)	Aiyl okmotu and local pasture user committees: all pastures, except for those located on forest fund lands	Annually to local community members Lease forbidden

Source: Own compilation on basis of the quoted legal documents

The first Land Code ZKKR from 1991, declared prior to Kyrgyzstan's independence, defined the Rayon (district) Councils of People's Deputies as responsible for the management and the allocation of pastures located within the borders of their own territories (art. 46 par. 8, art. 85). The councils were replaced after independence by the rayon administrations. The superior Oblast (regional) Councils of People's Deputies and, after 1991, the oblast administrations were responsible for pastures extended across rayon borders (art. 47 par. 4, art. 65). Lastly, Kyrgyzstan's Council of Ministers was competent for grasslands extended across oblast borders (art. 85). This was new because during Soviet time, the users – that is, notably state and collective enterprises – were responsible for the management of all pastures allocated to them independent of their location. With the new law, the pastures could be leased to state, collective as well as private users permanently or long-term for up to 25 years (art. 9).

A new management arrangement related to a pasture categorization from Soviet enterprises was introduced in 1995 by a presidential decree on the land and agrarian reform UPKR-MNRGPZAR. The determining factor here was the distance between the grasslands and settlements. Pastures near to settlements were to be managed by local 'committees for land and agrarian reforms' which were later transformed to the rural authorities, the so-called *aiyl okmotu*. The Ministry of Agriculture and Food had to manage 'remote pastures' located in greater distances from settlements than the first ones in agreement with oblast administrations. The rayon administrations were managing pastures of 'intensive use' that are located in between. Though, in many cases, the affiliation of the pastures to these categories and therefore the management responsibility remained unclear, as concrete threshold values of the distances were not defined, nor were allocations of the pastures to specific categories conducted officially.

Generally, the mentioned norms were significant for all pastures, exceptionally for those located on lands of the so-called state-owned forest fund. According to the Forestry Codes LKKR from 1993 and 1999, this area consists of territories covered by forests as well as areas that were allocated to the forestry sector for its purposes (LKKR 1993 art. 4; LKKR 1999 art. 7). In such cases, the management resided with the government and the SAEPFUGKR (LKKR 1999 art. 23–25). However, the rayon administrations achieved the right to allocate such pastures to interested people from the respective district and the oblast administrations would allocate to those from the respective region (LKKR 1999 art. 19, 20).

The ZKKR, amended in 1999, adopted the spatial pasture categorization of the UPKR-MNRGPZAR but replaced the Ministry of Agriculture and Food with the oblast administrations, making them responsible for 'remote pastures' again (art. 13 par. 2, art. 15 par. 2, art. 17 par. 1). With few exceptions, it also defined temporal leasing contracts generally as the only way to acquire pasture entitlements (art. 30 par. 3–5). This regulation was also codified by the law 'on management of agricultural land' ZKKR-UZSN in 2006 (art. 21).

The emergence of different socio-ecological pasture-related problems in the course of the first decade of independence, in this case mainly ecological ones, showed that these attempts to create a working legislation with clear management and allocation mechanisms remained ineffective. The repeatedly shifting regulations of management responsibilities hindered the establishment of work experience, routine and expertise within the responsible organizations. Additionally, they were not able to fulfil their duties due to being underequipped with human and material resources, as well as deficiencies of information and cross-linked knowledge that is essential for this job. These reasons led to insufficient implementation of the mentioned legal regulations and, therefore to a situation, where the 'use regime operate in parallel to, rather than in accordance with, the law' (Undeland 2005, 22). As a result, for more than a decade, no well-balanced management, allocation and utilization practices were working on the ground.

This requirement led to the decision to develop a comprehensive and reliable legal base for pasture relations embracing management, allocation and utilization matters. Assisted by the IBRD, this resulted in the government's resolution no. 360 'on pasture land lease and use' PPPAIP in 2002. Nevertheless, according to the key category 'spatial location' in relation to settlements and land funds, the already existing mosaic of diverse management responsibilities persisted (par. 10, 15, 39). Two formal procedures were designed to allocate pasture entitlements in the form of leases for up to 10 years: via auctions and via provisions to local communities for communal purposes and to economically vulnerable people for their individual needs (par. 4, 7). But again, the high expectations were not fulfilled. First, the division of management responsibilities according to the pasture's spatial category proved to be impractical due to the same reasons as previously. Second, the designated allocation procedures proved to be unrealistic for several reasons: The auction process designed was complicated and therefore caused high temporal and organizational costs for all participants. It also proved to be impractical as mobile stock breeders were to perform several successful acquisitions at the same time to obtain usage rights for all pastures required for a complete annual grazing loop. Therefore, the market-based auctions were seldom performed. Additionally, the ambitious approach of pasture allocation to communities and individuals, following the motive of social responsibility, was hardly applied. However, informal allocation practices and payments in which interpersonal power relations play an important role remained the custom (Undeland 2005, 31–35). As a result, economic strong actors achieved better pasture accesses than the poor. Short-term utilization forms oriented towards extraction maximization increased because of the legal uncertainty of resource access obtained in informal ways and the inability of people to sue. Therefore, spatially uneven and maladjusted utilization patterns and degradation processes persisted even after the turn of the millennium.

With the currently valid law 'on pastures' ZOP established in 2009 as a consequence of the failure of the previous norms, radical changes in the management and allocation practices had been introduced. Nowadays, local administrations and committees appointed by the pasture users themselves are formally responsible for the sustainable pasture management and allocation, but again not for those grasslands located on the forest fund for that still the SAEPFUGKR is responsible (art. 1, 4 par. 1). According to ZOP, utilization rights can now be acquired annually by local community members obtaining so-called 'pasture tickets'; however, leasing is explicitly forbidden (art. 2, 5, 15). From this obstacle, two questions arise: How will local structures and actors implement their duty against the background of being structurally underequipped with resources and knowledge? How can the approach be ecologically sustainable and provide legally secured planning reliability when it envisages the reviving of usage rights annually, it forbids long-term leases and informal allocation and utilization practices are so common?

The top-down-initiated introduction and the genesis of the hitherto existing pasture-related regulations can be subsumed as a 'trial and error' approach of combining Soviet understandings, as the state property and spatial categories of pastures, with new principles as initially the decentralized administration and market mechanisms and later the participative resource management within new legal norms. Thereby, the strategy of Kyrgyzstan's political leadership looks like a repeatedly 'incomplete, ad hoc arrangement dealing' (Undeland 2005, 22) with the challenges of the post-Soviet transformation. Consequently, the formal regulations were inconsistent, discontinuous and fragmentary in character, as well as inappropriate to specific local conditions. Besides, they hindered the actors' ability to design sustainable and long-time working management and utilization regimes so far. The users did not dispose over planning reliability and therefore often appropriated resources at short notice. In doing so, they were oriented towards extraction maximization without sufficient consideration for the long-term impacts of their actions. Thereby, the legislation has contributed to pasture-related rivalries, unequal allocations and degradations in situ. This will be shown by drawing a sketch of a socio-ecological pasture challenge in the walnut-fruit forest region in consideration of the legal framework developments in the course of the post-Soviet period.

7.5 A Pasture-Related Challenge in the Walnut-Fruit Forest Region

The research area is located on the northern edge of the Fergana Basin at the south-facing slopes of Tian Shan's Fergana Range in the administrative entity named Bazar Korgon Rayon (Fig. 7.2). The walnut-fruit forests growing there at elevations between 1,100 and 2,000 m asl are a regional distinctive feature. They cover around 30,000 ha and consist of a great variety of woody species, particularly walnut (*Juglans regia*), and several wild forms of fruit species (Gottschling et al. 2005; Griza et al. 2008, 46). Nowadays, more than 50,000 people live close to these forests. Their livelihood strategies depend in part on local land and forest resources (Schmidt 2006; Schmidt and Doerre 2011, 289).

Closely connected with this regional peculiarity, in the otherwise sparsely wooded country is the remarkable variety of legal land categories relevant for pasture management and allocation procedures: Large pasture stakes are located on forest fund lands. In this respect the LKKR has to be considered along with other legal norms. Relatively small areas are legally defined as pastures located 'near to settlements' and 'in remote distance'. The main part is constituted by pastures of 'intensive use'. This means that after 1991, additional to the SAEPFUGKR and the local forestries as its executive branches, the public administrations at the local, district and regional levels have had to play a central role in the management and legal regulation of pasture relations. One of the most prominent post-Soviet pasture challenges, though not the only, of the rayon is related to the pasture system *Kara Art*. Despite ecological problems partly induced by maladjusted uses, an inter-local conflict over access and utilization forms ignited here (cf. Fig. 7.2). Hence, the region is especially well suited to an examination of the effectiveness of Kyrgyzstan's legal pasture regulations.

During Soviet times, most parts of Bazar Korgon's pastures were utilized through state and collective farms based in the lowlands and foothills of the Fergana Range. As a part of the forest fund, though not directly utilizable for forestry purposes, *Kara Art* was allocated in the 1950s on the basis of a long-term leasing contract to the collective farm *Engels* based in Sovetsk. Its herders mostly living in Uch Bulak used the area for dairy cattle grazing during the summer period. After a division of the farm in 1977, the upper pasture section named *Sha Murat*, not belonging to the forest fund, was allocated to the newly established enterprise *60 years of October*, which specialized in sheep husbandry (SKBK 1958; KIRGIZGIPROZEM 1983a; GAOZh 1997a, b). The pastures were used and managed exclusively through these farms.

In 1991, according to the ZKKR, the rayon administration became responsible for the management of 'intensively used' pastures. However, this body was unable to execute its duties in resource allocation according to legal requirements and in supervising the pasture utilization. So, after the dissolution of both farms, the former herdsmen informally took over the pastures for private purposes whilst the inhabitants of the nearby settlements of Zhai Terek and Arslanbob also gradually took possession of the territory in search for new income opportunities: Whereas the inhabitants of Zhai Terek were sufficiently provided with farmland, there existed an immediate need for pastures. In contrast, in Arslanbob, a shortage of arable farm land was observable (cf. Fig. 7.2). As the areas of *Kara Art* should serve both demands, this was the beginning of a rivalry between cultivation and animal husbandry locally unknown before. In the course of the UPKR-MNRGPZAR, the rayon administration delegated the responsibility for 'intensively used' pastures that were used before by Soviet state and collective farms to the respective local authorities. The pasture *Sha Murat* was placed at the disposal of the Kenesh authority, former Sovetsk. *Kara Art* as a part of the forest fund was reallocated to the state-owned forestry *Arstanbap Ata* (GOP 1997). Although the rayon administration was designated by the LKKR to allocate 'intensively used' forest pastures, its influence relating to this matter became insignificant. In contrast, within the weakened but still state-owned forestry structures, individual actors such as managers and rangers became powerful gatekeepers controlling pasture access and started to exploit their positions for individual purposes. Informal allocation practices and maladjusted uses increased. The secession of the forest district *Zhai Terek* in 2000 of which *Kara Art* is a part, incisively influenced the evolution of the rivalry between cultivation and animal husbandry to an inter-local conflict between the inhabitants of both settlements who demanded *Kara Art* as pasture or cropland, respectively (GLSKR and GUL 2004, 11–12; Mamaraimov 2007). Despite the important role of the non-place-based rayon administration in the allocation of 'intensively used' forest pastures as intended by the PPPAID, the newly established forest enterprise revealed as the only serious stakeholder within the area of pasture management and allocation. In this respect, the PPPAID proved to be in fact a 'paper tiger'. The new ZOP remained meaningless here as the area belongs to the forest fund.

What I observed during my field studies is that despite their own interests, the employees of the new established enterprise seem to be more loyal to the inhabitants of their settlement than to their neighbours from Arslanbob. I learned that the pasture users from Arslanbob feel discriminated against and penalized by the rangers of the *Zhai Terek* Forestry. Pasture access based on temporal leasing contracts as put forward by the ZKKR and PPPAID was an exception, as told to me by the respondents.

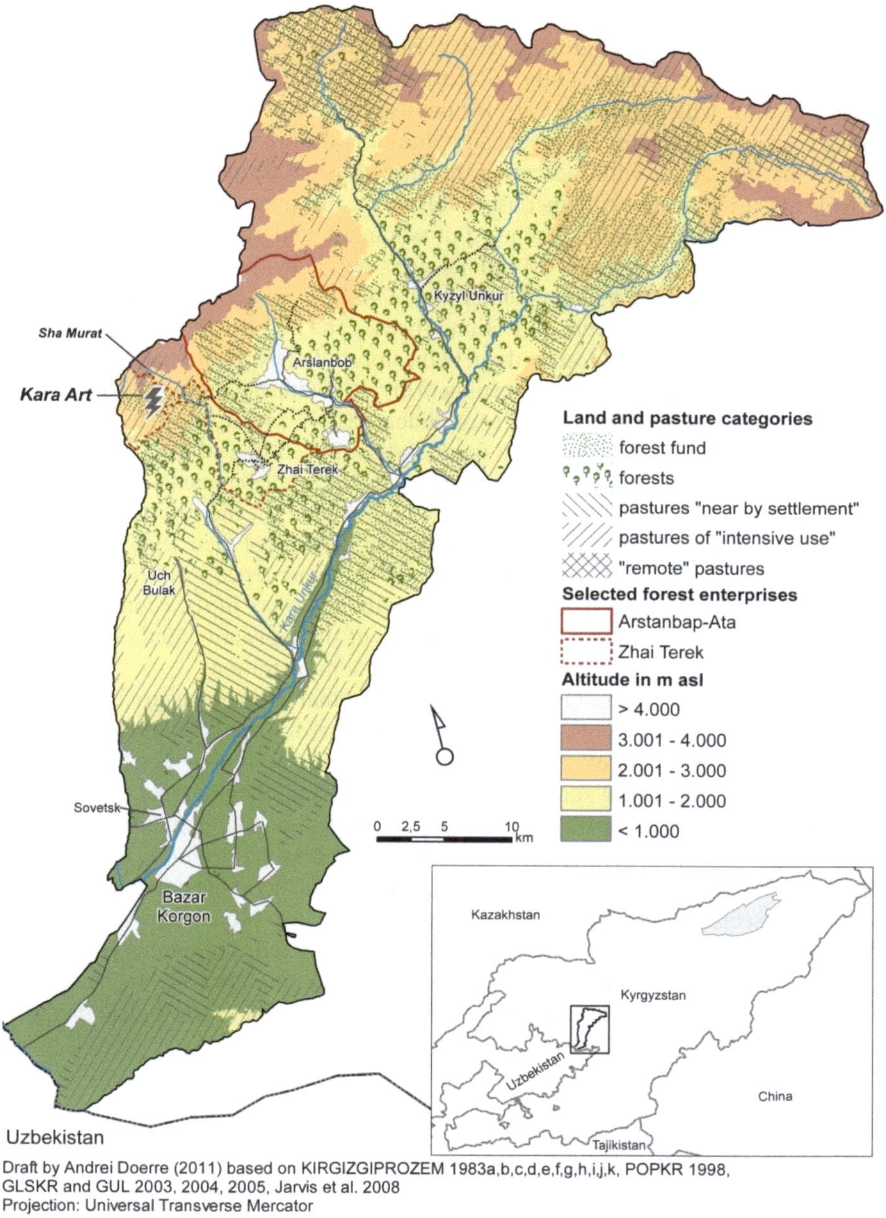

Fig. 7.2 Land categories relevant for pasture management and allocation in Bazar Korgon Rayon (Design: Andrei Dörre)

According to the management of the forest enterprise *Zhai Terek*, in 2007, 64 pasture users were registered in *Kara Art*. But following my own observations, at least 80 users were present. Only 16 of them had a leasing contract and a 'forest ticket', both of which are necessary for the use of forest fund pastures. The documents contain information on the leasing duration, the size, location and demarcation of the plot along with indicating the agreed usage form. The other 16 could provide at least a 'ticket'. All other users utilized the resource on the basis of informal agreements with the forestry staff most of whom are underpaid. In most cases, the paid fees did not correspond to the official sums settled by the rayon administration. This stands as proof that the forest enterprise *Zhai Terek* did not integrate the public administration into the pasture allocation process. The purpose of these measures is to define the amount of the fees without being controlled by a third party. Moreover, all income can be retained without needing to divide it with anyone else. Additionally, in many cases, the forestry enabled forbidden pasture practices such as farming, overstocking, goat keeping on forest fund lands or the cutting of living trees by not intervening in such cases as a countermove to informal payments. Therefore, a remarkable situation arose in which it became lucrative for the forest enterprise to allow certain harmful practices not in spite of but because of their legal ban. This is an important cause, if not the central one, for the emergence of socio-ecological pasture problems on *Kara Art*. It accrues from the interplay of the economic needs of the users and forestry staff, the unreliability of the legislation and enforcement authorities as well as the weakness of the public administrations.

The dilemma is evident: On the one hand, informal agreements are expressions of creative handlings of complicated and therefore hard to apply legal requirements and allocation procedures. They enable economic activities beside the top-down-initiated impractical standards. On the other hand, they hamper the management and control of pasture utilization. Additionally, such agreements allow only those actors who are able to pay for the service to access the resource. By doing so, they enforce the socio-economic stratification of the society. This practice also leads to tenure insecurity where it is impossible to sue for a land plot acquired informally. Additionally, ecologically harmful usages were promoted. In summary, it can be stated that the informal agreements of the given example are fostering resource–related social conflicts between the local population as well as ecological damages of the pasture.

7.6 Conclusion

Kyrgyzstan's pasture-related legislation after 1991 consists of formal norms that are characterized as top-down-initiated, external interventions which have caused several unintended effects at the local level. The failure of the regulations seems to be caused decisively by the attempt to formulate a generally valid and highly differentiated legal frame which neglects the characteristics of the country, which is distinguished by radical social transitions and at the same time is equipped with diverse, local-specific settings. The legislation was resting upon certain untenable presumptions such as the existence of efficient institutions, authorities and market mechanisms. It seems to be more conducive to reduce the national regulations down to the absolutely necessary frame conditions and to enable the social actors of the subordinated scales to find their own solutions that serve the respective socio-economic and ecological situations. These specific local and regional resource management, allocation and usage regulations should be legitimized through the participation of local people within the processes of decision-making and institution building. Where pasture-related challenges across national borders persist, trans-boundary strategies for problem mitigation could be advisable approaches. Such attempts can make a decisive contribution to the integrated and sustainable development of Kyrgyzstan's society in economic, social and ecological matters by balancing initially opposing interests.

The newest pasture law follows this approach by allocating pastures to local communities and delegating legal resource responsibilities to local structures. To achieve the desired goals, this strategy has to be accompanied by capacity-building measures and by financial support to empower the responsible local organizations and actors to fulfil their duties. This has happened until now insufficiently. Moreover, the ZOP has also to be applied for pastures located on forest fund lands, as in the given example. Whereas this measure means a strengthening of local communities, it also leads to a weakening of the SAEPFUGKR and the forestry enterprises. It is expected that these organizations will offer resistance against such actions. However, the solution to the challenges described seems to be impossible without such a radical step.

References

Akademiya Nauk Kirgizskoi Sovetskoi Socialisticheskoi Respubliki and Gosudarstvennyi Komitet Kirgizskoi Sovetskoi Socialisticheskoi Respubliki po delam izdatel'stv, poligrafii i knizhnoi torgovli (1982) Kirgizskaya Sovetskaya Socialisticheskaya Respublika. Encyklopediya (AN KSSR and GK KSSR IPK. KSSR. Encyclopedia). Glavnaya Redakciya Kirgizskoi Sovetskoi Enciklopedii, Frunze

Asian Development Bank [ADB] (2010) Central Asian atlas of natural resources. ADB, Manila

Baibagushev E (2011) Recent changes in pastoral systems. Case study on Kyrgyzstan. In: Kreutzmann H, Abdulalishoev K, Zhaohui L, Richter J (eds) Pastoralism and rangeland management in mountain areas in the context of climate and global change. Deutsche Gesellschaft für Internationale Zusammenarbeit, Bonn, pp 102–118

Beyer J (2010) Ethnonationalismus in Kirgistan. Die Ereignisse im Juni 2010. Zentralasien-analysen 31–32:11–141

Brylski P, Schillhorn-van Veen T, Eliste P (2001) Kyrgyz Republic. Mountain rangeland and forest sector note. Environmentally and socially sustainable development working paper 33. 10 Sept 2001. World Bank, Washington, DC

Central Intelligence Agency [CIA] (2011) The world factbook 2010. Kyrgyzstan. https://www.cia.gov/library/publications/the-world-factbook/geos/kg.html. Accessed 7 June 2011

Fergana.ru (2010) Uzbekistan i Kyrgyzstan postroyat zagrazhdeniya na konfliktnykh uchastkakh gosgranicy i dogovoryatsya po pastbishcham (Uzbekistan and Kyrgyzstan will erect barriers on disputed border sections and will reach an agreement on pastures). 2 June 2010. http://www.ferghana.ru/news.php?id=14855. Accessed 14 June 2010

Glavnoe upravlenie geodezii i kartografii (1987) Atlas Kirgizskoi Soveckoi Socialisticheskoi Respubliki (GUGK 1987. Atlas of the KSSR). GUGK, Moskva

Gosudarstvennaya Lesnaya Sluzhba Kyrgyzskoi Respubliki and Glavnoe Upravlenie Lesoustrojstva (2003) Proekt organizacii i razvitiya. Kyzyl-Ungurskii leskhoz (GLSKR and GUL Project of the organization and development. Kyzyl-Ungur Forestry). Bishkek

Gosudarstvennaya Lesnaya Sluzhba Kyrgyzskoi Respubliki and Glavnoe Upravlenie Lesoustrojstva (2004) Proekt organizacii i razvitiya leskhoza "Arstanbap-Ata" (GLSKR and GUL. Project of the organization and development of the Arstanbap-Ata Forestry). Bishkek

Gosudarstvennaya Lesnaya Sluzhba Kyrgyzskoi Respubliki and Glavnoe Upravlenie Lesoustrojstva (2005) Proekt organizacii i razvitiya i lesochozyajstvennaya vedomost' Zhai Terekskogo lesnichestva (GLSKR and GUL. Project of the organization and development as well as forestry register of the Zhai Terek Forestry). Bishkek

Gosudarstvennyi Arkhiv Oblasti Zhalalabad (1997a) Istoricheskaya spravka na kolkhoz im. "60 let oktyabrya" Sovetskogo sel'soveta, Bazar-Kurganskogo raiona. Fond 452, opis' 1 (GAOZh. Historical information on the collective farm 60 years of October of the settlement Soviet Sovetsk, Bazar Korgon rayon. Stock 452, inventory 1)

Gosudarstvennyi Arkhiv Oblasti Zhalalabad (1997b) Istoricheskaya spravka na kolkhoz im. Engel'sa Bazar-Kurganskogo raiona Sovetskogo sel'soveta. Fond 434, opis' 1 (Historical information on the collective farm *Engels* of the Bazar Korgon rayon, settlement Soviet Sovetsk. Stock 434, inventory 1)

Gottschling H, Amatov I, Lazkov G (2005) Zur Ökologie und Flora der Walnuß-Wildobst-Wälder in Süd-Kirgisistan. Archiv für Naturschutz und Landschaftsforschung 44:85–130

Granicy otgonykh pastbishch. Postanovlenie N 236 B [GOP] 2.7.1997. Glava raionnoi administracii Bazar Korgon (GOP. Decree no. 236 B. 2.7.1997. Head of the rayon administration Bazar Korgon)

Griza E, Venglovskii B, Sarymsakov Z, Karraro G (2008) Tipologiya lesov Kyrgyzskoi Respubliki. Prakticheskii orientirovannyi dokument dlya polevoi ocenki lesnykh nasazhdenii i ikh pravil'noe upravlenie (Typology of the forests of the Kyrgyz Republic. A practical instrument for the empirical assessment and sound management of the forests). Intercooperation, Bishkek

International Bank for Reconstruction and Development [IBRD] (2001) Kyrgyz Republic. Country assistance evaluation. Report 23278. IBRD, Washington, DC. http://lnweb90.worldbank.org/oed/oeddoclib.nsf/DocUNIDViewForJavaSearch/079E0FAF183B4BBC85256B2F006AEF1A/$file/kyrgyz.pdf. Accessed 18 Oct 2010

International Crisis Group [ICG] (2002) Central Asia: Border disputes and conflict potential (Asia Report 33) 4.4.2002. Osh/Brussels. http://www.crisisgroup.org/~/media/Files/asia/central-asia/Central%20Asia%20Border%20Disputes%20and%20Conflict%20Potential.ashx. Accessed 3 March 2011

International Crisis Group [ICG] (2010) The Pogroms in Kyrgyzstan (Asia Report 193) 23.8.2010. Bishkek/Brussels. http://www.crisisgroup.org/~/media/Files/asia/central-asia/kyrgyzstan/193%20The%20Pogroms%20in%20Kyrgyzstan.ashx. Accessed 24 Aug 2010

Isakov KI (1974) Opyt uluchsheniya i ispol'zovaniya gornykh pastbishch Kirgizii (Improving and utilization practices of Kyrgyzstan's mountain pastures). Trudy Kirgizskogo Nauchno-issledovatel'nogo Instituta Zhivotnovodstva i veterinarii 21:3–14

Jarvis A, Reuter HI, Nelson A, Guevara E (2008) Hole-filled seamless Shuttle Radar Topographic Mission 90m [SRTM 90] data V4. Digital Elevation Database v 4.1. Tiles X50-Y4, X50-Y5, X51-Y4, X51-Y5, X52-Y4, X52-Y5. International Centre for Tropical Agriculture. http://srtm.csi.cgiar.org/SELECTION/inputCoord.asp. Accessed 20 Oct 2009

Johnson DL, Lewis LA (1995) Land degradation: creation and destruction. Blackwell Publishers, Cambridge

Kasybekov A (2009) Spletni zagorodili dorogu (Rumours closed the road). Vechernii Bishkek 125 (9793) 8.7.2009. http://members.vb.kg/2009/07/08/panorama/7_print.html. Accessed 24 July 2010

Kazybekov E (2007) Zhajloo – zhalpy eldik bailyk (Summer pastures – social prosperity). Beles 6/7(875–76):5

Kirgizskii Gosudarstvennyi Proektnyi Institut po Zemleustroistvu Kirgizskoi Socialisticheskoi Respubliki (1983a) Karta Bazar-Kurganskogo Rayona Oshskoi oblasti Kirgizskoi SSR. Masshtab 1: 50,000 (KIRGIZGIPROZEM. Map of the Bazar Korgon rayon of the Osh oblast' of the KSSR. Scale 1: 50,000)

Kirgizskii Gosudarstvennyi Proektnyi Institut po Zemleustroistvu Kirgizskoi Socialisticheskoi Respubliki (1983b,c,d,e,f,g,h,i,j,k) Kormovo-Botanicheskaya Karta. Gosudarstvennoe Kolkhoznoe Ob"edinenie "Zhivprom", Kolkhoz imeni Frunze, Sovkhoz "Sajdykum", Kolkhoz "Taldy Bulak", Kolkhoz imeni 22. Parts"ezd, Kolkhoz imeni Lenina, Kolkhoz imeni Dzerzhinskogo, Kolkhoz imeni Engel'sa, Kolkhoz "Komsomol", Kolkhoz "60 let Oktyabrya" Bazar-Kurganskogo raiona Oshskoi oblasti. Masshtab 1: 50,000 (Fodder-botanical map. State Collective Farm Association Zhivprom, Collective Farm Frunze, State Farm Saidykum, Collective Farm Taldy Bulak, Collective Farm 22. Party congress, Collective Farm Lenin, Collective Farm Dzerzhinskii, Collective Farm Engels, Collective Farm Komsomol, Collective Farm 60 years of October of the Bazar Korgon rayon of the Osh oblast'. Scale 1: 50,000)

Kreutzmann H (2004) Mittelasien: politische Entwicklung, Grenzkonflikte und Ausbau der Verkehrsinfrastruktur. Geographische Rundschau 56(10):4–9

Lesnoi kodeks Kyrgyzskoi Republiki N 1198-XII. 7 maya 1993 (LKKR no. 1198-XII. 7 May 1993)

Lesnoi kodeks Kyrgyzskoi Respubliki N 66. 8 iyuljy 1999 goda (LKKR no. 66. 8 July 1999)

Ludi E (2003) Sustainable pasture management in Kyrgyzstan and Tajikistan: development needs and recommendations. Mt Res Dev 23(2):119–123

Mamaraimov A (2007) Taerga Aer Likmi? (With cunning on the arranged?). Didor 2(225) 29.1.2007:1, 4

Melvin N (2011) Promoting a stable and multiethnic Kyrgyzstan: overcoming the causes and legacies of violence. Central Eurasia Project. Occasional papers series 3. Open Society Foundations, New York

Mikhailov G (2010) Uzbekskie beteery uhodyat iz Kirgizii. Mezhdunacional'nyi konflikt v prigranich'e dvukh stran ischerpan (Uzbek armoured tracked vehicles are leaving Kyrgyzstan. The ethnic conflict in the border lands of both countries is solved). Nezavisimaya Gazeta 3 June 2010. http://www.ng.ru/cis/2010–06–03/6_kirgizia.html. Accessed 23 July 2010

Nacional'nyi Statisticheskii Komitet Kyrgyzskoi Respubliki (2009) Kyrgyzstan v cifrakh 2009 (NSKKR. Kyrgyzstan in Figures 2009). NSKKR, Bishkek

Penkina L (2004) Estestvennye pastbishcha i etnokulturnye tradicii (Natural pastures and ethno cultural traditions). Bishkek. ftp://ftp.fao.org/docrep/nonfao/lead/x6400r/x6400r00.pdf. Accessed 24 March 2011

Perechen' otgonnykh pastbishch Kyrgyzskoi Respubliki. Utverzhden postanovleniem Pravitel'stva Kyrgyzsko Respubliki N 775. 30 noyabrya. 1998 (POPKR. Confirmed by decree of the Government of the KR no. 775. 30 November 1998)

Polozhenie "O poryadke predostavleniya v arendu i ispol'zovaniya pastbishch"N 360. 4.6.2002 goda (PPPAIP no. 360. 4 June 2002)

Schmidt P (2001) The scientific world and the farmer's reality: agricultural research and extension in Kyrgyzstan. Mt Res Dev 21(2):109–112

Schmidt M (2006) Transformation der Livelihood Strategies im ländlichen Kirgistan. Verlorene Sicherheiten und neue Herausforderungen. Occasional Papers Geographie 32. Zentrum für Entwicklungsländerforschung, Freie Universitaet Berlin, Berlin

Schmidt M, Doerre A (2011) Changing meanings of Kyrgyzstan's nut forests from colonial to post-Soviet times. Area 43(3):288–296

Schoch N, Steimann B, Thieme S (2010) Migration and animal husbandry: competing or complementary livelihood strategies. Evidence from Kyrgyzstan. Nat Resour Forum 34(3):211–221

Shamsiev B, Katsu S, Dixon A, Voegele J (2007) Kyrgyz Republic. Livestock sector review: embracing the new challenges. The World Bank, Washington, DC. http://www-wds.worldbank.org/external/default/WDSContentServer/WDSP/IB/2007/03/14/000090341_20070314160221/Rendered/PDF/390260KG0Lives1iew0P09028701PUBLIC1.pdf. Accessed 19 Nov 2007

Skhematicheskaya karta Bazar Kurganskogo raiona Dzhalal Abadskoi oblast' Kir. SSR. Predstavlennaya k godovomu zemel'nomu otchetu po forme N22 za 1958 g. Masshtab 1:100,000 (SKBK. Attached to the annual land use report according to form no. 22 for 1958. Scale 1:100,000)

State Agency on Environment Protection and Forestry under the Government of the Kyrgyz Republic [SAEPFUGKR], United Nations Development Programme in the Kyrgyz Republic [UNDPKR] (2007) Kyrgyzstan. Environment and natural resources for sustainable development. UNDP, Bishkek

Steimann B (2008) Niemand hier respektiert meine Grenzen. Konflikte zwischen Hirten und Goldsuchern auf Kirgistans Weiden. Neue Zürcher Zeitung 16./17.2.2008:6

The Times of Central Asia [TCA] (2010) Kyrgyz protesters block road from Uzbek enclave. The Times of Central Asia 3.6.2010:1, 5

Tishkov V (1995) 'Don't kill me, I'm a Kyrgyz!': an anthropological analysis of violence in the Osh ethnic conflict. J Peace Res 32(2):133–149

Trilling D (2010) Inter-ethnic tension rattles Bishkek. http://www.eurasianet.org/node/60891. Accessed 8 March 2011

Ukaz Prezidenta Kyrgyzskoi Respubliki "O merakh po dal'neishemu razvitiyu i gosudarstvennoi podderzhke zemel'noi i agrarnoi reformy v Kyrgyzskoi Respublike" N 297. 3 noyabrya 1995 goda (UPKR-MNRGPZAR no. 297. 3 Nov 1995)

Undeland A (2005) Kyrgyz livestock study. pasture management and use. n.pl. World Bank, Washington, DC

United Nations Environment Programme [UNEP], United Nations Development Programme [UNDP], North Atlantic Treaty Organisation [NATO] and Organization for Security and Co-Operation in Europe [OSCE] (2005) Environment and security. Transforming risks into cooperation. Central Asia. Ferghana/Osh/Khujand area. Vienna/Geneva/Bratislava/Brussels. http://www.envsec.org/centasia/pub/ferghana-report-engb.pdf. Accessed 15 Aug 2010

Urumbaev M (2004) Pastbishche razdora (Pastures of discord). In: Vechernii Bishkek 15.4.2004. http://members.vb.kg/2004/04/15/panorama/17_print.html. Accessed 30 Nov 2010

Zakon Kyrgyzskoi Respubliki "O pastbishchakh" N 30. 26 yanvarya 2009 goda (ZOP no. 30. 26 Jan 2009)

Zakon Kyrgyzskoi Respubliki"Ob upravlenii zemlyami sel'skokhozyajstvennogo naznacheniya"N 4. 11 yanvarya 2001 goda (ZKKR-UZSN no. 4. 11 Jan 2001)

Zemel'nyi kodeks Kyrgyzskoi respubliki N 123. 9 iyul' 2003 goda (ZKKR no. 123. 9 July 2003)

Zemel'nyi kodeks Kyrgyzskoi respubliki N 45. 2 iyun' 1999 goda (ZKKR no. 45. 2 June 1999)

Zemel'nyi kodeks Respubliki Kyrgyzstan N 574-XII. 19 aprel' 1991 goda (ZKRK no. 574-XII. 19 April 1991)

Zhuk O (2010) Vina Bakieva v Aksyinskikh sobytiyakh ochevidna (Bakiev's guilt for the Aksyevents is evident). Delo 39 (833). http://delo.kg/index.php?option=com_docman&task= cat_view&gid = 43&Itemid = 76. Accessed 3 Dec 2010

Chapter 8
Conflicting Strategies for Contested Resources: Pastoralists' Responses to Uncertainty in Post-socialist Rural Kyrgyzstan

Bernd Steimann

Abstract Based on a case study in rural Kyrgyzstan, this chapter explores the multiple constraints which shape agro-pastoral practices today and how local households respond to them. To do so, it distinguishes between different forms of uncertainty, that is, social and political, livelihoods, knowledge and ecological uncertainties. Results show that Kyrgyz agro-pastoralists must respond to much more than ecological uncertainties only, and that the extent of these uncertainties differs a lot between the rich and the poor. Responding to these uncertainties, wealthier households often refer to different sets of norms and practices to justify their claim on pastures and other resources – a practice typically described as 'forum shopping'. At the same time, rich and poor households alike constantly try to diversify their cash income sources, often beyond agro-pastoral production. The example of herders taking paid employment in a foreign mining company on their communal pastures shows that such diversification can rapidly become precarious in the sense that newly adopted income sources may contradict with pastoral practices and may seriously threaten subsistence production in the long run.

Keywords Agro-pastoralism • Pastures • Uncertainty • Livelihoods • Diversification • Forum shopping • Kyrgyzstan

8.1 Introduction

Pastoral production systems are often interpreted as a direct coping response to ecological uncertainty, because their intrinsic mobility allows people to respond flexibly to droughts, winter fodder shortage and general resource scarcity. Herders

B. Steimann (✉)
Human Geography, Department of Geography, University of Zurich,
Winterthurerstr. 190, CH-8057 Zürich, Switzerland
e-mail: bernd.steimann@helvetas.org

H. Kreutzmann (ed.), *Pastoral practices in High Asia*, Advances in Asian Human-Environmental Research, DOI 10.1007/978-94-007-3846-1_8,

react to environmental trends and shocks by exploiting seasonal differences in forage availability through migration and/or by adapting flock sizes (Scoones 1995; Hesse and MacGregor 2006; Milner-Gulland et al. 2006). In the context of post-socialist Central Asia, however, ecological uncertainty alone seems insufficient to explain current agro-pastoral practices and their often close links with other forms of income generation. Over the last two decades, political, economic and social transformation processes have fundamentally altered the prospects of the rural population to make a living from animal husbandry. In the case of Kyrgyzstan, a rapid privatization of the agrarian sector and the collapse of productivity after 1991, as well as profound yet incomplete administrative reforms, have given rise to a number of challenges which shape agro-pastoral practices today. More recently, the 'traditional' use of pastures has increasingly become challenged by a constantly expanding mining industry (Zozulinsky 2007).

Building on a case study of a rural village in Central Kyrgyzstan, this article therefore asks for the different forms of uncertainty agro-pastoralists are confronted with today and how local socio-economic disparities influence people's coping responses. So far, several studies examined how rural Kyrgyz households cope with transformation in general and how they strive to diversify their livelihoods (cf. Ronsijn 2006; Shigaeva et al. 2007). However, few scholars have systematically differentiated the various transformation-related challenges, and have linked them to the concrete behaviour of agro-pastoral households. In addition, there is yet little knowledge on the growing competition between agro-pastoralism and other forms of pasture use such as the mining industry. The article therefore first identifies different forms of uncertainty which appear relevant for Kyrgyz agro-pastoralists, that is, livelihood, knowledge, ecological, and social and political uncertainties (Mehta et al. 1999, 2001). It then refers to the notions of forum shopping (Meinzen-Dick and Pradhan 2001) and livelihoods diversification (cf. De Haan and Zoomers 2003; Scoones 2009) to analyse people's differentiated coping responses. Thus, the purpose of this chapter is to contribute to a better understanding of current agro-pastoral practices from a socio-economic point of view, as well as to shed light on existing negotiations and tensions between different stakeholders concerning the access to and the use of pasture resources.

8.2 Research Approach and Methods

8.2.1 Transformation and Uncertainty

Transformation research has repeatedly related processes of post-socialist transformation to the rise of uncertainty and has identified two main causes. First, the collapse of key institutions of the socialist system – state-sponsored health care, education, social security and guaranteed employment – has led to the loss of former safeguards and routines and thus to a severe destabilization of post-socialist societies. In many cases, rapid decentralization policies have also fostered legal and political uncertainty, especially at the local level, where human and financial resources to adopt the new rights and responsibilities are often absent (Hopfmann 1997; Herbers 2006). Second, the privatization of former state property has fundamentally redefined property rights over resources. Because 'property relations exist not between persons and things but between persons in respect of things' (Hann 2006, 19), this has fundamentally altered the social relations between people. In the absence of sufficient political and legal security, this alteration has often given rise to new conflicts over resources between competing individuals and user groups (Kandiyoti 2002; Sikor 2002; Bichsel et al. 2010). Overall, post-socialist transformation has made interactions between institutions, policies and actors at different levels less predictable, thus increasing transaction costs and uncertainty for all those involved. Uncertainty can thus be seen as both an outcome, as well as a major driving force of transformation processes.

Authors like Mehta et al. (1999, 2001) and Little et al. (2001) have further developed the linkages between human conduct and uncertainty and have contrasted the latter with the notion of risk. They argue that actors can deal with risk by calculating alternative outcomes or probabilities and can avoid or at least minimize it if economic and social costs allow. By contrast, uncertainty describes a situation characterized by indeterminacies, making it impossible to calculate probabilities. Yet since uncertainty can affect all spheres of human conduct, the concept requires further analytical distinction. Mehta et al. (1999, 2001) differentiate between livelihood, knowledge, social and political, and ecological uncertainties. These categories can however not always be clearly distinguished from one another, and people experience uncertainty very differently, depending on their wealth, gender, social or political affiliation as well as their skills and knowledge.

Livelihood uncertainties refer to the generally unstable and unpredictable conditions in the economic sphere, including people's often insecure access to natural resources, capital and labour and commodity markets. In the absence of reliable rules and regularities, the behaviour of other actors is difficult to predict, as are the outcomes of what they do (Meinzen-Dick and Pradhan 2001). *Knowledge uncertainties* exist because each individual relies on different sets of information that are, furthermore, located within particular institutional settings. Thus, people's knowledge about institutional arrangements and the behaviour of others is always incomplete and necessarily gives rise to uncertainties (Mehta et al. 2001). *Social and political uncertainties* relate to altered power relations and forms of cooperation at various levels, as well as to social developments such as rural outmigration. For instance, decentralization policies may, at least temporarily, add to political uncertainties, since it redistributes responsibilities and powers between various actors (ibid.). *Ecological uncertainties* relate to the unpredictable and variable nature of ecosystems with which people interact. Since the effects amongst variables within an ecosystem are often complex and intermittent, non-equilibrial dynamics are often the norm. This is particularly true for dryland ecosystems, where resource management is often highly complex and requires flexible and adaptive strategies. Consequently, pastoralism is usually interpreted as a direct coping response to ecological uncertainty, since mobile herders can respond flexibly to droughts and general resource scarcity (Scoones 1995; Hesse and MacGregor 2006).

8.2.2 *Responding to Uncertainty: Forum Shopping and Livelihoods Diversification*

People often respond to uncertainty by widening their range of options, that is, by increasing the flexibility of their ways to achieve certain objectives. One widely adopted means of doing so is 'forum shopping' (Meinzen-Dick and Pradhan 2001, 12). In most social settings, various kinds of law such as state law, religious law and customary law coexist and overlap. This coexistence and interaction of various cognitive and normative orders is usually called legal pluralism (cf. Griffiths 1986; von Benda-Beckmann 2002). In an uncertain environment, actors can thus try to hold on to their resources by manoeuvring between various legitimating principles and standards of measure. Thus, forum shopping can also exacerbate knowledge uncertainties: If others refer to multiple normative and cognitive orders to legitimize their claims, predicting their behaviour becomes difficult.

Another form of increasing flexibility is livelihood diversification. Recent actor-oriented research suggests that under constantly uncertain circumstances, people often seek to secure or improve their capital base by diversifying their livelihood sources. Frequent diversification strategies in rural areas include farm and off-farm labour, the establishment of a small private business or migration for labour. Over the last decade, livelihood studies have repeatedly established a global tendency amongst rural households to increasingly diversify their livelihoods beyond the agricultural sector and the local level, thus developing a very diverse, often multi-local portfolio (De Haan and Zoomers 2003; Scoones 2009). But diversification does not necessarily improve a household's economic situation. Whilst the relatively rich can usually afford to plan ahead to build on complementarities in a strategic manner, households in financial distress are often pushed into precarious short-term diversification. Though this may help them to secure a living in the short run, it often exacerbates future impoverishment, because the newly adopted activities compete with existing ones (Barrett et al. 2001; Little et al. 2001; Ashley et al. 2003).

8.2.3 *Methods*

The case presented here stems from Jergetal, one of three villages constituting a community (*ayil okmotu*) located in Naryn *oblast* (province), about 25 km from the *oblast* capital at an elevation of 2,150 m asl. Research was carried out over a total of 10 months, in the form of a mixed-methods, three-stage case study. First, a quantitative household survey was carried out in April and May 2007 which covered 397 households and provided the basis for household selection. Second, 28 households representing different wealth categories were visited for semi-structured in-depth interviews, held from September to December 2007. Third, another round of qualitative in-depth interviews was held from July to September 2008 with representatives of 21 herding households on the spring, summer and autumn pastures. Throughout the whole research process, additional interviews and focus group discussions were held with various experts, government and non-government representatives from the local to the national level.

8.3 Agricultural Privatization at the National and Local Level

Soon after gaining formal independence in August 1991, Kyrgyzstan faced a deepening national crisis in food procurement. Restructuring the agricultural sector was therefore a primary goal of the Kyrgyz government, who quickly adopted structural reform measures promoted by the international donor community through conditional loans and grants. Working on the basis of the Washington Consensus, international policy advisors promoted a swift, market-led transition consisting of an immediate liberalization of prices, the privatization of most state property and the withdrawal of the state from the economic sphere. Agrarian reforms gained momentum in late 1993, when, in view of the rapid impoverishment of the rural population, de-collectivization was declared compulsory for all *kolkhozes* and *sovkhozes* (collective and state farms; Bloch et al. 1996, 13). Except 25% of arable land that was reserved for a state land redistribution fund, all arable land within a farm had to be equally distributed as land-use shares to all current and former farm workers and their family members. Livestock, machines and built infrastructure had to be distributed as private property, whilst pastures remained in state ownership. By 1996, about 82% of all *kolkhozes* and *sovkhozes* had been privatized and transformed into more than 32,000 small peasant farms, plus a few cooperatives and joint-stock companies. In 1999, a new Land Code finally converted the land-use shares into secure ownership rights (Abazov 1999, 218; Jones 2003, 264). Although land and livestock shares were calculated on an individual basis, they were distributed by household. Thus, the household emerged as the new main production unit in Kyrgyz agriculture and had to bear all production-related risks. The result was a dramatic collapse of agricultural output. Between 1990 and 1996, the number of sheep and goats fell from 10 million to 3.7 million heads, mainly due to fodder shortage, disease and mass slaughter. Since then, flock numbers have increased again but have not yet reached pre-independence levels (Farrington 2005; Undeland 2005). Crop and fodder yields dropped to a similar extent after 1993, not only due to the partitioning of arable land but also because of a lack of cash investment, fertilizers and working machinery. Today, only a quarter of all arable land is still in use for grain production (Christensen and Pomfret 2007, 39; Mamytova and Mambetalieva 2008).

In the case of Jergetal, the division of livestock and land reserves as well as most machinery and barns started in December 1993. Every individual received 0.27 ha of irrigated arable land and three sheep. Larger households also received a cow or a horse.

Barns were handed over to the leaders of local tribal groups [Kyrg. *uruu*] for further distribution. By contrast, agricultural machines remained with the *ayil okmotu*, from where groups or individuals could lease them for a period of 5 years. After that, the lessee had the right to buy the machine from the *ayil okmotu*. The whole privatization process was carried out by a local commission consisting of about 40–50 people, including former *kolkhoz* elites, local elders and other well-respected people from the village. Many *kolkhoz* agronomists joined the commission because they were best qualified to carry out the often complicated calculation and demarcation of land parcels. By late 1994, the distribution at village level was more or less complete.

8.4 Local Disparities and Uncertainties

8.4.1 Local Disparities

Jergetal village has 675 ha of arable land, most of it irrigated, and the community has been allotted a total of 91,597 ha of pastures (Jergetal Ayil Okmotu 2007). However, despite the seemingly equal distribution of assets in the mid-1990s, the current distribution of livestock and arable land at household level is highly uneven. Survey results show that 13% of all households have no own animals at all, 53% own small flocks of 5.1–50 sheep equivalent and 3% own more than 350 sheep equivalent (Table 8.1). Thus, more than two-thirds of all households own less than 50 sheep equivalent, mostly sheep, one to two milking cows and/or a horse. This is usually just enough to cover subsistence needs. At the same time, households with more livestock usually also own more arable land per capita than households with little or no livestock.

To a considerable degree, these disparities reflect the long-term effects of the rapid privatization process. On the one hand, the early distribution of assets in Jergetal was anything but transparent and benefited mostly those who had already been in control of key resources before. On the other hand, the collapse of the state economy and the subsequent attempts to decentralize the administration caused many new forms of uncertainties with which rural households must cope today.

Table 8.1 Average per capita land ownership of households grouped by livestock holdings, Jergetal 2007 [$n=397$] (own survey)

Household type	*n*	%	Sheep equivalent per household	Average private arable land per capita [ha]
Large farms	10	3	>350.0	0.35
Mid-sized farms	104	26	50.1–350.0	0.34
Small farms	210	53	5.1–50.0	0.27
Very small farms	20	5	0.1–5.0	0.16
No livestock	52	13	No livestock	0.19
Total	397	100	–	0.28

8.4.2 Livelihoods Uncertainties

For most rural people, the collapse of their *kolkhoz* resulted in the concrete loss of regular wages, adequate pensions and subsidized commodities. Today, rural income opportunities are rare, and old-age pensions and child allowances are hardly ever sufficient to survive. Commodity prices have been subject to massive – mostly upwards – fluctuations in recent years. Unfortunately, the privatization programme, which was intended to counter these trends pre-emptively by endowing households and individuals with private means of production and the right to access the emerging free market, failed to guarantee equal starting conditions for everyone.

In the case of Jergetal, a large share of livestock already disappeared before the distribution, when many herders and representatives of the *kolkhoz* elite illicitly appropriated animals from the *kolkhoz* flocks. When the distribution finally began, the remaining animals were sick and weak, so that most of the sheep and goats which were allotted to the households of ordinary workers soon died, either from starvation or disease. As for arable land, some households received their shares in one plot close to the village, whilst others received several small plots in different and often distant locations. High transport costs then prevented many from cultivating their fields, so that the overall productivity decreased. Today, many of the asset poor who cannot afford to cultivate all their land even no longer consider remote fallow land as their own.

Also the management of pastures became problematic. The 2002 'Regulations on the Procedure for Providing Pastures for Lease and Use' defined three different types of pastures, making the *ayil okmotu* responsible for the allocation of winter pastures near to the village, the *rayon* administration for the intermediate (spring and autumn) pastures and the *oblast* administration for the remote (summer) pastures (Undeland 2005, 51). Consequently, herders and livestock owners practising a three-pasture seasonal cycle had to negotiate their use rights – in the form of lease contracts for intermediate and remote pastures – with three different administrative entities, usually located in three different places. In Jergetal, the situation became even more complicated by overlapping responsibilities between the *rayon* and the State Forest Department, which owns a large share of intermediate pastures (Fig. 8.1). Herders who wanted to rent these pastures therefore had to address both authorities, who applied different terms and conditions. Unfortunately, the new law 'On Pastures' passed by the Kyrgyz Parliament in 2009, which abolished the administrative tripartition, did not address these overlapping responsibilities of local communities and the State Forest Department. Finally, the perimeter of a foreign gold mine established in 2006 on Jergetal's intermediate pastures constitutes an additional layer of authority, whose origin is generally unclear. Most local respondents vaguely assumed that 'the state' had issued the licence, but neither the *ayil okmotu* nor the pasture specialist at *rayon* level had any saying in the issue. Overlapping authorities and claims thus cause considerable confusion and uncertainty for local agro-pastoralists.

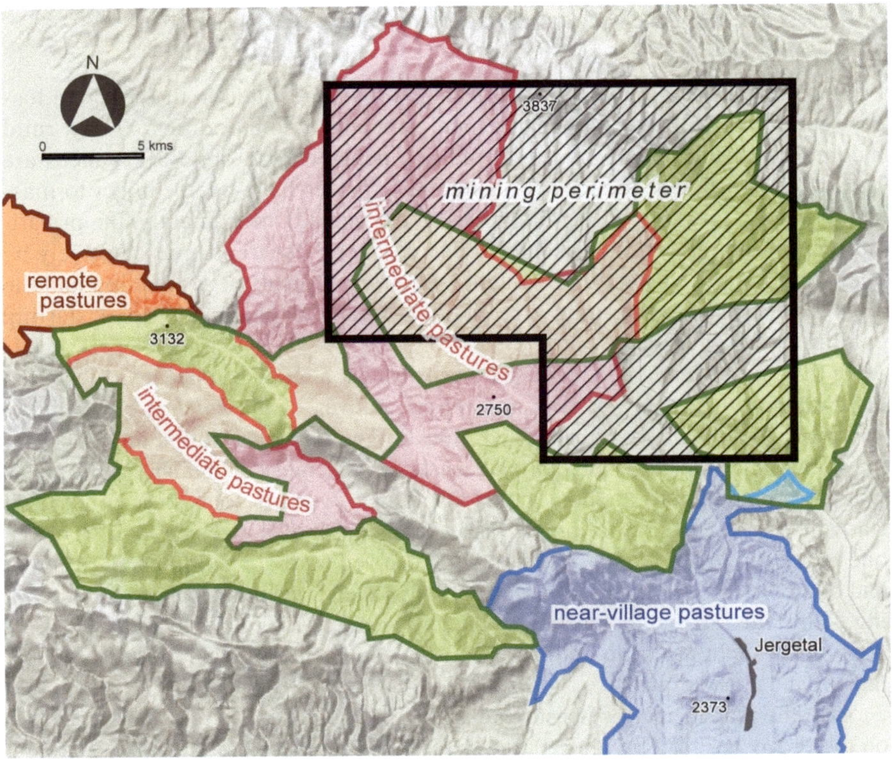

Fig. 8.1 Pasture area allotted to Jergetal village and under the authority of the communal (*blue*), the *rayon* (*pink*) and the *oblast* administration (*red*); under the authority of the State Forest Department (*green*); and the gold mine's perimeter (*black*) (Design by Bernd Steimann, based on various maps)

8.4.3 Knowledge Uncertainties

Labour was highly divided in the *kolkhoz* system, and so was professional knowledge about agro-pastoral production. Even then, the unequal distribution of knowledge allowed some to accumulate more resources and wealth than others. Due to secured wages and subsidized goods, however, those with very limited skills and knowledge could still make a living, even if it was difficult.

But after 1991, comprehensive knowledge regarding agro-pastoral production became crucial to make a living. Many of those who knew about the *kolkhoz*' imminent collapse and who were in control of its resources diverted healthy animals to their own stables or sold machines before they could be distributed to the public. Later on, their professional knowledge allowed them to quickly build up their own herds, whilst households of ordinary workers often struggled to do so:

> We had a lot of goats at that time (…), but we didn't have a clue about the pastures, and the fodder was made ready by the *kolkhoz*. (Smallholder, Jergetal 2007)

Thus, in terms of its procedure as well as its outcome, the privatization process mainly benefited those who already knew how to cultivate land and/or keep animals. Today, adequate knowledge is particularly important for those who wish to practise agro-pastoral production beyond mere subsistence. For instance, many of those working as herders today either herded for the *kolkhoz* in the past or inherited the necessary knowledge from their parents. Since herding animals is one of the few local income opportunities, such knowledge has become critical to people's livelihood security.

Another form of knowledge uncertainty concerns people's incomplete knowledge about formal rules and procedures, which is particularly apparent in relation to pasture management. Many herders have not the faintest idea about the pasture legislation and the involved organizations and institutions. This also increases their knowledge uncertainty about the behaviour of others.

8.4.4 Social and Political Uncertainties

Social and political uncertainties refer, amongst others, to altered power relations and forms of cooperation at various levels. In Jergetal, people's enthusiasm for economic cooperation beyond the own household experienced a number of setbacks after 1991. The neoliberal privatization paradigm clamoured the superiority of private property and entrepreneurship whilst associating collective action with 'outdated' socialist ideals. Thus, there was no support for small-scale cooperatives, and those who nevertheless tried to farm jointly after privatization soon failed. Another setback came when a half-hearted attempt by the Kyrgyz government to foster local cooperation in 2006–2007 drove people's trust in each other and in state institutions to an all-time low. In the long run, this has not only curtailed local opportunities for economic synergies but has also increased people's uncertainty about the behaviour of others:

> In the past, they [neighbours] supported each other, but now we cannot even ask for bread. (...) Every family is individual now (...) everybody takes care of his own business. (Smallholder, Jergetal 2007)

Uncertainties also arise when people must deal with state representatives. Many public servants are unable to fulfil their duties, either because they lack the necessary skills and knowledge or because they have no funds to invest. This often results in a high turnover of public servants – the communal land-use specialist in Jergetal was replaced three times during the research for this study – and in very little cooperation between the communal and the *rayon* level, as well as amongst the various *rayon* departments. Meaningful coordination of public services thus becomes very difficult.

8.4.5 Ecological Uncertainties

The collapse of the state-led production and the subsequent privatization have undoubtedly aggravated people's vulnerability towards climatic variations.

In Jergetal, a series of exceptionally dry years after 2005 resulted in low yields, forcing many households to buy fodder instead of selling it. When combined with chronic trouble getting hold of good seeds, sufficient irrigation water and other inputs, farming has thus become too unpredictable for many:

> I have 1.5 hectares of land (…). In total, the costs of cultivation would amount to about 15,000 Som [US $363]. If the harvest is bad, I'll go bankrupt. The risk is very high and that is why I don't do it. You have to pay for irrigation, plowing, transport, seeds – for everything. And then if you can't harvest in the end… If you have animals you'll get young ones, you can always sell them (…). (Smallholder, Jergetal 2007)

As a result, many households have increasingly turned to animal husbandry. However, also pasture productivity has been badly affected in recent years, mostly due to the intensive use of village-adjacent pastures. In addition, the foreign mining company on the intermediate pastures presents another threat to the environment, whose long-term effects are as yet difficult to predict. Indeed, even before this mining company started to operate in 2006, different small companies were present on the village's intermediate pastures. They all left behind large ditches, pits and pools, turning once fertile pastures into barren landscapes. Many local herders now fear that the new mining company will also abstain from renaturalizing abandoned pits, thus further diminishing pasture productivity in the long run. At the time of research, nobody knew exactly how long the company intended to stay and in which direction it planned to extend its activities. In view of the mine's large perimeter, which includes a considerable part of the village's intermediate pastures (Fig. 8.1), this constitutes another momentum of ecological uncertainty for local pastoralists.

8.5 Pastoralists' Responses to Uncertainty

8.5.1 Forum Shopping

Evidence from Jergetal suggests that wealth in the form of livestock has a considerable influence on how people respond to uncertainties and, vice versa, how the intended and unintended consequences of these responses affect people's livelihoods. Thus, wealthier households draw more often than others on both formal and informal norms and practices to legitimize their claims to arable land and pastures. In the early 1990s, several mid-sized and large farm households began to use communal land for fodder cultivation without asking for communal permission. Having invested much time and money into re-cultivating the plot and being aware that not many other households would have the requisite means to do the same, many of these farmers considered the land their property. However, when the communal authorities noticed their behaviour, the farmers usually agreed to sign a lease agreement that legalized their claim ex post. Similarly, many wealthier households spent considerable time and money to conclude a pasture lease agreement with the *rayon* administration to legitimize their exclusive access to intermediate pastures and thus defend them from other herders. However, as soon as they had obtained the contract, many of them stopped paying the annual lease fee or reverted to informal payments. Being aware of the state's inability to enforce the law and to sanction offenders, they can afford to manoeuvre between acceptance and rejection of formal institutions. By contrast, less wealthy herders hardly ever concluded a lease contract. Some of them are completely unaware of the lease system, but many are just unable to pay for all the fees and documents. Others again simply reject the lease system as something unnecessary, knowing that the state will hardly ever sanction such misdemeanour.

The examples show that comparably wealthy households can often afford to use formal rules and regulations when they come in handy in order to secure their livelihoods, but they also recombine these rules with other, less formal strategies and routine behaviour. The practices observed thus highlight the central importance of people's wealth for how they deal with uncertainties. Those able to pay for the cultivation of faraway land or the conclusion of a pasture lease contract can draw on several sources to legitimize their rights and can thus establish stronger property rights. This again highlights the significance of the socio-economic disparities outlined above for people's access to natural resources – irrespective of the formal property rights they were endowed with in the course of the agrarian reforms of the 1990s.

It would however be wrong to assume that only the wealthy practice forum shopping. Instead, less wealthy households also manoeuvre between different frames of reference, although their range of possibilities is not so large. One example of this is the behaviour of local herders who participate in the annual communal meeting to negotiate the annual herding fees with the communal authorities but adjust the fees to suit their own needs a few days later. There are several explanations for why herders can afford to reject the communal authority. On the one hand, their key position in the functioning of animal husbandry at local level endows them with considerable negotiating power, since people who cannot herd themselves depend on their services. On the other hand, most herders know that the communal authorities have few means to sanction their behaviour or else they put up with potential fines. A consequence of these varying practices is that no local consensus on the appropriate use and management of pastures has yet emerged, making it difficult for herders and livestock owners to develop a long-term perspective regarding animal husbandry.

8.5.2 *Precarious Diversification*

Arable land and livestock are essential for people's livelihoods in Jergetal, but they mainly serve subsistence needs. Survey results from spring 2007 show that only 43% of all households are able to generate a noteworthy cash income from agro-pastoral production, be it through the sale of animals, meat, skins, wool and milk products or by herding other people's animals. However, whilst most mid-sized and all large farm households have one or two agro-pastoral cash incomes, only about one-third of all small farm households can do so. For the latter, herding, that is, taking care of other people's animals in return for payment in cash or kind, has become an increasingly attractive way to generate a seasonal cash income, since local livestock numbers have increased in recent years. At the time of research, more than 15 local households worked as herders, mostly from early May to late September.

Nevertheless, most local households remain unable to make a living from animal husbandry and must depend on other, often highly limited income opportunities. Land cultivation is important in terms of fodder cultivation but does hardly ever directly contribute to household cash incomes. This is so because access to land and markets is often difficult, and input costs for irrigation, seeds and machinery are high. The few regularly paid jobs are usually with state institutions, where monthly wages range between 1,800 and 4,500 KGS (40–100 USD). Similarly, opportunities for paid (non-)agricultural and (non-)farm labour are very limited. Whilst a few comparably wealthy families can afford to hire paid workers for labour-intensive tasks, the less wealthy often support each other through *ashar*, a traditional practice of unpaid mutual help amongst kin and neighbours. Due to this shortage of local employment opportunities, social support by the state has gained striking importance as a source of cash income. Two-thirds of all households depend either on old-age pensions and/or child allowances, although the amounts paid are very low at 650–2,000 KGS (15–46 USD) per month in the case of pensions and 75–300 KGS (1.7–7 USD) per month in the case of child allowances (own survey and NSC 2007).

However, local income patterns began to change considerably with the emergence of a Chinese-Russian mining company on the scene in 2006. Unlike other, smaller mining enterprises in the region, the management had obtained all necessary licences from the national and the provincial level to secure a 5-year mining right for a perimeter of 21,458 ha on the village's intermediate pastures (Fig. 8.1). The company soon began to improve the old existing roads from the next highway to the mineral deposits. Although local people generally appreciated this investment, they soon realized that the mine began to spoil their drinking water by contaminating several brooks with dangerous chemicals. When local protests increased, the Russian manager decided to pay the *ayil okmotu* a first instalment of 100,000 KGS (2,300 USD) in compensation for the damage. After that, the management began to employ local people as miners, offering them 15-day shifts and daily wages of up to 1,000 KGS (23 USD) – an exceptionally good salary for rural Kyrgyzstan. By summer 2008, around 70 people from the village had taken up employment in the mine, also including herders living close to the mine during the summer.

Apparently, ordinary mine workers neither had a work contract nor any kind of insurance. They also did not know whether they would still be employed at the end of their shift. In addition, nobody exactly knew how long the mining company would operate for. Despite these precarious working conditions, however, not only the less wealthy had taken this opportunity to improve their household income.

Even households with comparably large flocks tried to send as many household members as possible to the mine:

> My two brothers already work in the goldmine. If we could find somebody to look after our animals, I would go there as well. (...) But I am the only one now who can look after our animals, so I stay here on the jailoo. (Herder with a mid-sized farm, Jergetal 2008)

At the same time, an increasing number of herding families spending summer close to the mine began to sell milk, butter and yogurt to the mining company. This allows them to cut down transport costs, whilst the mine pays up to one quarter more than the traders near the highway. One herding household that installed its yurt only a few hundred metres from the next open pit even started to exchange milk products for electricity. Another household was even found to have concluded a seasonal milk supply agreement.

In short, the mine offers manifold economic incentives in the form of local employment, comparably good remuneration and trading opportunities. Thus, if a household does *not* send someone to work in the mine, it is less because of concerns about potential negative long-term effects for pastoral production, rather than due to a lack of workforce. This does not mean, however, that people are unaware of the obvious contradiction between pastoral production and mining activities. Instead, many who work in the mine continue to complain about the ongoing destruction of pastures, which repeatedly forces them to find a new place for their yurt, about the lack of renaturation measures and the use of dangerous chemicals.

8.6 Conclusions

Building on a case study from Central Kyrgyzstan, this chapter examined how different forms of uncertainty affect agro-pastoral production and how local households respond to these uncertainties. It is an attempt to go beyond the usual interpretation of pastoralism as a response to ecological uncertainties (Scoones 1995) but instead to look at it as a practice embedded in the complex processes of post-socialist transformation.

Without doubt, ecological uncertainties such as droughts and fodder shortages have remained amongst the foremost problems of Kyrgyz pastoralists. In fact, the state's withdrawal from agrarian production and the subsequent arrival of private investors such as mining companies have probably even aggravated people's vulnerability towards environmental constraints. At the same time, however, the multiple and often parallel transformation processes following the collapse of the Soviet Union have caused a variety of other uncertainties which affect agro-pastoral production as well. Thus, the intricate and inefficient pasture legislation has forced many households to circumvent the law by 'informally' practising seasonal rotation and has even fostered local conflicts (livelihoods uncertainties). Yet many people also lack the necessary skills and knowledge to practise animal husbandry in a profitable and sustainable way (knowledge uncertainties). At the same time, people's will to cooperate with each other beyond the own household has reached an all-time low, so that predicting the behaviour of others has become increasingly difficult (social uncertainties). Thus, agro-pastoral practices in post-socialist rural Kyrgyzstan do not just respond to ecological imponderables but are also shaped by an often adverse political, social and economic environment.

However, analysis has also shown that people respond very differently to uncertainties. The described practices of forum shopping show that uncertainty can even be beneficial to those with sufficient power, since it allows them to respond more flexibly to certain challenges and opportunities. At the same time, the less wealthy usually lack the necessary negotiating power to make use of different norms and rules. Nevertheless, for a majority of the rural Kyrgyz, there is hardly any alternative to the combination of agro-pastoralism and other, often precarious income sources such as working in a gold mine. It may be assumed, however, that in the long run, the outcomes of such diversification again differ for the rich and the poor. For many smallholders and households without livestock, working in the mine may seem like the only way to improve their standard of living without having to migrate to urban areas. Often in immediate need of cash but usually short of financial capital, they must make use of every income opportunity, even if it threatens to seriously worsen the resource base on which their own pastoral production will depend in the future. By contrast, wealthier households already have some savings in the form of livestock or cash. In addition, they are often in a stronger position to secure and defend their access to pastures. Thus, their need for an additional source of cash income may not be so immediate. Nevertheless, if compared to agro-pastoral production cycles and the related uncertainties, even wealthy households may consider the insecure yet well-paid employment in the mine an attractive income opportunity. The difference, however, is that they seem to be in a better position than their less wealthy neighbours to secure long-term access to any type of pastures. Once the mine has ploughed up the intermediate pastures, access to the remaining pastures will have to be negotiated afresh, with the same advantages for wealthier households.

The chapter thus shows how the distinction between different types of uncertainties can help to better understand pastoral behaviour in a post-socialist context. In addition, the described nexus of 'traditional' pasture use and the mining industry raises a number of concerns for the near future, not least in view of Kyrgyzstan's reportedly 'huge potential for further exploration and development of existing and unexplored [gold] deposits' (Zozulinksy 2007, 2). Since access to reliable financial institutions is often limited for the rural population, many people prefer to invest surplus sums of money in livestock, which can be converted into cash whenever needed (Schoch et al. 2010). It may therefore be assumed that many households will use the salary from the mine to buy more animals, to keep flock sizes stable by purchasing more winter fodder, or to cover day-to-day expenditures in cash, instead of converting livestock into cash first. This would inevitably lead to an increase of livestock numbers and, consequently, would further raise the pressure on pasture resources.

Acknowledgements This research received support from the National Centre of Competence in Research North-South (NCCR North-South), with financial assistance from the Swiss National

Science Foundation (SNSF) and the Swiss Agency for Development and Cooperation (SDC). I thank Gulzar Kuruchova, Ayzaada Bekboeva and Aisuluu Kurmanbay Kyzy for research assistance, as well as Christine Bichsel, Urs Geiser and Susan Thieme for commenting on earlier versions of this chapter. Responsibility for the ideas expressed herein remains mine alone.

References

Abazov R (1999) Policy of economic transition in Kyrgyzstan. Central Asian Surv 18(2):197–223

Ashley C, Start D, Slater R, Deshingkar P (2003) Understanding rural livelihoods in India: diversity, change and exclusion. ODI Policy Guidance Sheets, Overseas Development Institute, London

Barrett CB, Reardon T, Webb P (2001) Nonfarm income diversification and household livelihood strategies in rural Africa: concepts, dynamics, and policy implications. Working paper series by the Center for the Study of Inequality, Cornell University

Bichsel C, Fokou G, Ibraimova A, Kasymov U, Steimann B, Thieme S (2010) Natural resource institutions in transformation: the tragedy and glory of the private. In: Hurni H, Wiesmann U, with an international group of co-editors (eds) Global change and sustainable development: a synthesis of regional experiences from research partnerships. Perspectives of the Swiss National Centre of Competence in Research (NCCR) North-South, vol 5. Geographica Bernensia, Berne, pp 255–269

Bloch PC, Delehanty JM, Roth MJ (1996) Land and agrarian reform in the Kyrgyz Republic. LTC research papers. Land Tenure Center, University of Wisconsin-Madison

Christensen G, Pomfret R (2007) Distortions to agricultural incentives in the Kyrgyz Republic. Agricultural distortions working paper 4, World Bank's Development Research Group, Brussels and Adelaide

De Haan L, Zoomers A (2003) Development geography at the crossroads of livelihood and globalisation. Tijdschrift voor Economische en Sociale Geografie 94(3):350–362

Farrington JD (2005) De-development in Eastern Kyrgyzstan and persistence of semi-nomadic livestock herding. Nomadic Peoples 9(1/2):171–196

Griffiths J (1986) What is legal pluralism? J Legal Plur 24:1–50

Hann C (2006) 'Not the horse we wanted' – postsocialism, neoliberalism, and Eurasia. LIT Verlag, Münster

Herbers H (2006) Landreform und Existenzsicherung in Tadschikistan: Die Handlungsmacht der Akteure im Kontext der postsowjetischen Transformation. Fränkische geographische Gesellschaft, Erlangen

Hesse C, MacGregor J (2006) Pastoralism: drylands' invisible asset? Developing a framework for assessing the value of pastoralism in East Africa. iied. Issue paper 142. International Institute for Environment and Development, London

Hopfmann A (1997) Transformation und Zukunftsfähigkeit. Zur Kritik eines nachholenden Transformations- und Entwicklungsparadigmas. Utopie kreativ (78):40–49

Jones KD (2003) Land privatization and conflict in Central Asia: is Kyrgyzstan a model? In: Burghart D, Sabonis-Helf T (eds) In the tracks of Tamerlane: Central Asia's paths to the 21st century. Center for Technology and National Security Policy, Washington DC, pp 259–273

Kandiyoti D (2002) Inwieweit lassen sich die Analysen des Postsozialismus übertragen? Der Fall Zentralasien. In: Hann C (ed) Postsozialismus. Transformationsprozesse in Europa und Asien aus ethnologischer Perspektive. Campus, Frankfurt/New York, pp 345–371

Little PD, Smith K, Cellarius BA, Coppock L, Barrett CB (2001) Avoiding disaster: diversification and risk management among East African herders. Dev Change 21:401–433

Mamytova J, Mambetalieva G (2008) Kyrygzstan unveils food security plan. IWPR's Reporting Central Asia (526)

Mehta L, Leach M, Newell P, Scoones I, Sivaramakrishnan K, Way S-A (1999) Exploring understandings of institutions and uncertainty: new directions in natural resource management. IDS discussion paper 372. Institute of Development Studies, University of Sussex, Brighton, UK
Mehta L, Leach M, Scoones I (2001) Editorial: environmental governance in an uncertain world. IDS Bull 32(4):1–9
Meinzen-Dick R, Pradhan R (2001) Implications of legal pluralism for natural resource management. IDS Bull 32(4):10–17
Milner-Gulland EJ, Kerven C, Behnke R, Wright I, Smailov A (2006) A multi-agent system model of pastoralist behaviour in Kazakhstan. Ecol Complex 3:23–36
NSC (2007) Socioeconomic development of the oblast, 2002–2006. National Statistics Committee of the Kyrgyz Republic, Naryn Oblast Office, Naryn
Ronsijn W (2006) Coping during transition in rural areas. The case of post-soviet Southern Kyrgyzstan. Conflict Research Group working paper (4)
Schoch N, Steimann B, Thieme S (2010) Migration and animal husbandry: competing or complementary livelihood strategies evidence from Kyrgyzstan. Nat Resour Forum 34:211–221
Scoones I (1995) Living with uncertainty: new directions in pastoral development in Africa. Institute of Development Studies, London, pp 1–36
Scoones I (2009) Livelihoods perspectives and rural development. J Peasant Stud 36(1):171–196
Shigaeva J, Kollmair M, Niederer P, Maselli D (2007) Livelihoods in transition: changing land use strategies and ecological implications in a post-Soviet setting (Kyrgyzstan). Central Asian Surv 26(3):389–406
Sikor T (2002) The commons in transition. CEESA discussion paper no. 10
Undeland A (2005) Kyrgyz Republic livestock sector review – embracing new challenges. Europe and Central Asia Region, Environmentally and Socially Sustainable Development, World Bank
von Benda-Beckmann F (2002) Who's afraid of legal pluralism? J Legal Plur 47:37–83
Zozulinsky A (2007) Kyrgyzstan: mining industry overview. Business Information Service for the Newly Independent States (BISNIS). U.S. Commercial Service, Bishkek

Chapter 9
Pastoral People and Shepherding Practices in the Western Himalaya (Himachal Pradesh): A Historical Perspective

Chetan Singh

Abstract Shepherding in the Western Himalayan region of Himachal Pradesh (India) was carried out primarily by agro-pastoralist. They were an integral part of the socio-economic system of precolonial Himalayan states. Due to environmental and social factors, varied pastoral practices had evolved historically in different parts of Himachal. The establishment of British rule in India represented a fundamental break from the past. Colonial interests now dominated, and pastoral practices were restricted and altered in response to the newly introduced forest laws.

Keywords Gaddis • Forest Act • Himalaya • Kanet • Kinnauras • Kulu • Migration • Pastures • Trade

9.1 Pastoralism, the State and Himachal Himalaya

In large parts of Asia, nomadic pastoralism was often combined with sociopolitical forms of organization that may be considered tribal (Tapper 1991, 54). Pastoralism, nomadism and tribalism do not, however, invariably overlap. Tribal groups were not always pastoralists or nomads, and many were, in fact, sedentary agriculturists. Conversely, not all pastoralists were tribesmen. In India, the former could be an integral part of a hierarchical caste-oriented society even supporting a complex non-tribal state structure. The relationship between pastoralism, nomadism and tribalism has, therefore, been an extremely complicated one.

C. Singh (✉)
Department of History, Himachal Pradesh University,
Shimla 171005, India
e-mail: profcsingh@gmail.com

H. Kreutzmann (ed.), *Pastoral practices in High Asia*, Advances in Asian Human-Environmental Research, DOI 10.1007/978-94-007-3846-1_9,

Photo 9.1 Mobile pastoralism is a visible phenomenon on Himachal's roads

States dependent primarily on sedentary agriculture found it difficult to establish political and economic control over nomadic and tribal people. The long-term objective of such states was to convert pastoralists to settled agriculture. But enforced sedentarization could hardly be adopted by states whose revenue from settled agriculture, though crucial, was inadequate and where pastoralism contributed critically to the revenue. Western Tibet was one such region of the world. The medieval Tibetan state successfully harnessed resources generated by pastoralists and exercised control over them without compelling them to take to agriculture (Vitali 1996, 111).

Even pastoralists usually defined themselves in the context of a larger non-pastoral world—both socially and economically. Logically, therefore, pastoralism can be seen as 'a particular activity within regions where more diverse economic activities are pursued', and it uses spaces within a region that are unsuitable for agriculture (Galaty and Johnson 1990, 20; Salzman 2002, 260). Pastoralists have always engaged with economically diverse, non-pastoral societies. Without attempting to clearly define pastoralism, we might, for the present, describe it as a 'cultural economy' that was 'dependent on extensive systems of livestock husbandry' (Rodgers 1991, 201). This creates enough space for numerous diverse societies with pastoralist characteristics.

The supplementary economic activities and occupations of pastoral people can be partly explained by their interaction with sedentary society. Also important were their religious beliefs and customary practices that also contributed to their pastoralist identity. It is ecological factors, however, that were central to the practice of pastoralism and other occupational strategies associated with it. It is commonly argued that pastoralism is a social response to ecological conditions (Brown 1987).

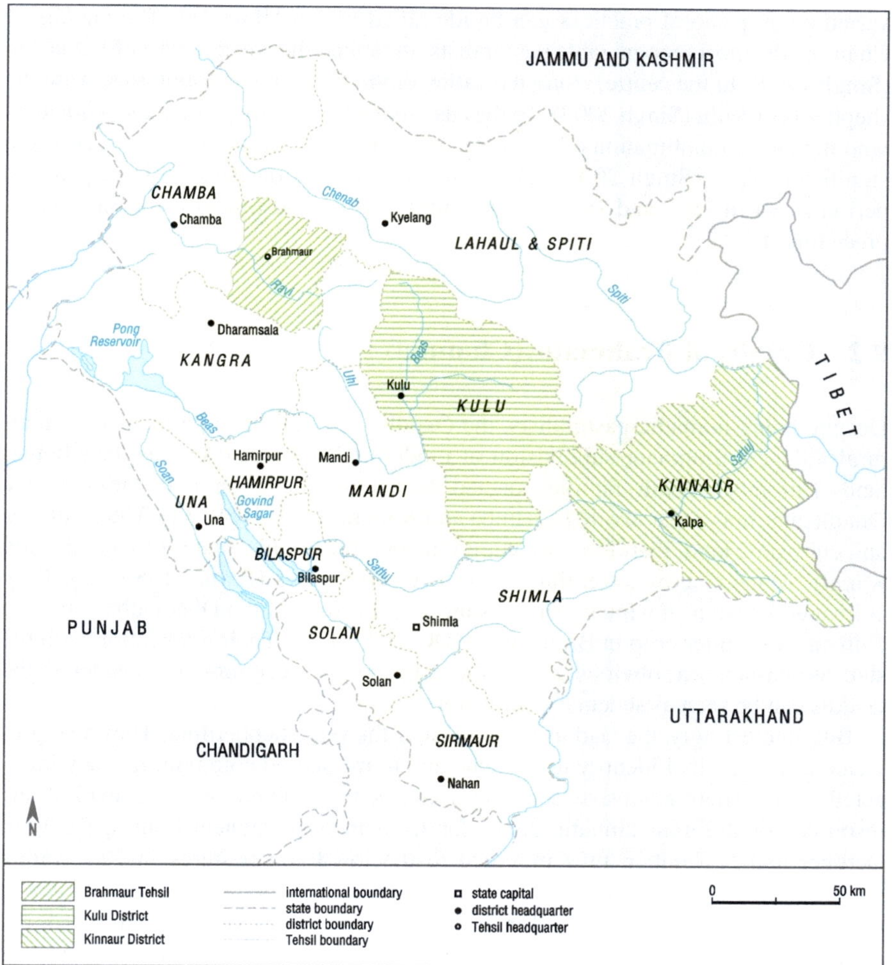

Source: topography based on Mutiah 1990

Cartography: Hilberer

Fig. 9.1 Map of Himachal Pradesh (Design: Hermann Kreutzmann)

The mountainous terrain of Himachal encouraged the emergence of an agro-pastoral economy (Photo 9.1). Within Himachal, however, the degree of dependence on pastoralism varied. Agricultural holdings were the primary unit of revenue assessment by the state. The landholding of the household was the basis for estimating the number of animals that it owned. For the peasant household, a varying combination of shepherding and cultivation was necessary. Opportunity permitting, the peasants preferred to keep 'large flocks of sheep and goats' instead of labouring on agriculture (Parmar 1959, 1 para., 5).

A noticeable diversity in agro-pastoral practices was found in Himachal. This is explained partly by geographical variations (Fig. 9.1) but more perhaps by forms of social organization. By ignoring small divergences, three large areas with noticeable variations in pastoral practices can be identified within Himachal. The Gaddis of Chamba, the most recognizable pastoralists, inhabited the western part of Himachal (Singh 1997). In the centre, along the valley of the Beas and its tributaries, were the shepherds of Kulu (Singh 2009). To the east, in the Satlej valley, were the Kinnauras who pursued a combination of economic activities wherein pastoralism occupied a significant place (Singh 2004). The following description pertains largely to the period between 1800 and 1950, but it would largely hold true also for the century preceding it.

9.2 Gaddis of Brahmaur (Chamba)

Despite being nomadic pastoralists, the Gaddis were linked to permanent agricultural villages. Brahmaur subdivision of modern Chamba district was their homeland. But most Gaddi families owned agricultural land on both sides of the Dhauladhar range that divided Chamba and Kangra (Lyall 1889, 38). Their villages and cultivated land, therefore, lay within the two precolonial states of Chamba and Kangra. Many Gaddis were, thus, subjects of both states (Barnes 1889, 42). They cultivated wheat as a winter crop in Kangra and traversed the Dhauladhar range to cultivate a summer crop in Brahmaur (KDG 1926, 177). Their association with the state as peasants was obvious. If only this aspect of their economy is considered, the Gaddis can be seen as sedentary cultivators.

But, interestingly, the Gaddis spent most of the year shepherding. They also projected a strong tribal identity to the state and to the settled communities they interacted with. Their extensive seasonal movements enabled them to exploit the resources of different climatic zones far from their permanent homes. Between October and November, they migrated to the low-lying pastures of Ravi valley within Chamba, or to Kangra, Nurpur and Pathankot. Some families stayed back in their Brahmaur villages to care for the cattle (CSG 1910, 203, 228). In April, the flocks were brought back to the home villages to manure the fields. For the summer months starting in June, they were placed under the care of a *mahlundi* (master shepherd) who took them to the alpine pastures. Whilst some flocks browsed the high grasslands of Churah and Brahmaur, others crossed the difficult passes to distant pastures in Pangi and Lahaul. By September, the flocks returned over the Pangi range to the pastures of the Ravi river catchment. October saw them grazing in the jungles of Bhattiyat, Nurpur and Pathankot (CSG 1910, 279).

A few Gaddis who migrated to lower areas in winter made a living ‘threshing rice’ and ‘serving in other people’s houses’ (CSG 1910, 203). The majority, however, herded sheep and some continued to do so in the alpine pastures throughout summer. Their constant migration made the Gaddi flocks dependent on the resources of the villages they passed through. A mutually acceptable system of exchange between the Gaddis and these villages was, therefore, established. The shepherds manured the village fields by resting their flocks on them for a fixed number of days. Sometimes a sacrificial sheep was given to the village immediately below a pasture to propitiate the local deity (Lyall 1889, 41; KDG 1918, III, 222). Villages located near river crossings and mountain passes occasionally charged a fee from the Gaddis. They drew this benefit from their strategic location, but the large number of sheep must also have put immense pressure on local resources. Bridges, usually made of grass ropes, needed constant repair and rebuilding (KDG 1918, III, 222–3; MSG 1908, 48). So the peasants, certainly, benefited from the migratory Gaddi flocks. As the flocks came into Kangra valley from Brahmaur in winter, they were received by peasants eager to keep them for some time to manure their fields. The shepherds were paid for this and made good money as they moved their sheep from one landholding to another (Barnes 1889, 30; Lyall 1889, 39, 42).

During the second half of the nineteenth century, the colonial government's dependence on timber increased. The supposed damage caused by migratory flocks to young trees became a pretext for enacting new forest conservation rules in 1891. The duration for which the Gaddis could stop in Kulu valley during their migrations to and from Lahaul was reduced. This decreased the availability of sheep manure in Kulu and the productivity of maize and rice fields fell sharply (KDG 1918, II, 89). However, it was not easy to dispossess the Gaddis of their traditional pastures. Many such pastures were in Chamba, whilst others were in Kangra, Lahaul and other places to which the Gaddis migrated in summer and winter. Their strongest inherited claims (*warisi*) were in Chamba and Kangra. A *warisi* was the hereditary right of a Gaddi family to graze its flocks in a specific forest area at certain times of the year. The right originated from a title (*patta*) granted by the Raja (Lyall 1889, 39). For this they paid fixed cash rent (Lyall 1889, 41; KDG 1926, 274). Though the nature of the *warisi* was not explicit, it was probably more in the nature of a 'managership' than an ownership (KDG 1926, 269; Lyall 1889, 39). In Lahaul, *warisis* were, nevertheless, gifted or sold amongst Gaddis, but the new user was required to rest his flock for a short time in the fields of the original claimant when they migrated between winter and summer pastures (KDG 1918, III, 221). The *waris*, or person claiming the *warisi*, was recognized as the 'master of the flock' (*mahlundi).* He was required to muster other shepherds and their flocks to graze the pasture. The *mahlundi* negotiated with the state officials and village communities on behalf of the other shepherds (Lyall 1889, 39; CSG 1910, 279 note). His efforts were compensated by the *mailani* or payment made by cultivators for sheep droppings.

The value and legitimacy of a *warisi* in pastures was considerable. British officials recognized it even whilst asserting complete governmental ownership of the forests. A *warisi* was admitted even where village communities controlled local forests (Barnes 1889, 19; KDG 1926, 262, 379). Partly at least the pastoralists' strength was their economic importance for the peasantry and the state—especially in precolonial times. It was their flocks that converted extensive forest resources into taxable commodities. *Banwaziri* (forest taxes)—especially grazing dues—were an important source of revenue for all Himalayan states.

A *trini* (grazing tax) was levied in most pastures. Migratory flocks of professional shepherds paid higher taxes than sheep owned by local peasants (Lyall 1889, 24; MSG 1908, 47–8). Precolonial Chamba state obtained much *trini* in the form of wool or sheep and goats. The income derived by government from grazing dues must have been considerable. When the British sought to close access to some forests of Dehra (Kangra) in 1874–1875 by terminating all users' rights, the peasants resisted till they were assured that apart from some other concessions, they would also receive a 'share of the grazing revenue collected from the Gaddis' (KDG 1926, 296).

Despite its recognizable tribal identity, however, Gaddi society was not egalitarian. Gaddi-Khatris claim a high status based on an oral tradition of being migrants from Panjab. Moreover, intricate social subdivisions amongst Gaddis indicate a hierarchy legitimized by Brahmanical ideas. Village menials and artisans, such as Kolis, Riharas, Lohars, Badhis, Sipis and Halis, who performed the most laborious tasks in the settled villages, were located at the bottom of the social ladder (Rose 1970, II, 256–7; CSG 1910, 138–9). They were not even acknowledged by their superiors as belonging to the Gaddi community. Claim to a superior Gaddis position was asserted by first, the Brahmins; second, the Khatris and Rajputs; and finally, the Thakurs and Rathis (CSG 1910, 137). Intermarriage between superior Gaddi groups was common, notwithstanding whether they wore the sacred thread or not (Ibbetson 1970, 205; Rose 1970, II, 256). However, the high caste status of these Gaddis did not translate into equal relations with corresponding castes in non-Gaddi society.

Brahmins and Rajputs attempted to overcome this barrier during the 1901 census by preferring to be enumerated under their respective caste names and not as Gaddis (CSG 1910, 137). This transition was, however, difficult to make, and the Gaddis remained quite distinct from the more Brahmanized societies with which they interacted. This was a distinction that Barnes, the first British revenue settlement officer of Kangra, had already emphasized when he wrote that 'the most remarkable race in the hills. In features, manners, dress and dialect, they differ essentially from all the rest of the population' (Barnes 1889, 42).

9.3 Pastoralism in Kulu

The precolonial principality of Kulu today covers two districts. Kulu district is spread over 5,503 km^2, whilst Lahaul-Spiti stretches across 13,835 km^2 of largely barren terrain. The Kulu area consisted of the Beas valley and its many tributaries. Its Outer Saraj area, however, lies along the Satlej river.

Hamilton, a colonial official of Panjab, argued that Kulu peasants could pay higher land revenue because they obtained additional income from selling the wool of their flocks. Agriculture, by itself, provided them 'little more than their food' (Diack 1898, 2 para., 4). Cash obtained from the sale of wool and *ghee* (clarified butter) helped the peasant family make market purchases and pay state taxes (KDG 1918, 155). Yet it was also noted that 'even the graziers with large flocks of sheep and goats' were landowners 'in a small way' (Diack 1898, 1 para., 2).

Agriculture was combined with transhumant pastoralism in Kulu. The method of shifting flocks from one place to another was very similar to that of the Gaddis earlier mentioned. Like the Gaddis, the Kulu pastoralists accessed different kinds of pastures at different times of the year. By the end of June, the flocks were:

> ... taken further up the hillsides to the *gahrs*, pastures in the forests at about 8 to 11 thousand feet elevations. The pastures, large open glades among the trees, are more properly called *thach*, which word is also applied to the level space in which a flock is penned for the night. In July when the rains have set in or are about to commence, the flocks are driven still higher up to the *nighars*, the sheep-runs on the grassy slopes above the limits of forest growth...The flocks remain in the *nighars*, till the end of the rainy season, about the middle of September, and are driven back to the *gahrs* where they graze till the cold becomes severe and drives them down first to the villages of their owners and thence to their winter quarters. (KDG 1918, 107; Diack 1898, 36 para., 63, 41 Appendix II)

The rajas of Kulu and the village communities recognized the rights of pastoralists in specific alpine pastures. British revenue officials later recorded these rights in the register of 'customary practices' (*wajib-ul-arz*) of Kulu subdivision. Each shepherd had a *ban* (grazing area), but if he failed to use if, others could do so for a night as they migrated to their pastures (Diack 1898, 41 Appendix II).

Large flocks were moved to warmer pastures outside Kulu in order to survive the winter. Kulu sheep and goats, therefore, migrated to neighbouring areas Mandi where pastures were leased from the Mandi Raja (Lyall 1889, 86 para., 96). Whilst many flocks of the main Kulu valley migrated to low-lying adjoining states, the pastoralists of Saraj *waziri* had access to sufficient winter pastures in lower areas within Saraj.

On the other hand, during summer the mountainous Shanshar and Kanawar areas of Kulu had alpine pastures that were open to flocks from outside, provided the needs of local sheep were met (Diack 1898, 41). Access, in summer, to the rain-free trans-Himalayan pastures of Lahaul and Spiti was also important for some shepherds of Kulu and for a number of Gaddis. Chandra valley in Lahaul was too dry for widespread cultivation. In summer, flocks of sheep and goats grazed in the higher uninhabited reaches where only grass grew (KDG 1918, 183, 221; Harcourt 1972, 7). Six flocks of Gaddi shepherds (from Chamba and Kangra) and one flock from Jagatsukh (Kulu) had the right to graze in Spiti during the early decades of the twentieth century (KDG 1918, 287).

Irrespective of the legality of their rights, it was estimated that apart from the Gaddi flocks, 'more than half the sheep and goats of the Kulu *tahsil* were driven to Lahaul-Spiti (Diack 1898, 36 para., 63). Transhumant pastoralism was central to the economy of Kulu valley that had a large number of sheep/goats. Inadequate pastureland in winter and in summer drew the Kulu shepherds into a continuous migratory movement.

However, shepherds of some parts of Kulu, especially of Rupi and Saraj, did not migrate in summer. They had sufficient alpine pastures not only for their own flocks but also for outsiders (Diack 1898, 36 para., 63, 41 Appendix II). Within Kulu, the higher reaches of the Parbati, Sainj and Tirthan river valleys had the best summer pastures. Whilst Saraj saw limited pastoral transhumance in both seasons (summer and winter), the Rupi flocks had to be shifted down in winter.

Shepherding provided revenue to the precolonial state. The practice of taxing the sheep and goats of the peasantry continued under the British. The tax allowed local flocks to gaze in all pastures, including the *nighars*, (high alpine pastures). Because peasants owned the flocks, this tax was considered part of land revenue. In 1891, however, an additional charge was imposed on the flocks. The pastures of different *kothis* (revenue districts) of Kulu were now differentiated. A single tax was levied on the home flocks of a single *kothi*. But flocks grazing outside their home *kothi*—in the high summer pastures of Lahaul, Rupi and Saraj—had to pay twice as much (Diack 1898, 37 para., 64). Kulu shepherds with flocks in the winter pastures of Mandi also paid high grazing rates (KDG 1918, 105).

After the Forest Act of 1865 was enacted, the colonial administration claimed ownership of the forests. Access to forest resources was now strictly regulated (Rangarajan 1996, 27, 30). The Forest Act of 1878 introduced even greater regulation with the primary objective of controlling timber extraction. Pastoralists now came to be regarded as particularly destructive of valuable timber in forests (Rangarajan 1996, 65, 69). For the shepherds of Kulu, the consequences were serious. About 60% of the total forest area of Kulu was classified as demarcated or reserved forest. Peasant rights were either limited or non-existent in such forests (KDG 1918, 120). In the Inner Saraj area of Kulu, the forest area categorized as 'reserved' was particularly large. This effectively meant that forests and pastures within reserved areas could no longer be converted to cultivated land. More importantly, access to pastures in such forests, too, was denied. As a result, livestock in Inner Saraj decreased noticeably (Singh 1952, 38 para., 52). Over the long term (1911–1945), cultivated area in Kulu increased, and the entire increase would logically have occurred in areas located outside the reserved forest. The strict forest laws, and fewer and firmly regulated pastures, compelled the migratory Kulu agro-pastoralists to pay greater attention to agriculture.

The imposition of new forest laws had another interesting consequence. Despite the seasonal migration of the Kulu and Gaddi flocks to Lahaul, many of the best pastures in the Chandra valley of Lahaul had remained quite unexploited before the establishment of British rule (KDG 1918, 221–2). The Kulu and Chamba flocks now increasingly utilized these pastures because the new forest regulations restricted access to many pastures in their home territory and to grazing grounds on migratory routes (KDG 1918, 115–16, 223). Lahaul had no reserved forests and offered the best summer grazing. By not increasing the grazing dues for Gaddi shepherds in Lahaul, the colonial officials probably hoped to attract them away from pastures situated within forests. Moreover for the British, the Lahaul area—so devoid of trees—could not serve any better purpose (KDG 1918, 223).

Even after the country won independence, forest officials continued to regard pastoralists as intruders into the forests and peasants as especially destructive of forest resources (Parmar 1959, 1 para., 7). An imagined increase in the number of their cattle was seen as the cause of soil erosion. It was more likely, however, that after being excluded from their old grazing grounds in reserved forests, the agro-pastoralists were compelled to resort to the few forests that remained unreserved (Parmar 1959, 2 para. 11). Foresters now used the deterioration of such forests to emphasize the over-exploitative nature of pastoral practices (Singh 1952, 8 para., 5).

For pastoralists, on the other hand, livestock did not merely carry economic value. It was of sociocultural significance in equal measure. The optimal and frugal utilization of natural resources was critical for their survival. Even before the emergence of 'scientific' forestry, traditional pastoral societies had 'evolved effective systems to manage common resources in the common long-term interests' (Warren 1995, 196). With the establishment of colonial rule, however, the resources were no longer available to them. The government adopted policies that encouraged the 'peasantization' and 'sedentarization' of pastoralists. This process has continued in independent India.

9.4 Trader Pastoralists of Kinnaur

Present-day district of Kinnaur occupies what was once the northeastern part of the princely state of Bashahr. It covers the upper reaches of the Satlej river basin and is bound by mountains ranging from 16,000 to 21,000 ft above sea level. Most of Kinnaur is an enormous mass of rocky spurs thrown out by the Great Himalayan Range. These spurs create steep and narrow valleys through which mountain streams flow rapidly to join the Satlej. The Baspa valley has some open spaces, though most of the cultivable land and villages in Kinnaur are located at elevated places on the slopes flanking the rivers.

The principality of Bashahr was divided into two territories: Kochi and Kinnaur. Kochi, in the south, was better cultivated and relatively open, whilst Kinnaur was extremely mountainous and sparsely cultivated. Gerard, who travelled through Kinnaur in 1818, noted that the grain produced in the villages of Kinnaur was 'insufficient for consumption' (Gerard 1993, 294). W. Murray, Deputy Superintendent of the Simla Hill States in the 1820s wrote that cultivated area was so limited that it looked like 'patches or steps of stairs up the slopes of the mountains' (Murray 1911, 288; Hamilton 1932, 32).

In short, people in Kinnaur could not have survived on agriculture alone. On the other hand, pastoralism or large-scale sheep rearing based entirely on the internal resources of Kinnaur was not viable either. Low-lying areas adjoining Kinnaur provided the additional grain and summer pastures. Without access to these external resources, life in the area would have been very difficult. This dependence on distant resources made Kinnaur similar to many other agro-pastoral regions of the world (Black-Michaud 1986; Molnar 1981, 33). The Kinnauras followed a combination of occupations. Apart from being peasants and pastoralists, they were also, as Murray reported, 'occupied in extensive commercial intercourse and trade; rearing vast flocks which form their chief dependence and trafficking into remote countries under great hardship and privation to gain a comfortable subsistence for their families at home' (Murray 1911, 278–9, 282). There existed an interweaving of agriculture, pastoralism and trade, and it is difficult to quantify the contribution that each of these made to the total economy.

More than a century later, not much appears to have changed. Hamilton, who prepared the forest working plan for Kinnaur, wrote:

> ...the people make up for the shortage of food by importing grain from the low hills and plains and a large proportion of the community is engaged in this as well as in other trade

> with Thibet; for which purpose sheep and goats are the chief means of transport. Sheep are also kept for the production of wool both for trading purposes and for the local manufacture of clothing. There can, therefore, be no question of the great importance of sheep and goats for the very existence of the Kanawaris… (Hamilton 1932, 32).

The pivotal position of sheep and goat rearing in the lives of Kinnauras is reflected by Parmar's report of 1959:

> The local people have come to believe that their economy is based on the rearing of sheep and goats…. The cultivated holdings are steep and… manuring is only possible by penning sheep and goats there. The terrain is so difficult that transport of foodstuffs and other necessities of life is not possible by other means of transport and that trade with Tibet is only possible with the sheep and goats. They, therefore, want that no limitations on their number and no restriction, whatsoever, of their grazing in the forests should be imposed (Parmar 1959, 58).

Parmar, a forest official, regarded migratory flocks as 'an evil' because they were 'allowed to enjoy unfettered freedom to graze'. He emphasized that their number needed 'to be brought down with a heavy hand' (Parmar 1959, 14).

In all Himalayan areas bordering the Tibetan plateau, sheep and goats have always served as pack animals for transporting goods and trade commodities (von Furer-Haimendorf 1988; Fisher 1987). Apart from being 'beasts of burden', they were also a regular source of wool and meat. In the Kinnaur villages near Tibet, yaks were also bred and were especially useful for carrying the heavy loads of traders travelling to markets in the Tibetan plateau (Fraser 1982, 263–4). It was important for a successful Kinnaura trader to possess a large flock, especially of pack animals. This required skilful husbandry, and the rearing of sheep and goats was a full-time occupation of many Kinnauras. There was also a great demand in Kinnaur for sheep bred in the adjacent Changtang area of Tibet (Moorcroft and Trebeck 1970, 316; Mitchell 1915, 21; von Furer-Haimendorf 1988, 193). They were particularly appreciated for their wool and their ability to carry large loads (Moorcroft and Trebeck 1970, 207, 316). The exact number of sheep and goats imported into Kinnaur is not available, but this could not have been negligible considering how important they were for the economy of the area (Moorcroft and Trebeck 1970, 21; von Furer-Haimendorf 1988, 193). The advice that Gerard, a nineteenth-century British official, received from a Kinnaura about travelling in the cold desert area is illustrative: 'load the sheep with the grain, finish it first, and then kill and eat the sheep: this is the way we travel on the inhabitable tracts; we never think of grain as long as we have plenty of sheep' (Gerard 1993, 149 note).

Kinnaur society had its share of complexities and hierarchies. Kanet agro-pastoralists were the largest and dominant section. Artisan groups such as the Domangs (smiths), Ores (carpenters) and Chamangs (weavers) made up the lower castes. The general term 'Koli' was often applied to the latter groups who also worked as labourers in the fields of the richer Kanet landowners. Unlike in more Brahmanized societies, there were no intermediary castes in Kinnaur. Intermarriage between upper and lower castes was impermissible, and untouchability was practised with regard to food and entry of the lower castes into the house of Kanets (Rose 1970, II, 446; Sanan and Swadi 1998, 54). Interestingly, Kanet clans found in Kinnaur were not found in other parts of Bashahr state. Even within Kinnaur, these clans dominated different *parganas* (administrative units) or different villages within *parganas* (Rose 1970, II, 446–7). In his late-nineteenth-century study, Rose had argued that the Kanets of Kinnaur were Jads, who were a sub-caste with which the Kanets of the lower ranges neither intermarried nor ate (Rose 1970, II, 458). However, the term Jad was used primarily for Kanets residing in the areas bordering Western Tibet and not for those in other parts of Kinnaur (SHSG 1911; Bashahr 22). Irrespective of their internal differences, the Kinnaura pastoralists were, according to local tradition, closely associated with the emergence and sustenance of Bashahr state.

Kinnauri and Tibetan societies probably had a few things in common. Muhammad Haidar Dughlat, a Mughal general who invaded western Tibet in the early sixteenth century, observed that its inhabitants consisted of two sections: the Yulpa were 'dwellers in villages' and the Champa were 'dwellers in the desert' (Dughlat 1972, 407). He also noted that a single flock of laden sheep trading 'in Hindustan and in the mountains of Hindustan' could number 10,000 (Dughlat 1972, 408). Though Kinnaur did not have such explicit divisions at the larger social level, the flock-owning families did, indeed, have two branches: the *kim mee* and the *arung mee* (Negi 2007). The *kim mee* resided permanently at their village home and cultivated the land. On the other hand, the *arung mee*—consisting of men and teenage boys—travelled the year round tending its flocks. This division of functions notwithstanding, the livestock and land of the family was collectively owned and represented a single economic unit. An additional uniting factor was the polyandrous system in which the *talang see* (the wife of the eldest brother who was the householder and part of the *kim mee*) was the common wife of the brothers (Rose 1970, III, 448).

The livestock of the family was also bifurcated: the *kim set* were kept permanently at the village whilst the *arung zet* were shifted seasonally over long distances (Negi 2007). Cows and oxen needed for agriculture were the *kim zet*. A small number of sheep could also be included amongst the *kim zet*. The number of *kim zet* was kept small because they had to be stall-fed through the long and harsh winter when pastures were under snow. Winter fodder for these animals was in the form of hay prepared from specific patches of land (*banjar ghasni*) near the village. In some areas, these hayfields had to be irrigated (Bhuj 1928, 23). Grazing grounds, called *panwis*, were set aside as pastureland. Customary practice required that these *panwis* be neither brought under cultivation nor allowed to become reafforested (Hamilton 1932, 11).

The migratory cycle of the Kinnaur flocks reflects the large area that they utilized. Between mid-May and end-October, the local flocks grazed in the alpine pastures of the Sutlej, Baspa and Bhaba valleys. Shepherds raised temporary shelters (*dogris*) in these higher pastures where 'they resided with their flocks for 4 or 5 months in the year' (Gerard 1993, 220). Some enterprising Kinnaura shepherds engaged in trade and moved with their flocks into Tibet in summer. Towards the end of October, the flocks began descending to the lower hills in Mandi, Bilaspur, Kulu, Suket, Sirmur, Simla and Solan and grazed there from mid-December to the end of April. However, almost a quarter of the flocks remained in Kinnaur, but shifted to low-lying pastures of the valleys that were almost snow-free. The sheep were shorn twice a year: in spring and autumn. Goats apparently seemed to require more care (Parmar 1959, 211).

Shepherds from different parts of Kinnaur herded their flocks to summer and winter pastures over which they had acquired a customary claim. Each flock followed a migratory route that had been established over a long period, and along which it had access to grazing areas (Parmar 1959, 54). As they moved between summer and winter pastures, the Kinnauras made money by using their pack animals to transport goods to and from villages that were virtually inaccessible (Parmar 1959, 51). It was noted in 1888 that wool was transported down to Rampur (the Bashahr capital) over paths that were so steep that sheep, instead of larger animals, had to be used for its carriage (Simla District Gazetteer 1888, 72).

All migratory flocks (*arung zet*) had two categories of animals. The first was the *laadi* (or *ladoo*) or pack animals, and the second were the *shai zet*, mainly female goats, ewes, lambs and kids (Parmar 1959, 211; Negi 2007). The *laadi* were always gelded males of the species. Rams and male goats kept for stud purposes, and the females of the species were never used as pack animals. On the migratory route, these two categories of animals travelled differently between camps. The *laadi* started at dawn and proceeded rapidly with their loads to the next camp so as to reach by midday. After being unloaded, they grazed for the rest of the day. The *shai zet*, on the other hand, moved slowly, browsing as they went along.

9.5 Conclusion

Pastoral practices in Himachal Pradesh were situated within a larger agro-pastoral context in which sedentary agriculture dominated. But agro-pastoral economies and societies evolved differently in different parts of Himachal. The Gaddis of western Himachal came nearest to being nomadic pastoralists. They migrated across an enormous range of territories and successfully asserted a distinct ethnic identity even within the principality of Chamba. In Kulu, the pastoralists were essentially peasants who utilized the natural resources of central Himachal to supplement their agricultural income. Kinnauras, in eastern Himachal, successfully combined several activities. They produced wool for sale, profited from trade between Tibet and Himachal and also engaged in the transportation of goods on payment. This was, of course, in addition to their role as peasants cultivating laboriously developed fields in their home villages. Irrespective of the divergences, however, one fact remains indisputable: there was no part of the western Himalayan region of Himachal where the pastoralist was not a familiar sight.

References

Barnes GC (1889) Report of the land revenue settlement of the Kangra District, Punjab. Report completed in 1850 and first published in 1855. Reprint. Civil and Military Gazette Press, Lahore

Bhuj PtC (1928) Tarmim Bandobast: tahsil Chini, 1928. Copied as typescript from the original unpublished version in 1970

Black-Michaud J (1986) Sheep and land: the economies of power in a tribal society. Cambridge University Press, Cambridge
Brown CW (1987) Ecology, trade and former Bhotia identity. In: Raha MK (ed) The Himalayan heritage. Gian Publishing House, Delhi, pp 125–137
Chamba State 1904 (1910) Punjab States Gazetteers, vol XXII A. The Civil and Military Press, Lahore
Diack AH (1898) Final report of the revised settlement of the Kulu Sub-division of the Kangra District. Civil and Military Gazette Press, Lahore
Dughlat MMH (1972) Tarikh-i-Rashidi. A history of the Mughals of Central Asia. Translated by Denison Ross E, Elias N (ed). Curzon Press, London
Fisher JF (1987) Trans-Himalayan traders. Economy, society and culture in Northwest Nepal. Motilal Banarsidass, Delhi
Fraser JB (1982) The Himala Mountains. First published 1820. Reprint. Neeraj Publishing House, Delhi
Galaty JG, Johnson DL (1990) Introduction: pastoral systems in global perspective. In: Galaty JG, Johnson DL (eds) The world of pastoralism. Herding systems in comparative perspective. The Guilford Press, New York, pp 1–31
Gerard A (1993) Account of Koonawur in the Himalaya. First published 1841. Reprint. Indus Publishing Co, New Delhi
Hamilton AFP (1932) Revised working plan for the Kanawar Forests (Sutlej Valley) of the Upper Bashahr Division, 1931–1960, vol I. Superintendent Government Printing, Lahore
Harcourt AFP (1972) Himalayan districts of Kooloo, Lahoul and Spiti, report completed in 1870. Reprint. Vivek Publishing House, Delhi
Ibbetson D (1970) Panjab castes. First published 1883. Reprint. Languages Department Panjab, Patiala
Kangra District 1924–5 (1926) Punjab District Gazetteers, vol VII. Part A. Superintendent, Government Printing, Punjab, Lahore
Kangra District, Parts II, III, and IV, 1917 (1918) Punjab District Gazetteers, vol XXX A. Superintendent Government Printing, Punjab, Lahore
Lyall JB (1889) Report of the land revenue settlement of Kangra District, Punjab, 1865–72. First published 1874. Reprint. Civil and Military Gazette Press, Lahore
Mandi State 1904 (1908) Punjab States Gazetteers, vol XII A. Civil and Military Press, Punjab Government, Lahore
Mitchell A (1915) Report on the administration of the Bashahr state, 1914–15. Liddel's Printing Works, Simla
Molnar A (1981) Economic strategies and ecological constraints: case of the Kham Magar of north west of Nepal. In: von Furer-Haimendorf C (ed) Asian highland societies: an anthropological perspective. Sterling Publishers Pvt. Ltd, New Delhi, pp 20–51
Moorcroft W, Trebeck G (1970) Travels in the Himalayan Provinces of Hindustan and the Punjab; in Ladakh and Kashmir; in Peshawar, Kabul, Kunduz and Bokhara 1819 to 1825. First edition 1837. Reprint. Languages Department, Patiala
Murray W (1911) Report from Lieutenant W. Murray, Deputy Superintendent, Hill States, to C. Elliot, Agent to the Governor General, Western Provinces, dated Camp Subathoo, 6th July 1824. In: Delhi residency and agency records, 1807–1857, vol I of Punjab Government Records, 8 vols in 10 parts. Superintendent Punjab Government Press, Lahore
Negi V (2007) Pashchimottar Himalaya ke Aranyak-Bhedpalak: Kinnaurke Sandarbh Mein (Samajik-Sanskritik Addhyayan) Titled in English as: The nomadic shepherds of North-West Himalayas: in the context of Kinnaur (A sociocultural study). Central Institute of Higher Tibetan Studies, Sarnath/Varanasi
Parmar BS (1959) Report on the grazing problems and policy of Himachal Pradesh. Forest Department, Himachal Pradesh, Shimla
Rangarajan M (1996) Fencing the forest: conservation and ecological change in India's central provinces, 1860–1914. Oxford University Press, New Delhi

Rodgers WA (1991) Environmental change and the evolution of pastoralism in South Asia. Stud Hist 7(2) n.s.:194–204
Rose HA (1970) A glossary of the tribes and castes of the Punjab and North-West Frontier Province. Based on the census report for Punjab, 1883 and census report for Punjab, 1892. First published, 1883, 3 vols. Reprint. Languages Department Panjab, Patiala
Salzman PC (2002) Pastoral nomads: some general observations based on research in Iran. J Anthropol Res 58(2):245–264
Sanan D, Swadi D (1998) Exploring Kinnaur and Spiti in the Trans-Himalaya. Indus Publishing Co, New Delhi
Simla District Gazetteer (1888–9) nd. Punjab Government, Calcutta
Simla Hill States 1910 (1911) Punjab States Gazetteers, vol VII. The Civil and Military Gazette Press, Punjab Government, Lahore
Singh B (1952) Final report of the fourth revised settlement of the Kulu Sub-division of the Kangra District, 1945–52. Controller of Printing and Stationery Punjab, Chandigarh
Singh C (1997) A strategy of interdependence: Gaddi, peasant and state in Himachal. In: Nathan D (ed) From tribe to caste. Indian Institute of Advanced Study, Shimla, pp 374–386
Singh C (2004) Between two worlds: the trader-pastoralists of Kinnaur. In: Ratnagar S, Heredia R (eds) Mobile and marginalized peoples: perspectives from the past. Manohar Publishers, New Delhi, pp 35–64
Singh C (2009) Pastoralism and the making of colonial modernity in Kulu, 1850–1952. Nomadic Peoples 13(2):65–83
Tapper R (1991) Anthropologists, historians and tribesmen. In: Khoury PS, Kostiner J (eds) Tribes and state formation in the middle east. I.B. Tauris & Co, London
Vitali R (ed) (1996) The Kingdom of Gu.ge Pu.hrang. According to mNga'.ris rgyal.rabs by Gu.ge mkhan.chen Ngag.dbang graps.pa. Dharamsala
von Furer-Haimendorf C (1988) Himalayan traders: life in highland Nepal. First published 1975. Reprint. Times Books International, New Delhi
Warren A (1995) Changing understanding of African pastoralism and the nature of environmental paradigms. Trans Inst Br Geogr New Ser 20(2):193–203

Chapter 10
State Policy and Local Performance: Pasture Use and Pastoral Practices in the Kumaon Himalaya

Christoph Bergmann, Martin Gerwin, Marcus Nüsser, and William S. Sax

Abstract In the Kumaon Himalaya, British colonial administrators as well as agents of the independent Indian Union intervened heavily in pasture use by adopting rationally governed and scientifically sanctioned development schemes. These measures mostly originated from outside and largely ignored local cultural logics through which a pastoral life also takes its form. We use the case of the Bhotiyas of the Kumaon Himalaya to explicate this interaction of state policy and local performance. On the one hand, we analyse recent development trends that occurred after India started to liberalise its market in the early 1990s. On the other hand, we describe a ritual practice through which the Bhotiyas channel emerging power relations and conflicts towards the outside of their migratory cycle. We conclude by suggesting an interdisciplinary perspective on pastoral practices in the Himalayan region.

Keywords Bhotiya pastoralists • State policy • Ritual practice • Interdisciplinary approach • Kumaon

C. Bergmann (✉) • W.S. Sax
Department of Social Anthropology, South Asia Institute, University of Heidelberg, Im Neuenheimer Feld 330, D-69120 Heidelberg, Germany
e-mail: c.bergmann@uni-heidelberg.de; william.sax@urz.uni-heidelberg.de

M. Gerwin • M. Nüsser
Department of Geography, South Asia Institute, University of Heidelberg, Im Neuenheimer Feld 330, D-69120 Heidelberg, Germany
e-mail: martin.gerwin@uni-heidelberg.de; marcus.nuesser@uni-heidelberg.de

H. Kreutzmann (ed.), *Pastoral practices in High Asia*, Advances in Asian Human-Environmental Research, DOI 10.1007/978-94-007-3846-1_10,

10.1 Introduction

Two broad approaches have so far shaped the discussion of how pasture use and pastoral practices in the Himalaya are affected by externally induced political and economic alterations. Some scholars hint at the importance of local cultural logics and plead for an understanding of daily routines from the inside (Gooch 2008). Others foreground the interrelations of pastoral communities with the adjoining lowland centres and show how they actively participate in the development process (Kreutzmann 2004). We intend to combine both positions by analysing how state policy and local performance interact on the ground where they are constantly shaped and reshaped in the context of pastoral life.

Our focus is on the so-called Bhotiyas of Kumaon, who practise combined mountain agriculture (Ehlers and Kreutzmann 2000) across different altitudinal belts, all of which are linked through seasonal migration (Nüsser 2006). Their mobility pattern favours the maintenance of livestock by means of reliable grazing resources including crop remnants from harvested fields. As former trans-Himalayan traders, their ethnicity and livelihood was traditionally associated with the Indo-Tibetan border, which was however sealed as a result of the India-China War in 1962. Till then, the Bhotiyas kept large flocks of sheep and goats as well as yak herds (Farooquee and Rao 2000). Trade caravans relied on these animals for transporting commodities across the high passes to Tibet. Yaks in particular were valued for their ability to cope with severe snow and difficult terrain conditions. After the cessation of trade, most of the yak herds were gradually decimated, and in the Gori Valley, as discussed below, they disappeared completely in the 1970s.

In this chapter, we expose that the development of pastoralism in the Kumaon region is highly variable, practised in various intensities over time and negotiated between local actors and external stakeholders. This is exemplified by the rather unexpected reactivation of seasonal migration and yak breeding during the 1990s when India started to liberalise its market and the Kumaon region was subsequently included into the 27th independent federal mountain state of Uttarakhand (Fig. 10.1). We further argue that pastoral production is dependent not only on shifting market relations and state policies but also on improvisational skills 'that permit continuous flow in human performance of all kinds' (Richards 1993, 63). These skills consist of embodied techniques and forms of knowledge, which are acquired and activated in settings of practical activity and engagement – when doing or making things – as well as in the interpretation of experience (Bourdieu 1977; Ingold 2000, 157–171). Rituals describe important cultural strategies for incorporating these practical interpretations as well as for invoking or acting on particular meanings, values and directives. Institutions, that is, the rules and norms that are effective for the use of biotic resources, provide an arena where daily procedures are formed and reproduced. Although ritualised and institutionalised practices have mostly been investigated separately, we suggest an interdisciplinary framework that scrutinises the mutual interplay of local performance and state policy in terms of negotiations over 'natural' and 'socio-cultural' resources and the power relations inscribed therein.

Sources: *Comprehensive Atlas of the World* (Eleventh Edition). London: Times; Generalni Stab 1:200 000; Georeferenced SRTM-Data; Landsat-Data
Draft and Cartography: M. Gerwin, M. Nüsser, N. Harm

Fig. 10.1 The Himalayan state of Uttarakhand

We base this suggestion upon various phases of field research that were conducted in Uttarakhand since 2004, including initial field trips to almost all the Bhotiya-inhabited valleys of Uttarakhand. Since 2007, we concentrated our work on the Gori and Darma Valleys of the Kumaon Himalayas, located to the east of Nanda Devi. The duration of field trips ranged between 1 and 15 months. The interdisciplinary data assessment for the present paper included participant observation, semi-structured interviews with local informants and government authorities, a questionnaire survey as well as translations and analyses of recorded audio and visual materials. Historical sources were surveyed in regional archives of Uttarakhand as well as in the Asia and Africa Department of the British Library.

10.2 State Policy and Local Performance

In the Indian Himalaya, pastoral practices have been regarded as maladaptive and backward since colonial times (Sharma et al. 2003; Agrawal and Saberwal 2004). State policies mostly aimed at the creation of landscapes corresponding to 'modern, scientific' paradigms, which increasingly restricted mobile forms of land use (Saberwal 1999). This resulted in a gradual distancing of pasture grounds and other resource deposits from local cultural practices, a typical result of Westernised development that has been critically questioned elsewhere (Braun and Wainwright 2001).

In addition, various geopolitical dynamics, of which the sealing of the Indo-Tibetan border in 1962 is only the most visible example, had profound effects on Himalayan pastoralists, particularly through the loss of additional income sources in trade as well as the massive expansion of infrastructural and administrative structures for military purposes. In this context, it becomes important to rethink the interrelation between lowland state actors and mountain communities (e.g. Kreutzmann 1996; van Schendel 2002). Upland regions and their inhabitants have often been described as passive and backward outsiders who are locked in the peripheries of a developing lowland state until they are accommodated into mainstream society. Recent studies indicate, however, that such groups actively participate in negotiations over their life world and deliver non-uniform responses to external interventions (Shneiderman 2010).

While market orientation and globalisation have created various problems for pastoralists, such as the privatisation and commercialisation of community-regulated resources, the Indian Government has also promoted certain policy tools that enable such groups to effectively represent their interests and become collaborators in the policymaking process (Agrawal 2005). Most noticeable is the expansion of forest councils (*van panchayat*) into the valleys of the High Himalaya during the 1990s, when they became shaped for a greater economic viability and environmental sustainability by the state government and various NGOs who increasingly entered the stage. The first councils of this sort in Kumaon were legally recognised in 1931, which makes them one of the oldest surviving examples of formally approved agreements between state authorities and local user groups worldwide (Agrawal 2005, 21). On the one hand, expert groups tend to look at these institutional frameworks as functionally designed and rationally governed units, whose possible outcomes can be calculated by means of quantifiable parameters, such as area size or numbers of users (Agrawal and Chhatre 2006; Jodha 2007). On the other hand, actual practices on the ground level are driven by local cultural logics in addition to an assumed transcultural rationality. This means to acknowledge that religious beliefs, social values and ritual practices also contribute to people's ability of coordinating seasonal movements and achieving a kind of resonance between the nutritional needs of animals, fluctuating weather conditions, conflicting labour tasks and the currents of social life.

10.3 The Bhotiyas in the Kumaon Himalaya

The languages spoken by the Bhotiyas belong to the Tibeto-Burman family (Grierson 1909; Willis 2007). However, most people are multilingual and also fluent in Hindi, Pahari[1] and Nepali. Though it has become the official scheduled tribe designation, the Bhotiyas are not an ethnically or culturally homogenous group. The British colonial administration had popularised this term as a uniform administrative category for all 'Tibetan-like' pastoral groups involved in trans-Himalayan trade (Brown 1992). However, various ethnonyms are used in common parlance.[2] The two major groups in Kumaon's Pithoragarh district are the Johari and Rang. The Johari reside in the Gori Valley, whose upper parts are referred to as Johar and Ralam. The Rang live further east and split into subdivisions that are also named according to the valleys they inhabit, namely, the Darmani (Darma Valley), Chaudansi (Chaudans Valley) and Byansi (Byans Valley). Our analysis gives special emphasis to the Johari and Darmani, whose lower settlements are located around Munsiari in case of the former and around Dharchula in case of the latter. Both locations are former trade posts that have developed into densely populated and urbanised centres of the district. In general, low-caste groups (today officially recognised as scheduled castes) were, and to some extent still are, closely associated with a Bhotiya household, for whose members they conduct various agricultural tasks.

The Bhotiyas' migratory pattern spans over several ecological zones of the Kumaon Himalaya, each with specific environmental potentials and limitations for crop farming, forest and pasture use (Nüsser 2006). The narrow belt of the Outer Himalaya (*Bhabar*) arises out of the northern parts of the Gangetic Plains (*Terai*). Up to an altitude of approximately 1,000 m, sub-humid tropical Sal forests (*Shorea robusta*) are found. The Lesser Himalaya follows as a 70–100 km broad belt in a northerly direction and covers an altitude between 1,500 and 3,000 m. In lower parts, this zone is extensively covered by *Pinus roxburghii* forests, which often exhibit a herbaceous understorey sustained by intentional burnings. The zone of the High Himalaya is about 30–50 km wide. Glaciated mountain peaks, of which some exceed an altitude of 7,000 m, dominate the scenery. Its narrow transversal valleys give rise to montane forests where evergreen oaks (*Quercus semecarpifolia*, *Qu. floribunda* and *Qu. leucotrichophora*) alternate with areas dominated by conifers (*Abies spectabilis*, *Cupressus torulosa*) or deciduous trees (*Alnus nepalensis*, *Aesculus indica* and *Acer* spp.). Above 3,600 m, high altitude grasslands are found, which prevail into the Tibetan Himalaya that forms the northernmost part of Kumaon. With relatively wide and shallow valley bottoms, these localities are covered with meadows and dwarf shrubs, which are widely used as summer pastures (*bugyal*). High passes leading onto the Tibetan Plateau had facilitated a flourishing cross-border trade until the war between India and China created a new periphery in 1962.

10.4 From an Old to a New Periphery: Pastoral Life in Transition

The beginning of British colonial rule in Kumaon started 1815 with the defeat of the Gurkhas, a ruling power from Nepal that had governed the area from 1790 onwards. At this time, the Bhotiya trade was restricted to the barter of locally needed products, such as grain, salt and cloth, whilst the more lucrative long-distance trade of *pashmina* wool took place further West through middle men in Ladakh (Raper 1812, 497–498, 530; Moorcroft 1818, 399–400). Seasonal migration during the winter months was directed towards the montane forests and grasslands at the margins of the High Himalaya (Traill 1832, 11–12; Walton 1911, 69). However, some traders also visited important market places and trade festivals in the Lesser Himalaya (Traill 1828, 193–195; Walton 1911, 209–213).

Right from the beginning, the British administration was aware that an involvement in the trans-Himalayan trade was crucial for advancing colonial ambitions and expanding their influence to market and resources of Tibet and Central Asia (Moorcroft and Trebeck 1837, xvii). In the following decades, they substantially reduced the taxes for the Bhotiyas, who in turn invested heavily in their sheep and goat livestock (Atkinson 1884, 143–151). In the course of the nineteenth century, this 'old periphery' of British sovereign territory gradually evolved into an 'integrated borderland' that was characterised by a drastically increased volume of trans-Himalayan trade (Walton 1911, 68–69) (Fig. 10.2).[3]

The ritualised trade partnerships the Bhotiyas maintained with their Tibetan partners, which we already described elsewhere (Bergmann et al. 2008), are a historical example to illustrate how local performances contribute to the flourishing of a pastoral production system. The Bhotiyas had popularised the half-truth that the Tibetans regard commensal relations as a necessary requirement for commerce and that this demands an ignorance of the important Hindu matters of touch pollution and beef eating. Besides reasons of logistical feasibility and language competence, this effectively kept away possible Pahari competitors even more effectively and thus strengthened the trader's monopoly.

The increase in trade volume, however, was also linked to major dynamics of the colonial economy in the Gangetic Plains, where wool mills had started an industrialised production during the 1860s (Roy 2003, 271–272). The Bhotiyas had started to bargain directly with middlemen of these mills in Haldwani and Tanakpur. The flocks of sheep and goats were taken along the lengthened road and grazed in forests and pasture grounds around these new market towns (Goudge 1903; Forest Department 1898). Moreover, the migratory groups negotiated customary agreements with sedentary farmers in order to graze their animals on harvested fields along the route (Forest Department 1889).

However, the Bhotiyas' seasonal migration was also influenced by more restrictive policies of the British administration that were implemented in order to meet increasing demand for timber, to be used in intensified railway construction. After the Indian Forest Act was established in 1878, large tracts of the *Terai* and *Bhabar* forests in the Himalayan foothills were demarcated as Reserved Forests (Guha 1989, 44). During the 1890s, the British introduced further regulations to directly act on migratory patterns, such as the installation of officially sanctioned grazing grounds (*parao*) in the Lesser Himalaya. While stays at these intermediate spots were temporally restricted to a maximum of 3 days, the allotted grazing grounds between the Gangetic Plains and the Outer Himalaya remained accessible from December to February with a fee of six *annas*[4] per animal (Forest Department 1898).

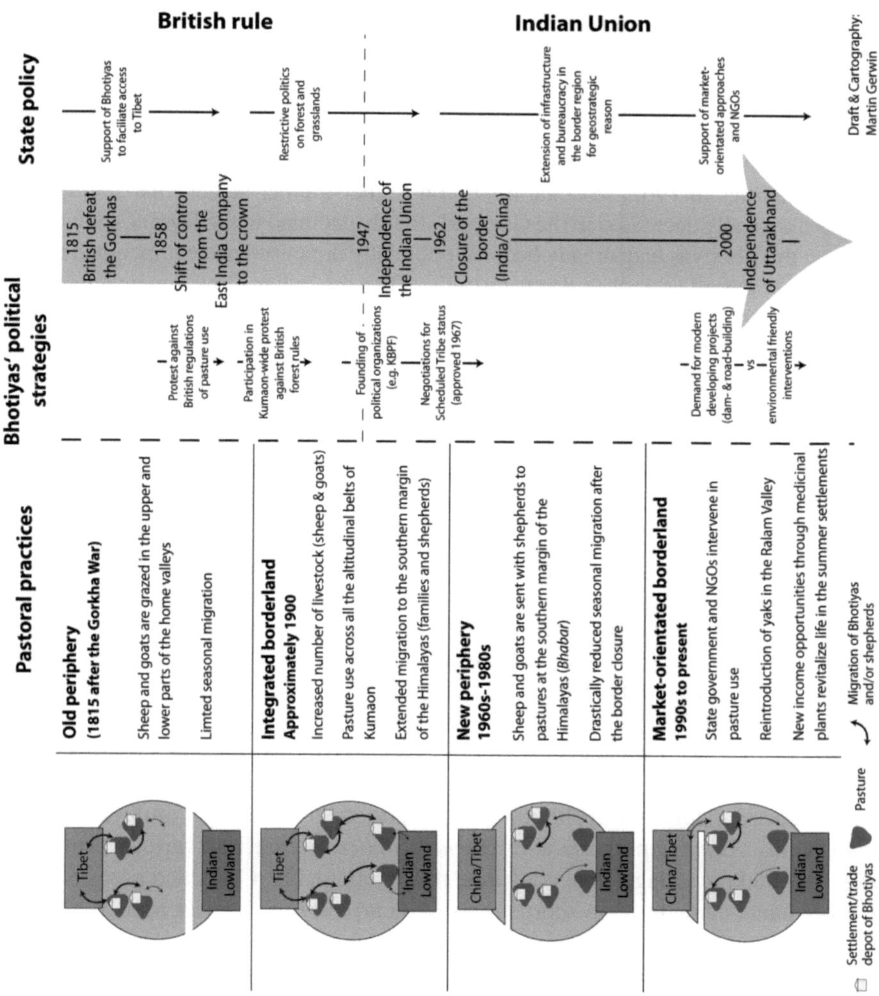

Fig. 10.2 The historical dimension of pastoral practices

The imposed rules triggered local discontent against British forestry, which ultimately led to an enduring rebellion of the entire Kumaoni population in the 1920s (Guha 1989; Agrawal 2005). Within this highly politicised conflict, commercial activities with the Tibetans began gradually to dry up. Tibetan wool was successively replaced by imports from Europe and Australia, and, furthermore, Tibetan salt was replaced by cheaper substitutes from the coastal areas of India (Roy 2003, 258–259). In the course of these changes, the main commercial routes shifted eastwards, linking Sikkim with the ports of Bengal (Foreign Department 1947). Older Johari frequently report that the most visible change was a dramatic increase in field cropping conducted by the members of a Bhotiya household themselves and not by appointed tenant farmers, as it had often been the case before.

Overshadowed by the closure of the Indo-Chinese border following the war between the two countries in 1962, seasonal migration to the summer settlements and alpine pastures drastically decreased in the Gori and slightly declined in the Darma Valley.[5] At that time, the Bhotiyas had already become politically organised and campaigned to be officially recognised as a scheduled tribe (ST). When the Indian Government approved this demand in 1967, the community benefited from quota access to universities, government services and the legislature, along with other privileges.

One of the most far-reaching forms of state intervention within this 'new periphery' was a large military road-building programme, which opened up the lower parts of all the Bhotiya valleys in Uttarakhand by providing infrastructure (Rawat and Sharma 1997). Furthermore, the Indian government promoted land reforms, the building of schools and the implementation of a decentralised administrative structure (Nautiyal et al. 2003). These policies advantaged Munsiari and Dharchula, which developed into densely populated centres of the valleys. In these new centres, non-agricultural employment became increasingly important, and Bhotiya groups were able to keep and expand their property in the growing bazaars.

All this was accompanied by a decrease of sheep and goats, and an increase in the number of cattle (KBPF 1947; GoU 2003), which are the preferred animals for all-year agriculture in the middle sections of the valleys. Around Munsiari, for instance, the herbaceous understorey of montane oak forests became progressively exploited for pasturing cattle, supplemented by stall-feeding during the winter months. These changes in livelihood also supported an increase in multifunctional trees (mainly *Alnus nepalensis* and *Aesculus indica*) as well as various fruit tree species around the villages that constitute this municipality. Grassland products, on the other hand, came to be in short supply and currently are heavily traded in local auctions that are arranged by the respective *van panchayat*.

These quite heterogeneous councils also include other tribal as well as non-tribal Pahari members. However, many Bhotiyas were in a better economic condition and even managed to enhance their standing through these institutions. Strategies include the reliance on scheduled tribe quota for accessing the ruling committees as well as their general visibility as a powerful group through self-founded development associations and cultural clubs as well as their presence in (local) scientific historiography and regional politics. These officially sanctioned institutional structures play a major role in the 'market-oriented borderland' (Fig. 10.2), which once again altered people's scope for action on Kumaon's mountain pastures significantly.

10.5 Adjusting Pasture Use and Pastoral Practice to a Market-Oriented Borderland

India planned its market liberalisation during the 1980s, and these plans were put into practice after the collapse of the Soviet Union and the subsequent orientation towards the USA, as well as international trade organisations in 1991 (Corbridge 2009). Following several rounds of negotiations on the border issue, the passes Shipki La in Himachal Pradesh and Lipu Lekh in Kumaon were reopened for trade in 1992 (Vasan 2006). Since 2008, China is India's largest trading partner with a volume of USD 51.8 billion (CII 2010).[6] All this encouraged a more confident approach of the Indian Government towards its controversial and partly disputed border with China, and this confidence increasingly has guided the actions of both policymakers and private investors (Mohan 2007). The removal of the so-called 'Inner Line' regulation in most parts of the Bhotiya valleys in the early 1990s is one sign of this process. Until then, strict controls on access due to security concerns inhibited economic development, including the nascent tourist industry (Statesman 1981; cf. Aggarwal 2004, 57–102 for Ladakh), and local residents frequently complained about this.

These interventions on the national scale interacted with various regional processes, of which the creation of Uttarakhand as an independent federal state in 2000 is the most important. Until then, the high mountain border region of Kumaon had formed the outermost part of the most densely populated Indian state, Uttar Pradesh, whose government regularly ignored demands coming from the periphery. The new government aimed at combining market-based approaches with a mountain specific policy, of which hydro-energy, tourism and medicinal plant extraction evolved as the most important building blocks (Garhwal Post 2009). Moreover, the whole state profits from the 'Border Area Development Program' that promotes investments in infrastructure, education and agriculture (Tribune 2009).

In the midst of these reconfigurations, a slight revitalisation of seasonal migration towards the upper settlements is noticeable in both valleys. In the Gori Valley, the number of families staying in the summer settlements has approximately increased by 30% since 2004. The majority of migrating people remain in their summer settlements for crop production and herb cultivation.[7] Most villagers possess only a small number of sheep and goats, or none at all. However, there are around 35 flocks with a maximum size of 400 animals that are pastured by Bhotiya shepherds in the upper valley during the summer months. While some work on their own behalf, non-migrating Bhotiyas, who are settled around Munsiari, employ the majority of them. The shepherds normally maintain strong kinship-based relations with residents of the upper villages through which access to pasture grounds is informally negotiated and granted.

During recent years, however, an increasing number of non-Bhotiya herds have also begun to enter the valley. The respective owners are from the lower parts of the Gori Valley, from the southern Pindari Valley and from Himachal Pradesh. These external user groups form a valuable income source for the *van panchayat* that now manage the alpine pastures surrounding the high altitude settlements. They levy a grazing fee of Indian Rupees (INR), 3–5 per sheep or goat and up to INR 20 per horse

Photo 10.1 A yak herder and his animals in the Ralam Valley at 3,700 m (Photograph © Marcus Nüsser September 26, 2008)

(one USD equals approx. INR 45). Most of these councils have only recently been (re-)established with the help of various non-governmental organisations (NGOs), which started to operate in the area from the 1990s onwards. The rise of these new actors in the Gori Valley forms part of an Indian-wide trend to support and facilitate the development of backward areas more efficiently and cost effectively (Townsend et al. 2004). In a similar vein, the Joint Forest Management Program of the 1990s, which was co-financed by the World Bank, heavily influenced the *van panchayat* in Kumaon. Through a focus on issues like participation, empowerment and gender equality, an internationally recognised set of communal resource regulations became gradually standardised (GoU 2001). But even though these new measures officially realised a greater integration of local users, they are still widely criticised by environmental activists and NGO workers. One reason, for instance, is an increasing bureaucratisation of organisational structures and annual implementation plans, which are seen as being far beyond the educational level of most local users (Sarin 2005).

The regional livestock sector has also been transformed by distinct NGO activities and development interventions. The example of the reintroduction of yaks in the Ralam Valley may serve to illustrate this case. After cessation of trade, most of the yak herds were decimated successively until the last herd disappeared in the 1970s in the Gori Valley. A local NGO organised the revival of yak breeding in 1996 after bringing a nucleus herd of 20 pure-bred yaks from Tibet. Till present, the Bhotiya herders support this endeavour (Photo 10.1) and promote inter-species breeding of

Photo 10.2 Ploughing of fields with yak hybrids in the summer settlement of Nagling in the Darma Valley (Photograph © Christoph Bergmann, May 12, 2008)

yaks with cows for milk production and ploughing, as well as using them as pack animals. The presence of these yak hybrids has the potential to influence earlier upward migration in spring (FES 2004, 19). In the Darma Valley, yaks and their hybrid crossbreeds are kept even without NGO support since the border closure (Photo 10.2). Villagers continue to bring these animals from the adjoining region of Nepal, where a constant supply from Tibet is still secure. Existing kinship ties often facilitate these transactions.

The reintroduction of yaks in the Ralam Valley and the maintenance of yak keeping in the Darma Valley show that various forms of pastoral practice are still alive and supported. Moreover, the increased (illegal) collection of aromatic and medicinal plants has brought new pressure on the alpine grasslands, as well as confronting local users and policymakers with new opportunities and challenges. Besides *Aconitum heterophyllum* (*atis*) and *Picrorhiza kurroa* (*kutki*), which are collected between August and November, especially *Cordyceps sinensis* (*kira*), a fungus that grows on a caterpillar larva, is of particular interest to local gatherers since approximately the last decade. This fungus is highly appreciated as a tonic, aphrodisiac agent and status symbol amongst the Chinese upper class, where large quantities of it are sold (Winkler 2008). It is extensively gathered from May to July and then sold at highly fluctuating rates ranging between INR 200,000 and 700,000 (approx. USD 3,200–15,000) per kg to middlemen in the markets of Munsiari and Dharchula, with an estimated yearly trade volume of 150–250 kg in the former and 250–350 kg in the latter location. The collection of this lucrative item widely contributes to the household income of both Bhotiya and non-Bhotiya residents, ranging from INR 15,000 (approx. USD 340) to more than INR 75,000 (approx. USD 1,685) in one season. The Uttarakhand Government had only slowly framed these activities through administrative procedures. Only in September 2010 did it officially delegate authority to the *van panchayat* over the collection and distribution of this lucrative item (GoU 2010).

From 2000 onwards, there had been numerous discussions in the Bhotiya villages about how to regulate these new forms of use of alpine grazing grounds. Most villages, for instance, adopted strict rules to enclose their grasslands and keep out external user groups. Moreover, officials from the Forest Department interacted more regularly with these groups and aimed at the implementation of specific working procedures, such as the rotational use of pasture tracts to enhance sustainability. The official support of the *van panchayat* marks strong continuities to the endeavours that were initiated during the 1990s through various NGOs. While this indicates a stronger presence of officially sanctioned schemes to regulate pasture use, these developments are not uniform. Without strong support from NGO workers and rather low efforts on the side of the Forest Department to rework and control the extraction of forest and grassland resources, *van panchayat* are far less important in the Darma Valley than in the neighbouring Gori Valley. Most of the alpine pastures are still used by local herds, although these are normally taken care of by shepherds from elsewhere. In what follows we expose a further socio-cultural facet of pastoral practices through which the complex tasks of livestock keeping become integrated with further activities, such as crop cultivation and downward migration.

10.6 The Ritualised Coordination of Seasonal Migration

In the course of their annual migratory cycle, residents of Darma Valley intervene most clearly in a ritual for Su-Ringding, the god of the land, which happens in mid-September, when most shepherds have come down from the alpine pastures to graze their animals on harvested fields and grasslands in the valley bottom. At that time, a variety of conflicting tasks are pending such as crop processing, wool shearing and various preparations for downward migration (*kuncha*[8]). Flock owners count their animals in order to pay employed shepherds, and also the low-caste people are reimbursed in cash or kind for their agricultural labour. While the diverging demands and loyalties of the various groups often give rise to conflict, the ritual foregrounds the ongoing nature of these engagements and brings forth a kind of resonance with the seasonal rhythms of their dwelt-in surroundings.

The ceremony starts early in the morning, when male villagers offer a small quantity of grain from the new harvest at the deity's shrine. In the meantime, the women of every household prepare some food from local grains and pulses. After a joint meal with one's own family, people begin to stroll around in the village reciprocating further food and drinks, a joyful activity that goes on for some hours. Eventually, everyone comes together at a place near the shrine, where a black goat is killed with a single stroke through its neck. Diluted with water, the animal's blood is directed along the sloped footpath towards the lower village boundary. This way of killing is exceptional, as people normally rip out the animal's heart through a little cut in the abdomen without any blood running out of the animal's body. Whenever it is beheaded during such a festive activity, this is not considered as a sacrifice for a deity but as an act that should satisfy and ward off evil spirits and demons. Having received their share, these beings are expected to stay away from the village where they are prone to spread unrest, conflict and disease.

Having cut up the meat, people start to prepare the further ceremony. Some flour, edibles as well as liquor from every household are collected in front of the shrine. Furthermore, every family holds ready a small mat made of straw (called *kayo*) on which some of these materials will be put at the end of the ceremony. For the time being, however, everyone eagerly awaits the arrival of the village elder who is supposed to conduct the further ceremony at the time of nightfall. This person then sits down in front of a wooden bowl or some other vessel that is filled with parts of the offerings, a small piece of the goatskin as well as some meat. When he starts to perform what is called *myilu tumo* ('throwing away'), low-caste drummers accompany him. The performance is constantly interrupted by loud comments of the audience's male section. All along they litter flour and distribute liquor amongst each other whilst the village elder – holding a wooden stick with a burnt tip, with which he fiercely mixes the ingredients of the vessel – recites the following story.[9]

> The gods and goddesses of my village are quite ordinary.
> The villagers, also, are somehow not different.
> Looking at their flocks, they seem rather poor.
> Also their crops do not grow very well.
> Among their sons not a single one is son-like.
> Among their daughters not a single one is daughter-like.
> None of their belongings seems to be special.
> But in the village in front, everything is better.
> Good houses, fertile fields, handsome sons, easily graspable daughters, huge flocks, and many fine goods are abundant.
> If you must get jealous, have envy or malice, then save it for those people! And if you have to go, then leave in that direction.
> But hey, those from the lower villages do the same, and they also send their people to the next village.
> And as this continues, the last of our villages will finally recite these lines for the whole of our beloved Darma region.
> Bad spirits and demons keep looking upon the gods and goddesses, and we who dwell in peace, prosperity, happiness, love, and with all virtues in the fourteen villages nestled in this Himalayan valley, this quiet heaven, this golden land, with jealousy, they always try to disturb and spread unrest.
> This envy, jealousy and malice, all bad words and conspiracies, all dangerous diseases, bad ideas and thoughts, tiredness, the origins of all kinds of violent outbreaks, power fights and ill effects we now send back from here.
> Today all these evil impacts will be shed into the flowing river to reach the far-away and deep waters.

But if this evil packet indeed goes down with the rivers into the deep waters, it will clash with those huge elephant-like beings residing there.
These giant creatures, which before crashing into this evil packet would have been quite amused, will of course ask, 'What in the world is this?'
These powerful beings will be affected by those evil influences and in a sudden burst of anger they will throw this damned packet high into the air. But guess what will happen to it? It will be carried along by the wind and again reach our beloved Himalayan glaciers.
All of a sudden this evil packet will fall out of the sky and land on one of our high pastures where our yaks are grazing.
Think about how this packet will look like in such a place where one cannot find even a single blowfly.
Having seen this evil packet, also the yaks will get furious and produce loud noises with their noses and leap about on their four feet.
But finally these fearless beasts will come close and pick this packet up with their horns. With all their strength they will again throw it back into the deep rivers.

In the closing part of the performance, the evil package is verbally directed through every one of the summer settlements and several other stations lying on the path towards the winter settlements. Its final destination is a place called Bakrihath, which is located shortly below the last winter settlement of Jauljibi near the confluence of the Kali and Gori Rivers – the 'deep waters'. Jauljibi hosts an international trade mart for livestock and commodities that annually takes place in January. There the Rang territory ends and local aristocrats, presumably the 'elephant-like beings' mentioned in the ritual chant, had their seats and were supported by their most loyal subjects in former times. Worth mentioning, for example, are the Rajwars[10] from Askot, local aristocrats who held proprietary rights in the village lands and harshly restricted the Darmanis' access to cultivatable fields in these lower parts. Darma shepherds complain that their position beyond the valley is much more insecure due to more unreliable social alliances, thefts of livestock, corrupt forest guards and the like. In a heightened atmosphere, the village elder proceeds as follows:

Give voice, make abuse
Throw it away, I say!
Give it away, I say!
The wood taken from the glowing fire is not going?
Far from it, certainly the wood taken from the glowing fire is going!
It is going from Sipu – the village of Syang Sai – hey!

[After every verse, ending with the name of the village and the tutelary deity of its residents, the audience begins to dance and shout]

We have done the worship for Su today.
Throw the evil packet away, I said!
After six months, after 160 days,
you neither have to say yes to someone,
nor do you have to do something.
If someone still does, then a six-inch nail goes right into his head
The wood taken from the glowing fire is going from Tidang – the village of Chung Sai – hey!

We have done the worship for Su today
After six months, after 160 days

you neither have to say yes to someone,
nor do you have to do something.
Admittedly our houses look like corn mills,
but our fields are rather tiny!
And our girls are just like a tokar.
And our boys are just like a takuva.[11]
The wood from the glowing fire is going from Darkar – the village of Riya Sai – hey!

Through variations of these lines and by naming every village along with its tutelary deity, the route towards Bakrihath is completed. One male member of each household then takes the small straw mat (*kayo*) on which some of the collected food and drinking materials are put. These offerings are either placed besides the shrine or on a nearby footpath from where crows are expected to pick them up and transmit their substance to the ancestor spirits (*syimi*). In the meanwhile, another man – normally a low-caste (scheduled caste) person whose daily task is to look after the villagers' cattle – approaches the scene and takes over the vessel and the wooden stick from the village elder. Equipped with branches of a prickly shrub as well as cow dung, the village youth drives him 3–5 to 7–9 times around the shrine and then finally towards the Darma River or one of its tributaries. They are expected to 'kill' the former, who yet always manages to escape and to throw away his baggage into the flowing water. The shouting crowd on his heels then desists from the initial objective and also gets rid of its 'weaponry' on the riverbank. When everyone has returned back to the village, the person that was in charge of the evil packet is applauded for his bravery and receives the two front forelegs of the goat that was killed before. Afterwards, the whole village joyfully dances and sings until late at night.

In view of the Darmanis' pastoral practices, the ritual indicates two major points. On the one hand, visible distinctions between different kinds of livestock are drawn. Yaks are associated with the uppermost and purest pastures, flocks of sheep and goats with wealth, whilst cattle are linked to conflict-prone village life.[12] Only the latter have a human spokesperson, the cattle herder, who has to take away the 'evil packet', whereas yaks are agents themselves when they finally close the annual cycle by throwing the evil packet back into the river. All this is driven by various power relations, both in terms of social distinctions within the village context as well as towards external groupings. On the other hand, the ritual foregrounds that the integration of different demands in labour as well as the scheduling of divergent agro-pastoral tasks before downward migration is always characterised by quarrels and instability. Like people attend to the growth of their crop plants and the movements of their livestock, they also take care of the ongoing process of their social affairs by channelling arising conflicts from the summer settlements towards the outside of their migratory route.

Rituals, like the one described above, have so far received little attention by outside scholars. The Bhotiya elites, however, have nowadays great interest in documenting such affairs and make them accessible to a wider audience through publications, internet forums, and local museums. In this manner, they transform these socio-cultural resources of daily life into a heritage, whose increased visibility also gains attention in the policymaking arena.

10.7 Discussion and Conclusion

Environmental conditions entail certain potentials and limitations for pastoral land use, such as the duration of snow cover or the onset of the vegetation periods in different altitudinal belts. However, the utilisation of the grass areas is never homogeneous and static. A variety of socio-economic changes and political interferences have affected the Bhotiyas' land use and migratory patterns in recent times. Far from being passive victims, they were ready to deal with new stakeholders entering the market-oriented borderland. The reintroduction of yaks in the Ralam Valley shows that officially approved institutions can provide a common platform for government officials, NGO workers and pastoral communities, who become active partners in shaping everyday forms of pasture use. These institutional arrangements are becoming increasingly important for regulating medicinal plant collection and access to grazing lands in the upper Bhotiya valleys.

Forms of regulation are, however, subjected to negotiation processes in which socio-cultural resources such as ethnicity, embodied practices, techniques and forms of knowledge play a vital role. Socio-cultural resources are important because they enable people both to transform the world and to 'play their part from within in the world's transformation of itself' (Ingold 2011, 6). This is not meant to reactivate romantic ideas of peasant oneness with nature but to emphasise the significance of local cultural logics within the context of pastoral production – when dealing with animals, cultivating fields, migrating, engaging in a *van panchayat* or carrying out a ritual. In rituals, people encounter further 'recognising agents' (Shneiderman 2010, 307), particularly from the divine world, including tutelary village deities and evil spirits who reaffirm aspects of pastoral practices that interact with but are yet different from stately approved procedures. Their presence mediates past trajectories and power relations with the locally situated activities and extra-regional influences that characterise current pastoral life. The attribution of agency to the yaks and 'elephant-like beings', which most plausibly refer to former regional aristocrats, in the ritual chant are two examples for that.

Pasture use and pastoral practices in the Kumaon Himalaya thus unfold within a multidimensional continuum that includes ritualised practices, realised subsistence strategies and the vocabulary through which external stakeholders justify or reject the implementation of specific policies (Fig. 10.3). Pastoral life depends on both 'natural' and 'socio-cultural' resources, which together drive a given production system and also shape the power relations inscribed therein. An appreciation of the interplay of state policy and local performance in the formation of pastoral utilisation strategies could enhance the policymaking process and also facilitate a more attentive assistance to those who actually dwell on high grounds for making a living.

Acknowledgements We are very grateful to the people from the Gori and Darma Valley for their generous hospitality and for taking the time to participate in the research process. We are also obliged to our friends in the Uttarakhand Government and in several NGOs. For comments on an earlier draft of this paper, we appreciate the efforts of Hermann Kreutzmann. The responsibility for

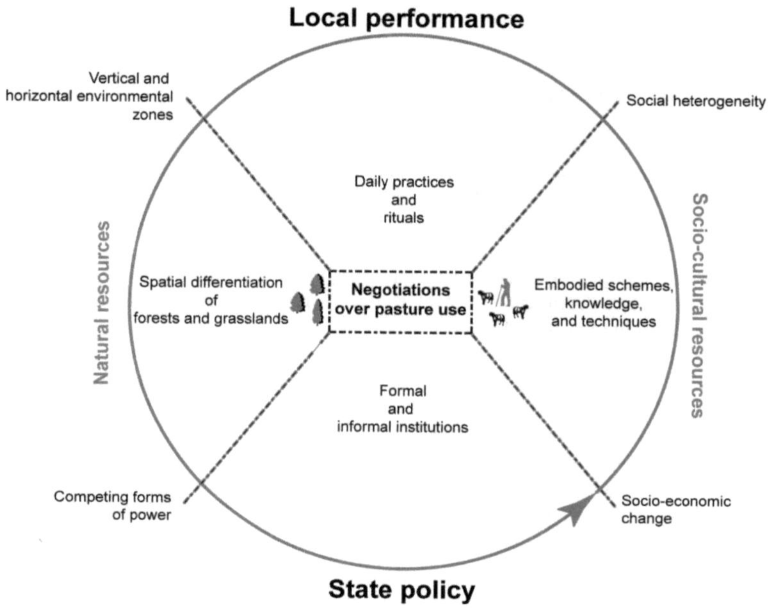

Fig. 10.3 Pasture use and pastoral practices between local performances and state policy

the final version rests completely with the authors. Finally, we are indebted to the Deutsche Forschungsgemeinschaft (DFG) for their generous grant of the ongoing project 'Changing Strategies of Resource Use: The "Bhotiyas" in the High Mountain Border Region of Uttarakhand, India'.

Notes

1. Pahari literally means 'of or belonging to the mountains' and commonly refers to the major Hindu hill populations in Nepal and India. These people speak Indo-European languages that are also classified by this term.
2. For the complexity of ethnic ascriptions amongst various Bhotiya groups compare Bergmann et al. (2008) for Uttarakhand, Nawa (2000) for Kumaon and far-western Nepal as well as Ramble (1997) for Nepal in general.
3. The quantity of imported wool to the Gori and Darma Valley rose from approximately 800 kg in 1,841 to more than 330,000 kg in the year 1901 (Goudge 1903).
4. Sixteen *annas* equalled one Indian Rupee (INR).
5. These differences are linked to varying land ownership conditions in the winter settlements of Gori and Darma Valley. Milam, the largest of all Bhotiya settlements in the Gori Valley, recorded a reduction of migrating families from 600 in the 1930s (Pant 1935, 240) to only 23 families in 1981 (GoI 1984). Even though this trend further consolidated to only 18 families in 2004 (Nüsser 2006), a slight revitalisation has taken place during the last years so that 22 families were counted in 2010. In the Darma Valley, seasonal migration reduced more slowly from a total of 2,674 people in 1961 to still 1,210 persons in 2001 (GoI 1966, 2003).
6. The Indian Government was keen to reopen the Lipu Lekh pass, as this allows a selected number of Indian citizens to make the pilgrimage to sacred Kailash in Tibet. Trade, however, plays a rather marginal role so far.

7. The main seasonal crops are buckwheat, barley, wheat, mustard, pulses, peas, potatoes and some other vegetables, such as cabbage. Cultivation takes place on traditional village fields, small vegetable gardens as well as on agricultural plots in ruins of old village houses. The latter are often irrigated and used for aromatic and medicinal plants, especially *Allium stracheyi* (*jambu*) and *Carum carvi* (*thoya*).
8. The terms of this section that are put in italics belong to the Tibeto-Burman Darma language.
9. Different versions of this ritual were recorded in September 2008 in village Bon, Son, Dugtu and Dantu in the upper Darma Valley. The description is based on the performance observed in village Bon. Parts of the story are accessible in the Darma language in Dhakriyal (2004, 253–238).
10. 'Rajwar' was the official title of the ruling powers in the Askot principality. They are considered as descendents from the medieval Katyuri kings of Kumaon, though their feudality became tributary to the Chand kingdom.
11. The *tokar* is the lower end, and the *takuva*, the upper stick of a spindle device, known as *takli*. The metaphor has clear sexual connotations, referring to male and female genitals.
12. This differs from a nomenclature reported in Stellrecht (1992), where sheep and goats are associated with high altitude pastures and ideas about purity. In the Dolpo region of western Nepal, however, yaks are the culturally most favoured livestock that also figures out prominently in rituals, even though sheep and goats are extensively reared due to their high reproductive rate and lower costs when compared to cattle (Bauer 2004, 25–38).

References

Aggarwal R (2004) Beyond lines of control: performance and politics on the disputed borders of Ladakh, India. Duke University Press, Durham

Agrawal A (2005) Environmentality. Technologies of government and the making of subjects. Duke University Press, Durham

Agrawal A, Chhatre A (2006) Explaining success on the commons. Community forest governance in the Indian Himalaya. World Dev 34:149–166

Agrawal A, Saberwal VK (2004) Whither South Asian pastoralism? An introduction. Nomad Peoples 8:36–53

Atkinson ET (1884 [1996]) The Himalayan gazetteer, vol II, part I. Natraj Publishers, Dehra Dun

Bauer KM (2004) High frontiers. Dolpo and the changing world of Himalayan pastoralists. Columbia University Press, New York

Bergmann C, Gerwin M, Nüsser M, Sax WS (2008) Living in a high mountain border region. The case of the 'Bhotiyas' of the Indo-Chinese border region. J Mt Sci 5:209–217

KBPF [Kumaon Bhotiya Peoples' Federation] (1947) The report of the Kumaon Bhotiya peoples' federation. Submitted before the minority sub-committee of the constituent assembly's advisory committee on 17th April 1947 [kindly handed over by Dr. S. S. Pangtey, September 2006]

Bourdieu P (1977) Outline of a theory of practice. Cambridge University Press, Cambridge

Braun B, Wainwright J (2001) Nature, poststructuralism, and politics. In: Castree N, Braun B (eds) Social nature. Theory, practice, and politics. Blackwell, Malden, pp 41–63

Brown CW (1992) What we call 'Bhotiyas' are in reality not Bhotiyas: perspectives of British colonial conceptions. In: Joshi MP, Fanger AC, Brown CW (eds) Himalaya: past and present, vol II. Shree Almora Book Depot, Almora, pp 147–172

CII [Confederation of Indian Industry] (2010) India-China economic & commercial relations. http://www.indiachina.org/resources/India_China_Economic_and_Commercial_Relations.htm. Accessed 17 May 2011

Corbridge S (2009) The political economy of development in India since independence. In: Brass PR (ed) Routledge handbook of South Asian politics. India, Pakistan, Bangladesh, Sri Lanka, and Nepal. Routledge, Abingdon, pp 305–321

Dhakriyal DS (2004) Himalayi sauka saanskritik dharohar, vol I (in Hindi, The cultural heritage of the Himalayan Shauka). Takshila Prakashan, New Delhi
Ehlers E, Kreutzmann H (2000) High mountain ecology and economy. Potential and constraints. In: Ehlers E, Kreutzmann H (eds) High mountain pastoralism in northern Pakistan, Erdkundliches Wissen 132. Steiner, Stuttgart, pp 9–36
Farooquee NA, Rao KS (2000) Conservation and utilization strategies of indigenous cattle and livestock among the transhumant pastoralists of Kumaun Himalaya (India). J Environ Syst 27:317–329
FES [Foundation for Ecological Security] (2004) Annual report 2003–2004. Foundation for Ecological Security, Anand
Foreign Department (1947) Annual report of the British trade agent, Gartok. PS/12/4209 Wool Trade 25th Feb 1942 – 30th June 1947. India Office Records of the British Library, London
Forest Department (1889) Damage caused to forests and grazing land by sheep and goats. File 59/1880–1893, Box 118. Regional Archives, Naini Tal
Forest Department (1898) Grazing of Bhutia sheep and goats in the forests of the Kumaon Civil Division. File 12/1896–1897, Box 62. Regional Archive, Nainital
Garhwal Post (2009) Nishank briefs IFAD President on state's development. http://www.garhwal-post.com/article/Dehradun/4248/. Accessed 30 Jan 2011
GoI (1984) Census of India, 1981. Uttar Pradesh. District census handbook (13): Pithoragarh. Director of Census Operation, Uttar Pradesh
GoI (2003) Census of India, 2001. District Pithoragarh, Uttaranchal. Office of the Registrar General and Census Commissioner, Delhi
GoI [Government of India] (1966) Census of India, 1961. Manager of Publication, Delhi
Gooch P (2008) Feet following hooves. In: Ingold T, Vergunst JL (eds) Ways of walking: ethnography and practice on foot. Ashgate, Aldershot, pp 67–80
GoU (2003) Livestock survey for the Tehsil Munsiari, 2001. Veterinary Department of the District Pithoragarh, Pithoragarh (unpublished)
GoU (2010) Government order, 98-PA/8–3[2]. Forest Department of Uttarakhand, Dehra Dun (unpublished, in Hindi)
GoU [Government of Uttaranchal/Uttarakhand] (2001) The Uttaranchal panchayati forest rules – 2001. Forest Department, Dehra Dun
Goudge JE (1903) Almora. Final report on the assessment of the Almora district and the hill pattis of the Naini Tal district. V/27/314/643. India Office Records of the British Library, London
Grierson GA (1909) Linguistic survey of India, vol III, Tibeto-Burman Family. Superintendent of Government Press, Calcutta
Guha R (1989) The unquiet woods. Ecological change and peasant resistance in the Himalaya. Oxford University Press, Delhi
Ingold T (2000) The perception of the environment: essays on livelihood, dwelling and skill. Routledge, New York
Ingold T (2011) Being alive: essays on movement, knowledge and description. Routledge, New York
Jodha NS (2007) Mountain commons. Changing space and status at community levels in the Himalaya. J Mt Sci 4:124–135
Kreutzmann H (1996) Ethnizität im Entwicklungsprozess. Die Wakhi in Hochasien. Reimer, Berlin
Kreutzmann H (2004) Pastoral practices and their transformation in the North-Western Karakoram. Nomad Peoples 8:54–88
Mohan CR (2007) Soft borders and cooperative frontiers: India's changing territorial diplomacy towards Pakistan and China. Strateg Anal 31:1–23
Moorcroft W (1818) A journey to lake Manasarovar in Undes, a province of little Tibet. Asiatic Res 12:380–536
Moorcroft W, Trebeck G (1837 [1976]) Travels in the Himalayan provinces of Hindustan and the Panjab, vol II. Faran, Lahore
Nautiyal S, Rao KS, Maikhuri RK, Saxena KG (2003) Transhumant pastoralism in the Nanda Devi Biosphere Reserve, India. A case study in the buffer zone. Mt Res Dev 23:255–262
Nawa K (2000) Ethnic categories and their usages in Byans, far western Nepal. European Bulletin Himalayan Res 18:36–57

Nüsser M (2006) Ressourcennutzung und nachhaltige Entwicklung im Kumaon-Himalaya (Indien). Geographische Rundschau 58:14–22

Pant SD (1935) The social economy of the Himalaya. Based on a survey in the Kumaon Himalayas. George Allen and Unwin, London

Ramble C (1997) Tibetan pride of place: or, why Nepal's Bhotiyas are not an ethnic group. In: Gellner DN, Pfaff-Czarnecka J, Whelpton J (eds) Nationalism and ethnicity in a Hindu kingdom. Harwood Academic Publishers, Amsterdam, pp 349–414

Raper FV (1812) Narrative of a survey for the purpose of discovering the source of the Ganges. Asiatic Res 11:446–564

Rawat DS, Sharma S (1997) The development of a road network and its impact on the growth of infrastructure. A study of Almora district in the Central Himalaya. Mt Res Dev 17:117–126

Richards P (1993) Cultivation: knowledge or performance? In: Hobart M (ed) An anthropological critique of development: the growth of ignorance. Routledge, London, pp 61–78

Roy T (2003) Changes in wool production and usage in colonial India. Mod Asian Stud 37(2):257–286

Saberwal VK (1999) Pastoral politics: shepherds, bureaucrats, and conservation in the Western Himalaya. Oxford University Press, Delhi

Sarin M (2005) Laws, lore and logjams. Critical issues in Indian forest conservation. International Institute for Environment and Development, London

Sharma VP, Köhler-Rollefson I, Morton J (2003) Pastoralism in India. A scoping study. DFID, New Delhi

Shneiderman S (2010) Are the Central Himalayas in Zomia? Some scholarly and political consideration across space and time. J Global Hist 5:289–312

Statesman (1981) The inner line legacy. Trend towards isolation, 28.12.1981. Newspaper archive of the History Department at the South Asia Institute, Heidelberg

Stellrecht I (1992) Umweltwahrnehmung und vertikale Klassifikation im Hunza-Tal (Karakorum). Geographische Rundschau 44(7–8):426–434

Townsend JG, Porter G, Mawdsley E (2004) Creating spaces of resistance: development NGOs and their clients in Ghana, India and Mexico. Antipode 36(5):871–889

Traill GW (1828) Statistical sketch of Kamaon. Asiatic Res 16:137–234

Traill GW (1832) Statistical report on the Bhotia mehals of Kamaon. Asiatic Res 17:1–50

Tribune (2009) Border roads. China ahead of India by miles. 15.10.2009. http://www.tribuneindia.com/2009/20091015/dplus.htm#1. Accessed 24 May 2011

van Schendel W (2002) Geographies of knowing, geographies of ignorance. Jumping scale in Southeast Asia. Environ Plan D Soc Space 20:647–668

Vasan S (2006) Indo-Tibetan border trade in Himachal Pradesh. China Rep 42:41–56

Walton HG (1911) Almora: a gazetteer, vol XXXV of the district gazetteers of the United Provinces of Agra and Oudh. Government Press, Allahabad

Willis CM (2007) A descriptive grammar of Darma: an endangered Tibeto-Burman language. Dissertation presented to the faculty of the graduate school of liberal arts at the university of Texas at Austin

Winkler D (2008) Yartsa Gunbu (Cordyceps sinensis) and the fungal commodification of the rural economy in Tibet AR. Econ Bot 62:291–305

Chapter 11
The Changing Role of Hunting and Wildlife in Pastoral Communities of Northern Tibet

Toni Huber

Abstract Hunting and wild animals have long been part of pastoralist life across the Tibetan Plateau, and especially in the northern Changtang region. Most recent research on Changtang hunting has focussed upon economic aspects in relation to conservation issues, wildlife ecology and status, human-wildlife conflicts and modern development. In contrast, the present study emphasizes social and cultural features of subsistence hunting practice and establishes some historical depth with which to contextualize data from recent decades. This chapter offers a rare diachronic perspective on hunting in a case study area located in the north-west of the Tibet Autonomous Region (China) and utilizes ethnohistorical evidence from throughout the twentieth century and contemporary ethnographic data from repeat fieldwork visits to the area. The results demonstrate that hunting in Changtang areas is best conceived of as a dynamic arena of practice. A subsistence hunting pattern for the region is described in relation to local ecological factors which seasonally determine hunting activity. This pattern is then viewed in relation to two historical periods of regional-level social and economic transition: a pre-modern wealth division between local pastoralist groups and the modern Communist period of collectivization into pastoralist communes. In conclusion, a range of local attitudes towards wildlife are examined in an attempt to open alternatives to the predominant economic, conservation and development-centred discussions of hunting and wild animals in Changtang pastoral communities.

Keywords Hunting • Pastoralists • Wildlife • Changtang • Tibetan Plateau

T. Huber (✉)
Central Asian Seminar, Institute of Asian and African Studies, Humboldt University, Unter den Linden 6, D - 10099 Berlin, Germany
e-mail: toni.huber@cms.hu-berlin.de

H. Kreutzmann (ed.), *Pastoral practices in High Asia*, Advances in Asian Human-Environmental Research, DOI 10.1007/978-94-007-3846-1_11,

11.1 Preamble

We must begin by noting that until the early 2000s, hunting was widely practised for a variety of reasons by pastoralists in the research case study area, although it had already become technically illegal some years earlier under Chinese state laws covering nature conservation and endangered species protection. A concerted crackdown by government officials culminating during 2002–2003 resulted in the confiscation and destruction of all hunting equipment in private ownership, with ongoing policing and penalties for illegal hunting now systematically implemented. Follow-up fieldwork during 2010 revealed that whilst subsistence hunting has now ceased, occasional destruction of predators and illegal poaching by a few pastoralist hunters was rumoured to be still occurring at remote sites. Thus, in sections below discussing recent observations, the present tense will be used since hunting does continue in a very limited and clandestine manner in the research area as of the time of writing.

11.2 Introduction: Local Hunting and the Case Study Area

Of the possible relations which pastoralists are known to maintain with wildlife in different societies, including hunting, domestication, temporary taming as pets and protection, we only have comprehensive evidence for Tibetan pastoralists hunting wild animals.[1] Hunting has traditionally served four principal goals when practised in Tibetan pastoral areas:

1. Supply of additional animal protein and fat for consumption by both humans and certain of their domestic animals
2. As a source of wild animal parts, especially hides, hair and horns (Photo 11.1) that are converted for use as material culture items
3. As a source of high-value wild animal products, especially organs, blood, flesh, horns and fine wool from particular species for trade
4. For control of predator and pest animals that kill or disturb livestock or that are perceived as grazing competitors for domestic herds

Nearly all of these goals of hunting have continued, to one degree or another, to be relevant until very recently in remote Changtang pastoralist communities. Viewed in economic terms, in pre-modern times, all four goals were chiefly pursued in order to maintain a subsistence economy rather than generate any surplus or commercial profit. However, during the modern (post-1960) period, there have been significant changes in the relative importance of these goals for pastoralists, due to specific economic and political demands to be discussed below.

Changtang pastoralists commonly practise hunting with a mixture of old and modern technologies. Steel leg traps, breech-loading and semi-automatic firearms and motor vehicles can be found in use by hunters, depending upon their levels of

Photo 11.1 Wild yak horn used as a milking pail in northern Gertse (Photograph © Toni Huber, 2003)

wealth or access to such items. However, many hunters still also use various traditional technologies (described in Huber 2005), often in combination with modern methods. Homemade muzzle-loading matchlock guns (and much more rarely, simple breech-loading rifles) have been in widespread use (Photo 11.2). An ancient style of leghold trap called *khogtse*[2] is still made from braided grass and animal hair with spikes of antelope horn (Photo 11.3).[3] The use of a combination of firearms and *khogtse* traps (and bows and arrows) for hunting in the region was first recorded in 1874 (Trotter 1915:165). A large-scale 'road trap' or type of game drive structure locally known as *dzaekha*[4] are also set up to catch migrating Tibetan antelope (*Pantholops hodgsoni*), although their use has recently died out due to official restrictions upon antelope hunting. Hunters also construct simple blinds or hides by digging shallow pits and erecting low stone walls within or behind which to lay in wait for animals with loaded firearms, whilst food bates are sometimes employed to attract particular species during winter. Dogs are almost always used to hunt species of wild sheep in rocky areas.

The case study area covered by this chapter extends over large parts of northern Gertse County and adjacent western Nima County (ca. 32°—34° N, 82° 30′—86° E) in the north-western region of the Tibetan Plateau known as the Changtang ('Northern Plains'). This is dry, cold alpine steppe country punctuated by a few low mountain ranges and saline lakes. Local pastoralists graze livestock on plains and

Photo 11.2 Hunting with a matchlock gun in northern Gertse (Photograph © Toni Huber, 2003)

Photo 11.3 Retired hunter with a *khogtse* leghold trap in northern Gertse (Photograph © Toni Huber, 2003)

Photo 11.4 Sheep herding at 4,870 m in the south of the case study area (Photograph © Toni Huber, 2002)

rolling hills that are sparsely vegetated by *Stipa* and *Stipa-Carex* zones and various dwarf herbs and scrubs.[5] Pastoralists in the case study area camp or dwell at widely scattered sites between 4,700 and 5,000 m; this represents both the northern and upper altitudinal limit for permanent human settlement on the Tibetan Plateau. Many of their encampments are now reachable via simple vehicle tracks but are remote in the sense that they are often located 100–200 km from the nearest county town. The local form of pastoralism depends largely upon sheep and goat herding (Photo 11.4), with only a small number of yak. A few horses and dogs are maintained as working animals. Pastoralism on the northern Changtang is marginal compared to many wetter, warmer and lower regions to the south and east of the Plateau. Whilst herd sizes are relatively small and animal product yields are typically modest by most regional standards, many Tibetans from outside the case study area maintain that its livestock often have tastier meat and fat, and that wool/hair quality (especially from goats) is high.

Above latitude 33° in the case study area, most human activity ceases or is strictly seasonal, and from this point on, a vast wilderness region extends for several 100 km northwards to the Kunlun Shan range and the southern margins of the Tarim Basin. The significance of this northern zone for pastoralists has long been its large herds of wild ungulates, especially the Tibetan antelope and the wild yak (*Bos grunniens*), which are both favoured game animals in the case study area.[6] Since 1993, all of

Gertse County north of latitude 33° and the whole of western Nima County north of latitude 32° have been included within the 298,000 km² Changtang National Nature Reserve, in which all hunting of wildlife is totally prohibited and nowadays strictly controlled.

In addition to direct field observations and discussions with active hunters throughout the case study area, part of the research method included 34 in-depth interviews specifically with elderly, retired hunters who ranged in ages from ca. 50–85 years. This data enabled some reconstruction of the pre-modern hunting culture prevailing in the region prior to the modern Chinese Communist administration of local pastoralist communities, which began around 1960. It further allowed for documentation of hunting during the period of collectivization into communes during the 1960s–1970s, as well as the subsequent economic and social reform period instituted by Deng Xiaoping and his allies post-1980. In addition to this data, we have an earlier set of direct observations (notes, maps and photographs) of local hunting left by Swedish explorer Sven Hedin (1909, 1913, 1922), who traversed the case study area twice during 1906 and 1908, and some of this will be compared with our ethnohistorical materials presented below.

11.3 Changtang Hunting: A History of Change

Tibetan hunting culture has always been dynamic and open to innovation. Here we can simply cite the introduction of firearms into Tibet during the seventeenth century and their subsequent universal acceptance as hunting weapons across the Tibetan Plateau, thus revolutionizing the ability to kill large wild ungulates. Beyond technological innovation, Tibetan hunting has long been transformed by changing political, socio-cultural and economic forces, not to mention ecological factors. Before discussing such issues in relation to recent history, I will first outline the basic subsistence hunting pattern of northern Changtang pastoralists as it is practised within local ecological and social contexts of the case study area. The account is drawn from direct observations, interviews and household surveys.

11.3.1 The Subsistence Hunting Pattern

On the micro-level of individual pastoral encampments or households in the case study area, hunting behaviour can constantly change due to the annual influence of ecological factors that in turn shape the domestic economy of Changtang herders. The principal factors are snowfall (and other forms of precipitation) and temperature, both of which can influence seasonal availability of grazing and hence livestock condition or even survival. Periodic heavy snows can lead directly to the starvation and freezing deaths of livestock of all ages and types, whilst sporadic and strong hail storms during lambing, kidding or calving seasons can also kill newborn animals.[7]

A second more seldom factor is the occasional outbreak of epidemic livestock diseases that can claim the lives of multiple animals at once. Finally, livestock can be lost each year to predation by wild carnivores. All such losses result in depletion of livestock available for domestic slaughter and consumption, a critical issue for families who maintain largely subsistence households supplying much of their own food. The traditional and relatively easy local means of compensating for this problem has been the hunting of preferred wild ungulates for additional food. The meat of wild yak, blue sheep, Tibetan argali sheep, Tibetan antelope and Tibetan gazelle is readily consumed by most Changtang pastoralists, although some also prefer to eat Tibetan wild ass.

In summary, the subsistence hunting pattern of northern Changtang pastoralists constitutes a food source supplement when pastoral production falls to a critical level or is in abeyance. Hunting of predators is aimed at curbing livestock loss and contributing towards sustainable pastoral production levels and must be appreciated as an integral aspect of the subsistence hunting pattern. Thus, levels of local subsistence hunting can be annually determined by weather conditions and events, and usually less so by disease outbreak and predator activity. All such ecological-type factors are highly variable and unpredictable.

In the case study area, a further element in determining the degree of dependency upon subsistence hunting over time has been general economic impoverishment of pastoralist households. Here ‘impoverishment’ means consistent inability to maintain pastoral production at levels that meet the annual subsistence needs of all household members. Whilst ecological-type factors can and do always shape the fortunes of pastoralists that lead to impoverishment, they are not the only reason for impoverishment to occur. In an environment where pastoralism is already potentially marginal for ecological reasons, successful pastoralists must possess both excellent skills in pastoral practices and sufficient household labour to enact them. My informants all stressed that, in addition to those suffering genuine ‘bad luck’ with ecological-type factors, poorly skilled or inexperienced pastoralists, and households consisting of few able-bodied persons, had always (in both pre-modern and modern times) been more likely to resort to, and regularly depend upon, subsistence hunting as a compensation.

Subsistence hunting was never the sole means of compensation for an impoverished or marginally producing pastoral household within the case study area. In both pre-modern times and in recent decades, other forms of compensation involving economic activities like trade or business ventures, natural produce harvesting and engagement in labour for others have been options. For a wide range of reasons, most marginal or impoverished pastoral households have not availed themselves of such options unless absolutely forced to. To give but one pre-modern example, prior to 1960, it was possible to harvest salt from the shores of local saline lakes and transport it to Indian Himalayan border trade markets for exchange or sale. However, to make this operation viable, a minimum of 400–500 sheep were required as pack animals for the salt, their wool was shorn and sold at the trade markets to raise additional profits, they carried goods on the return leg and some were killed as food supplies during the month-long journey. Marginal or impoverished households could never muster such numbers of sheep, and the absence of able-bodied household members for many months on the trade caravan further stressed the domestic production system, as well as leaving an encampment less able to cope with crises or less well defended in case of visits by livestock thieves. In addition, such caravans required members who knew the long-distance routes to the border and who had contacts and experience for successful trading. Elderly informants reported that trading caravans and other similar pre-modern economic options (contract herding or shearing for wealthier nomads, summer gold washing for Central Tibetans and so on) were all viewed as problematic compared to subsistence hunting. Hunting could easily be practised locally, it usually only required lower levels of effort over shorter periods of time, and it directly addressed food shortages. During the 1980s and 1990s, my informants who resorted to subsistence hunting did so instead of pursuing alternatives for quite similar reasons: lack of start-up capital or reasonable/secure lines of credit for business ventures, paid labour work being in far-distant locations, lack of skills (including literacy in Tibetan and Chinese) or self-confidence to take up other opportunities and simply because edible wild animals were locally available, relatively easy to kill and they were 'free' in terms of being an uncontrolled resource open for exploitation.

The general contours of the subsistence hunting pattern in the case study area appear to have been fairly consistent. Yet, we can also show that the intensity of, or level of dependence upon, local subsistence hunting has varied considerably across space and time, and that forms of hunting outside of the subsistence pattern have newly arisen, due to the circumstances of social and political history in this part of the Changtang.

11.3.2 A Pre-modern Underclass of Hunting Pastoralists

Prior to the 1950s, two distinct socio-economic classes of pastoralists occupied different zones of our case study area. At around latitude 32° of the central and western Changtang, a series of wide valleys connected by low passes runs east to west from Siling Tso across to Senge Khabab. Formerly the main pre-modern route for travel through the region, this transverse valley system now contains a modern highway and most of the main administrative centres. These broad 'southern' valleys often have extensive areas suitable for easy grazing. As one moves northwards, suitable areas of grazing vegetation are certainly available, albeit more localized and geographically scattered, thus more intensive herding is often required to utilize them.

In pre-modern times, it was not uncommon for a local pastoralist household to consist of six to eight or more persons. My elderly informants reported that, prior to Chinese occupation, wealthy pastoralist households with large herds (up to ca. 1,000+ animals) formerly monopolized these southern valleys and some of their lateral extensions for easier grazing throughout most of the year. The poorer households with few animals (ca. 50–150) usually spent at least half the year—including the coldest period from autumn to spring—at camps throughout the northern areas around latitude 33° and above. The latter half of this period represents the 'lean season' for pastoralists, when grazing is reduced by cold, animal body weight is lowered and food shortages can occur. However, autumn and early winter were also the best hunting seasons, when wild animals were fat and in good condition.

Hedin (1913, 1,185) already observed this basic division of local pastoralists into two groups during the early twentieth century: 'To the Changpas or "inhabitants of the north", who spend the winter in the north, the chase is the chief resource, and cattle-breeding is of secondary importance. The Tibetans in Gertse and Senkor, on the Bogtsang-tsangpo or in Naktsang, who own large herds, do not move northwards in winter, for with them hunting is an occasional occupation. The hunting tribes pursue the yak, the kiang [wild ass], and the antelope'. The 'hunting tribes' or 'Changpas' (*byang pa*, 'northerner') mentioned here refer to those impoverished or marginal pastoral households who survived by practising high levels of subsistence hunting throughout the cold season.

Hedin, who twice traversed the case study area during the cold season (November 1906 and February–March 1908), regularly encountered camps of these hunting pastoralists. He often remarked on the large ('astonishing') numbers of frozen wild animal carcasses that lined the insides of the tents.[8] One elderly informant recalled that as a 10-year-old child in this area, the inside of his family tent during winter was surrounded by a 'wall' (*rtsigs pa*) over 1 m high comprised of dozens of frozen antelope, gazelle and blue sheep carcasses, as well as the body parts of wild yak and wild ass. His father and uncles always killed much game from November on, when the animals were still fat and the temperatures low enough to keep the meat frozen. This was the typical situation every winter, and such stockpiles could keep the family fed for months on end. Hunted game meat of this sort was not only used to nourish the pastoralists themselves, it was also fed to herding and hunting dogs and regularly to all horses in a camp. Hedin (1909, 193) also witnessed this in the region: '...we saw them [Tibetan horses] run up to their masters for two large pieces of frozen antelope flesh, which they eagerly ate out of their hands like bread. They are just as fond of yak or sheep's flesh, and the Tibetans say that this diet makes them tough and hardy. We cannot help liking these small shaggy ponies, which live to no small extent on the offal of game'.

These hunting pastoralist groups usually moved back down to the southern valley systems around the beginning of summer (approximately early June). Their timing in departing southwards was set to coincide with a huge annual migration of wildlife moving in the other direction. Exactly during this season, adult female Tibetan antelope, accompanied by young female offspring, migrate en masse far into the northern wilderness zone where they visit regular calving grounds in order to give birth (Ridgeway 2004; Schaller 1998, 48–56). Each year, the pastoralists capitalized on this migration by setting up *dzaekha* game drive lines and *khogtse* leghold traps across the paths of regular antelope migration routes, thus enabling a final and easy game meat harvest before leaving the area. This special hunting period known as *marling* (*dmar gling*) has been described in detail elsewhere (Huber 2005). Because the trapping system was indiscriminate, many of the antelope killed during *marling* were pregnant females.

11.3.3 Communalization of the Hunt (1960s–1980s)

During the first decades of Chinese Communist administration on the Changtang, the local Tibetan socio-economic system of pastoralism was heavily modified by state interventions. This in turn transformed the pastoralists' hunting patterns and practices in a variety of ways. Initially, wealthy households were defined as 'class enemies' and had their large herds of livestock redistributed to the community, thus erasing former distinctions between the southern stockholding herders and the poorer groups who spent much of their time living from subsistence hunting in the north. Second, during the 1960s, all pastoralists were collectivized into people's communes. One central feature of communes was that the range of everyday pastoral activities most people normally performed became divided up into a series of specific types of labour and to which only certain persons were assigned. Commune members would then be allotted work-points based upon the type of task being performed. The work-points earned each day where tallied up by commune officials. Each pastoralist normally had one or two such areas of labour responsibility for which they earned work-points, and the various persons assigned to always perform the same type of labour were grouped together into working brigades.

Hunting was likewise collectivized as an aspect of pastoral labour under this system in Changtang pastoral communes. The groups who had to perform all the hunting within a commune formed units called 'hunting brigades' (*khyi ra sgrig 'dzugs* or *khyi ra ru khag*). These were normally comprised of five to six men selected by the commune leaders because of their skill in shooting and other hunting methods. Different techniques for hunting were traditionally employed to kill different types of game since each wild species has its own unique ecology and behavioural patterns that a hunter needs to understand and exploit to his advantage. Thus, hunting brigades were also organized and named on the basis of these techniques. For example, the 'dog hunting brigade' (*sha khyi sgrig 'dzugs*) specialized in hunting blue sheep and Tibetan argali sheep, the 'stalking brigade' (*'jab mda' sgrig 'dzugs*) specialized in wild yak, whilst the 'water [source] ambush brigade' (*chu sgug sgrig 'dzugs*) mainly hunted Tibetan antelope and Tibetan gazelle. Work in a hunting brigade earned comparatively high numbers of work-points because it directly contributed a valuable and often scarce resource—fresh animal protein and fat—which was shared with every member in the commune whenever the hunters killed game. Whilst they hunted year-round as necessary, the hunting brigades normally worked hardest from August to November, when game was fattest.

Thus, under the commune system, the local subsistence hunting pattern which had been a pre-modern economic mainstay of every impoverished individual pastoral household was now undertaken by a small and specialized group on behalf of the entire community. However, this new communal subsistence hunting soon became highly intensified due to both demands upon communes to produce more food and because of technological innovation.

Informants who were former hunting brigade members report that during the 1960s, a brigade never killed more than a half dozen game animals per week using their traditional muzzle-loading guns (all of which had been confiscated and become commune property) and trapping systems, whilst travelling to the hunting areas was undertaken on foot or by horse, and employed pack yaks or horse carts to transport the meat. By the 1970s, most hunting brigade members had been supplied by commune leaders with semi-automatic assault rifles or handheld submachine guns for hunting, of the types that were standard issue for People's Liberation Army troops of the day.[9] The effect of this new weapons technology upon harvesting of wildlife was dramatic. Informants estimate up to a tenfold increase in kill rates using this new type of gun, especially in the hunting of Tibetan wild ass that were not normally very shy of human approach and of blue sheep because their meat was favoured for eating. Soon trucks were being employed to transport huge numbers of wild ass carcasses killed by hunting brigades equipped with military-style weapons. Within a year or two of this killing regime, Tibetan wild ass became rarely seen and even locally extinct within several days travel around all communes in the case study area.

Former hunting brigade members reported that with these new weapons, one tended to actually hunt differently. If there was a group of animals in sight, hunters now never selected the healthiest or fattest to kill, nor picked them off one by one as in the past, but rather they kept on shooting without pause until every animal within range was either dead or laying dying on the ground. Some of my informants described herds of 20–30 wild asses being killed in this manner on occasion by well-armed hunting brigades, referring to it self-consciously—and with some obvious degree of shame—as 'slaughter of wildlife' (*ri dwags bshan*)[10] rather than 'hunting of wildlife' (*ri dwags rgyag*). One systemic reason behind such forms of unconstrained killing was that no work-points were awarded to hunting brigade members if they failed to kill any animals or hunted too few. Additionally, sometimes commune leaders instructed hunting brigades to kill every animal they encountered when out hunting; there was an official demand for high kill rates placed upon hunters and this for a variety of reasons.[11] Mainly this was due to additional food requirements during periods of poor pastoral production within communes, which was not uncommon since commune members often lacked motivation to work too hard. But, it was also because increasing numbers of administrators and officials began to settle in newly built and expanding government quarters in the county towns. Such officials often came from outside the districts, they neither produced nor could supply any of their own food requirements, and thus communes were charged with providing additional food for them, something achieved in part through extra hunting.

Whilst collectivization was heralded by Maoists in China as an era of great social revolution, when old cultural and social forms would all be swept away and replaced, all hunting brigades that I investigated maintained certain very traditional features.[12] They were exclusively male in composition, in keeping with the pre-modern (and still valid), pan-Tibetan hunting culture.[13] Moreover, no commune woman was ever allowed to touch the weapons and equipment of the hunting brigades for fear of creating bad luck during the hunt and pollution of the traditional *dralha* (*dgra lha*) deities associated with weapons and their users, something that was also believed to be dangerous to women themselves. This gender-based prohibition is also a standard part of life in any Changtang pastoralist's tent that I have visited. Finally, an interesting example of pan-Tibetan hunting ritual associated with the cult of the *dralha* was performed by hunting brigade members throughout the commune period. This rite involves smearing blood and/or fat from the newly killed game animal into the mouth of the gun barrel and along the gun's stock with one's finger. In the case study area, this simple rite is sometimes referred to as '[I the hunter] eat meat, [you the gun/*dralha*] drink blood' (*sha zo khrag thung*). It is a way of offering part of the kill to the weapon's and the shooter's *dralha* deity so that, satisfied, he will assist in hitting the target during the next hunt. Whilst all public ritual or religious activity was actually forbidden or highly disapproved of by the state and its officials during the commune period, the fact of hunting brigades carrying out their work in remote places far from settlements meant that they were beyond official scrutiny and could do what they pleased.

To conclude the previous sections, it can be noted that hunting observed in pastoralist communities in other regions of the Tibetan Plateau throughout the twentieth century has been described as economically 'optional', 'non-essential', 'peripheral' or 'not an important occupation'.[14] In the case of northern Changtang communities, such descriptions do not readily apply, because the subsistence contribution of hunting has too often been critical or decisive for maintaining pastoral households or collectives. This chapter also emphasizes that the role of hunting within local Tibetan pastoral economies could vary considerably due to seasonal factors at the local level, as well as social and political developments at the regional level.

11.3.4 The Advent of Commercial Hunting

Following the abandonment of communes within the case study area during the early 1980s, and the beginning of implementation of sweeping economic and social reforms under Deng Xiaoping and his supporters, the nature of hunting by Changtang pastoralists was once again transformed. A new system, sometimes called 'household responsibility', was introduced in which the entire spectrum of pastoral production reverted back to household management as had been the case during the pre-modern period. However, a major distinction from the pre-modern period was that now, every newly de-collectivized household unit was given a proportionally equal number of livestock, and pasture rights were also reallocated in a similar way. Until about the mid-1980s, this artificial economic homogeneity resulted in the subsistence hunting pattern being pursued fairly consistently in most pastoralist communities throughout the case study area; annual kill rates were all reported as being relatively low. And this was regardless of the latitude of dwelling.

This situation was not to continue for long. A new form of intensive commercial hunting of Tibetan antelope became widely and excessively practised throughout Changtang pastoralist areas since the mid-late 1980s. This was fuelled by a growing international market demand for *shahtoosh*—the fine, short wool of the antelope—as a luxury product. The ongoing consequences of commercial antelope hunting have had profound local effects, including an indefinite ban on all hunting, confiscation and destruction of all hunting equipment and increasing official stigmatization of the practice, which is criminalized and heavily punished in the breech. Nowadays, a long history of pastoralist subsistence hunting has effectively come to an end within the case study area.

Research related to the massive commercial over-hunting of antelope across the northern Changtang, and its major implications for conservation efforts, wildlife management and pastoralism has already been the focus of various studies (e.g. Fox et al. 2004; Naess et al. 2004; Schaller 1997, 2000). Apart from reiterating (Huber 2005) that commercial hunting by local pastoralists was something completely unknown until the 1980s, and that in the case of antelope, it was entirely generated by external demand—Changtang peoples never had their own uses for antelope wool—this well-documented topic will not be revisited herein. Rather, in conclusion, I will discuss the issue of local attitudes towards wildlife. This is not only of general relevance for understanding the practice of hunting. I would venture it is also intimately related to properly understanding the participation of some local pastoralists from my case study area in excessive commercial antelope hunting, as well as earlier involvements by local people in systematic, large-scale meat hunting for communes in the past.

11.4 Attitudes Towards Wildlife

Buddhism has often been associated with Tibetan attitudes towards nature, wildlife included.[15] Some observers have noted a link between Buddhist ideas and aversion to hunting amongst pastoralists.[16] Other evidence indicates that pastoralist belief in a class of local territorial gods may also influence attitudes towards wildlife and hunting.[17] Given such cultural records from various parts of the Tibetan Plateau, we might ask how these factors have been influential in our Changtang case study area. Furthermore, what are the main pastoralist attitudes towards wildlife, and have they changed along with transformations of local hunting over time?

11.4.1 A Dearth of Buddhism

Together with the great majority of Tibetans, Changtang pastoralists readily identify themselves as being Buddhist when asked. Whilst recognizing that Buddhism has recently become articulated with modern forms of national identity in Tibetan societies (and powerfully so, due to shared historical experience of Chinese occupation), we have to carefully qualify what being Buddhist actually means in the daily life of pastoralists in the case study area. The usual forms of institutionalized Tibetan

Buddhism that are so well known (mass monasticism, a highly literate scholastic tradition, politically and socially influential reincarnate lamas, popular pilgrimage centres and so on) are all absent in the case study area. The few tiny local monastic centres are all relatively recent foundations, and whilst clerical practitioners have circulated in the region, past and present, both their numbers and their religious and social rankings have typically been very low. If pastoralists practise as Buddhists at all, it is usually on the individual or family level of occasional engagement with popular ritual (pilgrimage, offerings and prayers, life cycle rites and so on). Thus, the actual presence of Buddhism as an organized religion is quite marginal in the northern Changtang.

The basics of Buddhist moral cosmology—which condemns any intentional harm of sentient beings and plots the negative post-mortem karmic consequences for those who do—are well known to adult pastoralists. However, any possible influence this might have on their behaviour is, according to my research data, only understandable in terms of local life histories. I found that from pre-modern times up until the late 1990s, the high majority of able-bodied male pastoralists had been involved in hunting as part of their regular domestic activities, thus they understood it with something normal and taken for granted. Hunting typically began in teenage years, with first gun use often being a sort of informal *rite de passage*, and ended for most men around 45–50 years of age, often due to deteriorating eyesight and other physical limitations. Local expectations of masculine behaviour, and engagement in a largely subsistence economy based upon manipulating animals, mean that Buddhist moral concepts are seldom reference points for the daily lives of men, who are highly pragmatic in fulfilling what life demands of them. It is only during middle age that men, having experienced death and illness in their social environment over time, begin to contemplate their own mortality more seriously and their possible post-mortem fates in terms of karma and rebirth. For this reason, it is quite typical that hunting becomes less attractive or personally problematic for middle-aged male pastoralists due to the Buddhist ideas they understand death and afterlife in terms of. A complementary study of the life histories of pastoralist hunters I conducted in Amdo (1999–2001), on the north-east of the Tibetan Plateau, revealed exactly the same pattern.

Thus, my data clearly reveal that Buddhism has played virtually no role in influencing attitudes and practices towards wild animals amongst male pastoralists as hunters,[18] and when this is known to occur, the men concerned have already reached an age when hunting is often no longer an option for them due to increasing physical frailties.

11.4.2 Territorial Deities

As for links between local territorial deities and attitudes towards wildlife and hunting, the principal belief amongst Tibetans is that the wild ungulates and carnivores in any area are considered the property (or 'herds') of such local gods. Thus, hunting game animals is potentially a theft which can be avenged by the deity (typically by illness, madness or misfortune) or, at very least, is viewed as a kind of permitted removal of game animals in dependence upon the purity and strength of moral and ritual relations between a local person and the god concerned. Whilst a few territorial gods are known from the south of the case study area, generally there is little or no interest in them on the part of the pastoralist population. In extensive interviews with over 30 local informants covering many ritual details of hunting, including direct questions relating to local territorial gods, no connection was made between hunting and belief in these gods. This result can be explained by the fact that all those pastoral populations historically inhabiting the southern transverse valley system and using areas to the north are descendents of immigrants who arrived in a series of waves several centuries ago from the far eastern Changtang[19] and who have not brought their former territorial gods with them, or at least not successfully reinstalled them in local landscapes.[20]

11.4.3 'Ownerless' Wildlife

My research suggests that Changtang pastoralists' attitudes to wildlife are based largely upon another set of assumptions that are unrelated to the universal and local religions just discussed. These assumptions are not necessarily readily articulated within any coherent or systematic doctrinal or ritual constellation. Whilst less obvious, careful attention to cultural practices and discourses can clearly reveal them. When I repeatedly observed how pastoralists hunted, killed and butchered wild animals, it was clear that they did so in 'cold blood' without any hesitation or compunction. In particular, the hunter's kill is completely devoid of any ritual activity performed for the sake of the dead animal.[21] This contrasts strongly with a whole variety of rites and behaviours that can attend the killing of domestic animals: prayers for the animal's positive rebirth might be uttered, the skulls or bones of domestic animals are memorialized and engraved with Buddhist ritual formulas for placement upon shrines, killing techniques which avoid actual bloodletting are used (e.g. strangulation) and killing one's own domestic animals is even avoided altogether by hiring a professional butcher. Why then this difference between a pastoralist killing of wild yak when hunting and the often ritualized killing of one of his own domestic yak or hunting a wild blue sheep compared with butchering a domestic sheep from one's own herd? Biologically, the animals are close to identical, and killing is killing after all.

As my informants explain it, the key distinction between wild and domestic species lies in concepts of ownership and property and how one relates to animals on this basis. A common and ancient Tibetan expression for wild animals is *semchen dagmey* (*sems can bdag med*), which literally means 'ownerless sentient beings'. With an 'ownerless' (*bdag med*) status, a wild animal is one which a person requires no permission to use, it does not fall within the category of property (it is not claimed, tamed or controlled by anyone), nor is it one which a human agent is compelled to take responsibility for (e.g. soteriologically, as in the ritualized killing of domestic animals). The attitude with which Changtang pastoralists have long harvested wild animals by hunting them is the same attitude that they take towards harvesting any other local resource base, such as salt deposits on a lakeshore or open summer pasturage on a mountain slope: wildlife is free, it is 'out there' in uncultivated and unsettled places, and available for harvest or use when required.

11.4.4 Wildlife as Enemy

The second significant point about pastoralists' attitudes to wildlife that can be discerned is a conceptual overlap between wild animals and human enemies or foes; both can be equally subject to the sequential acts of location, pursuit and killing. One seeks the whereabouts of a game animal/enemy, chases after it and kills it. Whilst it is common sense that a successful hunt (i.e. it must end with a kill) and a successful victory over an enemy (i.e. it must end in the complete neutralization of threat, most fully achieved by killing the threatening agent) are analogues, the actual fact of this conceptual overlap is explicitly evinced in a number of subtle aspects of the pastoralists' local culture. For one thing, the deities known as *dralha* (*dgra lha*, literally 'god of the enemy') are the key gods of both warfare and hunting. The *dralha* are in fact the only gods Tibetan hunters actually worship in relation to the act of hunting itself, just as men worship them when going into battle. The ritual complex of the *dralha* is identical for worship in both hunting and warfare. The *dralha* are believed to reside in two critical locations, in a man's weapons and battle/hunting equipment (including his horse) and on a man's right shoulder. They simultaneously protect the hunter/warrior from the prey/enemy's force and cunning (elusiveness) and conquer the prey/enemy by empowering a man's weapons with accuracy, penetrative effect and invincibility. All my informants, to a man, worshiped only their *dralh*a in relation to the hunt. The second instance in which we find conceptual overlap between wild animals and human enemies is in certain divination systems used on the Changtang for hunting. The traditional pastoralists practise scapulamancy using a sheep's scapula prior to a hunt in order to both locate the general direction in which the game animals will be found and to inquire whether there will be a kill or not (Q: will the effort of going out hunting be rewarded?). Exactly the same divination technique was employed in times past when human enemies had to be dealt with using conflict. In fact, when reading the scapula, the sites on the bone indicating information about 'enemy' (*dgra*) are the same as those for 'wild animals' (*ri dwags*), whilst the expression 'enemy prognostic' (*dgra phya*) can sometimes be used for both types of divination inquiries. Finally, I report what a 71-year-old informant told me during my 2010 visit to the case study area in answer to my question: 'When hunting, did you do anything special or particular upon killing a wild animal?' The old man responded without hesitation, 'I just used to say to myself *Dra rirag nyi nyatse kug.*[22]' This expression literally means 'Bend down the top of the neck of both enemies and wild animals'. This signifies that after achieving a kill, there is a final act to demonstrate the complete domination or utter subjugation of one's enemy/prey, which is beheading them (typically right at the skull's base) and then bending the neck stump so that it faces downwards in submission. The chanting of this expression upon killing is itself a kind of formula of aspiration for the nature of future encounters with any enemy/prey.

The notion that wild animals are the enemies of human beings is actually not at all out of place for pastoralists living in a northern Changtang environment, in which a whole range of wild carnivorous predators, including wolves, snow leopards, lynx, foxes and bears usually deplete valuable flocks and herds on an annual basis (Dawa Tsering et al. 2006). They literally steal the pastoralists' subsistence away from them. Equally troublesome is that large vegetarian ungulate, the wild yak bull, which can seek out herds of domestic yak in order to round up and drive off females to add to their breeding harems in the hills. A wild bull seeking domestic females will fight and injure or kill any domestic male who intervenes, and pastoralists suffer both loss of female and death of male animals in this manner. An enraged wild yak bull will also kill or injure any humans who intervene or who are perceived as a threat.[23] The killing of a local pastoralist by a wild yak bull occurred in my case study area just prior to the start of my research, and such deaths due to both wild yak and bear attacks are not uncommon right across the Changtang. All this more or less parallels a very similar and not so uncommon human menace prevalent in more lawless pre-modern times on the Changtang, raiding by bands of human livestock thieves, which could have all the same negative impacts on local pastoralist as wild animal predation and attack do. Wild animals are enemies, just as human beings are enemies.

11.4.5 Notions of Abundance

A final point on local attitudes towards wildlife concerns notions about the actual frequency or abundance of animals in the surrounding environment. As happened many times during my fieldwork, when I saw no local wildlife in an area where I knew hunting had been taking place, and asked where the game animals were, informants invariably indicated that they were in another valley or on mountains further away from the present location. When I visited these alternative areas and found no animals there either, the same answer was repeated by the locals. They in turn indicated an even more remote valley or mountain range as the present location of wildlife. The belief seemed to be that an absence of wildlife was only ever local, and that there was always a source of wildlife somewhere else not far away. Despite the obvious lack of wildlife in some areas due to decades of over-hunting, pastoralists' showed no perception of local or regional extinction of wildlife populations, nor that populations might in some way be finite and fragile.[24] Whilst working on the Changtang, Biologist George Schaller noted a revealing little narrative about the source of abundant wild animals, one that I myself heard from pastoralists during my research several times: 'It is widely believed that 100 animals will appear from the mysterious north, a place where few nomads have been, for every animal that is shot' (1998, 301). Hunting is therefore a bonus for conservation in local thinking. Such ideas about the spontaneous and rapid generation of wild animals in wilderness areas are in fact centuries old in Tibetan cultural history.[25]

11.4.6 Summary

If we try to understand how Changtang pastoralists thought and acted when participating in mass killings of wildlife for the commune system, or their reactions when smugglers and black marketeers offer them big money today for killing many antelope, or their repeated appeals to officials to relax hunting bans due to perceived grazing competition from wild animals or how they can readily kill a highly endangered snow leopard because it has slaughtered a few of their sheep, what explanations can we turn to? There is a tendency in many recent publications to present hunting practised by northern Changtang pastoralists almost solely in relation to productive systems and commercial profit. Hunting and attitudes towards wildlife appear solely as economic aspects of pastoralists' lives, inspired and directed by some form of local economic rationalism; if not explicit, this is at least the implicit assumption in the literature. Thinking and acting from economic perspectives certainly inform hunters' practice on the northern Changtang. However, the above investigation into pastoralists' attitudes reveals other aspects of social and cultural life on the Changtang that are highly relevant to addressing all of the types of questions we have just posed above. Pastoralists do view wildlife as an economic resource but largely in terms of their own local, cultural frameworks. Wildlife is considered a resource that is 'free' in the sense of having no responsible owner or 'free' from the burden of property rights. Wildlife, whether carnivores or certain ungulates, is also a resource which can take the form of a potentially threatening enemy, one that can undermine or destroy a person's economic base—or even take human life—in a range of ways merely by existing in the same neighbourhood and following its natural inclinations. And, no less significant, wildlife is a resource viewed as seemingly without limit in terms of its abundance. Hunting and the role of wildlife amongst Changtang pastoralists must be considered within this far more complex social and cultural field, one that also has various demonstrable roots extending back to past (sometimes ancient) patterns or models and historical experiences as well.

Acknowledgements Field research from 2002 to 2010 was funded by Victoria University (Wellington), Humboldt University (Berlin) and the Deutsche Forschungsgemeinschaft (Bonn) and conducted in conjunction with the Tibetan Academy of Social Sciences (Lhasa).

Notes

1. A few cases of temporary taming of some wild species, especially wild sheep, can be observed in the Changtang region. These animals are often found by hunters as orphans or strays and sometimes dedicated or donated to lamas, monasteries or pilgrimage shrines for care.
2. Tibetan words and text are rendered in simple phonetic form followed in parentheses or in footnotes by proper spellings using the Wylie system of Romanization. Certain local words (such as *dzaekha* or *Khogtse*) have no known or stable spellings.
3. The oldest examples of *khogtse* trap, perhaps more than 1,000 years old, have been excavated on the northern periphery of the Tibetan Plateau; see Stein 1921, vol. 2, 704, 767, 782; vol. 4,

plate LIV, item no. T. XV. A. i. 009. Also British Museum, Oriental Antiquities Department, OA MAS 796.

4. Although probably of prehistoric origins and related to other known alpine and sub-arctic game drive techniques (cf. Benedict 2005; Ingold 1980, 56–61; Popov 1966), *dzaekha* were first recorded in the northern Changtang during the late nineteenth and early twentieth centuries; Deasy 1901, 32 and Hedin 1913, vol. 3, 58. On recent *dzaekha* use, see Huber 2005 and Fox and Tsechoe Dorji N.d.
5. An overview of Changtang ecology is given in Schaller 1998, chapt. 2.
6. Other locally hunted wild species include blue sheep (*Pseudois nayaur*), Tibetan gazelle (*Procapra picticaudata*), Tibetan wild ass (*Equus kiang*) and Tibetan argali sheep (*Ovis ammon hodgsoni*). Occasionally, wolf (*Canis lupus*), two species of fox (*Vulpes vulpes*, *Vulpes ferrilata*) and Tibetan brown bear (*Ursus arctos*) are hunted as predator or pest animals. Although both snow leopard (*Uncia (Panthera) uncia*) and lynx (*Felis (Lynx) lynx*) are also hunted as predators in various Changtang areas, I obtained no reports of this for my fieldwork sites. For details of wild ungulates and carnivores of the Changtang, see Schaller 1998.
7. Lambing and kidding occur during late winter or early spring when weather is still cold and sometimes stormy, and the mortality rates are typically high.
8. See Hedin 1922, 93–95, 97, 113, 242–245, 264; Hedin 1913, vol. 1,179, 185–186.
9. In fact, the same guns were also intended for use by the 'local militia' (*yul dmag*) when not being used for hunting.
10. The verb *bshan ba* and noun *shan pa* are always used in relation to killing of domestic animals, and this expression is intentionally ambivalent as it mixes cultural categories.
11. Several informants reported that it was official policy to exterminate Tibetan wild ass in pastoral areas during the commune era. Government officials informed commune members that wild ass competed with livestock for valuable pasture resources and was thus a pest animal they must destroy at every opportunity. This conforms with well-known Maoist dogmas and practices of 'struggling against nature'; see Shapiro 2001.
12. The official Communist Party slogan of the day, 'smashing the four olds', was aimed at destruction and replacement of 'old customs, old culture, old habits, and old ideas'.
13. The same applies to domestic butchery; no pastoralist would ever knowingly eat meat from an animal killed by a woman.
14. See, for example, Goldstein and Beall (1990, 124, 127), Rinzin Thargyal and Huber (2007, 195) and Stubel (1958, 22).
15. See Huber 2004 for a review of legal protections for wildlife by pre-modern Buddhist states and religious institutions in Tibetan Plateau regions.
16. Goldstein and Beall 1990, 127; Stubel 1958, 22.
17. Goldstein and Beall 1990, 127; Rinzin Thargyal and Huber 2007, 106; Namkhai Norbu 1997, 48–49.
18. In response to a survey question asking Changtang pastoralists 'why should wildlife [such as predators] that cause conflict be protected?', only 2% of the 300 respondents choose to answer that 'Killing wildlife is against Buddhist teachings'; Dawa Tsering et al. 2006, 68.
19. See Karma Tshul khrims 2003b, 1–3, 21–23; Tshe ring rgyal po 2005; Tshe ring rgyal po 2006, 392–95; Trotter 1915, 165.
20. Here one can contrast the presence of a few mountain deities in places just outside of the case study area further to the south and which are located in the original home territory of the Sengkor Tsowa (*bSe'khor tsho ba*), a different population who claimed to have been settled in the area when the migrant Drongpa Changma Tsowa (*'Brong pa Byang ma tsho ba*) and Gertse Tsowa (*sGer rtse tsho ba*) populations arrived from the east. Missionary lamas from outside the region are reported as articulating connections between hunting and these mountain deities; Karma Tshul khrims 2003a, 30; Bellezza 2005, 101.
21. This is not to say the killing and butchering of hunted wild animals is non-ritualized in Tibetan contexts; it certainly is, although such rites as are performed relate directly to the human social order or to the *dralha* deities associated with a hunter's weapons.

22. *Dgra ri dwags gnyis kyi gnya' rtse bkugs.*
23. There is a very long Tibetan cultural history of representing wild yak as dangerous foes or enemies who must be destroyed, conquered or tamed by human heroes.
24. A similar lack of conservation awareness amongst pastoralists in the Aru Basin, immediately adjacent to my case study area, was noted by Fox et al. 2008, 10.
25. A popular fifteenth-century narrative biography of Tibetan Buddhism's most beloved saint, Milarepa, describes the rapid multiplication of a herd of Tibetan wild ass in the wilderness; Gtsang smyon He ru ka 1981, 597.

References

Bellezza JV (2005) Spirit-mediums, sacred mountains and related Bon textual traditions in upper Tibet. Brill, Leiden

Benedict JB (2005) Tundra game drives: an Arctic-Alpine comparison. Arctic Antarctic Alpine Res 37(4):425–434

Dawa Tsering, Farrington JD, Kelsang Norbu (2006) Human-wildlife conflict in the Chang Tang region of Tibet. WWF China – Tibet Program, Lhasa

Deasy HHP (1901) In Tibet and Chinese Turkestan: being the record of three years' exploration. T. Fisher Unwin, London

Fox JL, Tsechoe Dorji (N.d.) Tibetan antelope traditional hunting, its relation to antelope migration, and its rapid transformation in the western Chang Tang nature reserve. 15 pp. http://www.cwru.edu/affil/tibet/tibetanNomads/books.htm. Accessed 22 June 2011

Fox JL, Mathiesen P, Drolma Yangzom, Naess MW, Binrong Xu (2004) Modern wildlife conservation initiatives and the pastoralist/hunter nomads of northwestern Tibet. Rangifer 15(Special issue):17–27

Fox JL, Ciren Yangzong, Kelsang Dhondup, Tsechoe Dorji, Richard C (2008) Biodiversity conservation and pastoralism on the Northwest Tibetan Plateau (Byang thang): coexistence or conflict? J Int Asso Tibetan Stud 4:1–21. http://www.thlib.org?tid=T5558. Accessed 10 May 2011

Goldstein MC, Beall C (1990) Nomads of western Tibet: the survival of a way of life. Serindia, London

Gtsang smyon He ru ka alias Rus pa'i rgyan can (1981) Rnal 'byor gyi dbang phyug chen po mi la ras pa'i rnam mgur. Mtsho sngon Mi rigs dpe skrun khang, Xining

Hedin S (1909) Transhimalaya. Discoveries and adventures in Tibet, vol 1. Macmillan (Macmillan's Colonial Library edition), London

Hedin S (1913) Transhimalaya. Discoveries and adventures in Tibet, 3 vols. Macmillan, London

Hedin S (1922) Southern Tibet: discoveries in former times compared with my own researches in 1906–1908, vol 4. Kara-Korum and Chang-Tang. Lithographic Institute of the General Staff of the Swedish Army, Stockholm

Huber T (2004) The chase and the Dharma: the legal protection of wild animals in pre-modern Tibet. In: Knight J (ed) Wildlife in Asia: cultural perspectives. Routledge Curzon, London, pp 36–55

Huber T (2005) Antelope hunting in northern Tibet, cultural adaptation to wildlife ecology. In: Boesi A, Cardi F (eds) Wildlife and plants in traditional and modern Tibet: conceptions, exploitation, and conservation. Memorie della Società Italiana di Scienze Naturali e del Museo Civico di Storia Naturale di Milano 33(1):5–17

Ingold T (1980) Hunters, pastoralists and ranchers. Reindeer economies and their transformations. Cambridge University Press, Cambridge

Karma Tshul khrims [not attributed] (2003a) Lo bo sde chen dgon pa'i dkar chag kun gsal me long. N.p. 37 pp

Karma Tshul khrims [not attributed] (2003b) Stod mnga' ris sger rtse rdzong'brong pa byang ma'i lo rgyus. N.p. 137 pp

Naess M, Dondrup Lhagyal, Drolma Yangdzom, Mathiesen P, Fox JL (2004) Nomadic pastoralism in the Aru basin of Tibet's Chang Thang. Rangifer 15(Special issue):39–46
Namkhai Norbu (1997) Journey among the Tibetan Nomads. An account of a remote civilization. Library of Tibetan Works and Archives, Dharamsala
Popov AA (1966) The Nganasan. The material culture of the Tavgi Samoyeds, vol 56, Uralic and Altaic series. Indiana University Press, Bloomington-La Haye
Ridgeway R (2004) The big open. On foot across Tibet's Chang Tang. National Geographic, Washington, DC
Rinzin Thargyal, Huber T (eds) (2007) Nomads of Eastern Tibet. Social organization and economy of a pastoral estate in the Kingdom of Dege. Brill, Leiden
Schaller GB (1997) Tibet's hidden wilderness. Wildlife and Nomads of the Chang Tang Reserve. Harry N. Abrams, New York
Schaller GB (1998) Wildlife of the Tibetan Steppe. University of Chicago Press, Chicago
Schaller GB (2000) Wildlife conservation in the Chang Thang Reserve, Tibet. In: Ning W, Miller D, Zhi Lu, Springer J (eds) Tibet's biodiversity. Conservation and management. China Forestry Publishing House, Beijing, pp 21–28
Shapiro J (2001) Mao's War against Nature: politics and the environment in Revolutionary China. Cambridge University Press, Cambridge
Stein A (1921) Serindia. Detailed reports on explorations in Central Asia and westernmost China, vols 2, 4. Clarendon Press, Oxford
Stubel H (1958) The Mewu Fantzu. A Tibetan tribe of Kansu. HRAF Press, New Haven
Trotter H (1915) Account of Pandit Nain Singh's journey from Leh in Ladakh to Lhasa, and his return to India via Assam 1873–74–75. In: Burrard S.G (ed) Records of the survey of India, vol VIII (part 1): exploration in Tibet and neighbouring regions, 1865–1879. Office of the Trigonometrical Survey, Dehra Dun, pp 159–195
Tshe ring rgyal po (2005) Mnga' ris byang thang gi sger rtse tsho pa dang bse'khor tsho pa'i skor la dpyad pa. In: Gu ge tshe ring rgyal po'i tshe rtsom phyogs bsgrigs. Krung go'i Bod rig pa dpe skrun khang, Beijing, pp 267–278
Tshe ring rgyal po (2006) Mnga' ris chos 'byung gangs ljongs mdzes rgyan. Bod ljongs mi dmangs dpe skrun khang, Lhasa

Chapter 12
Implementation of Resettlement Programmes Amongst Pastoralist Communities in Eastern Tibet

Jarmila Ptackova

Abstract In the context of a major policy initiative by the Chinese state to better integrate western areas with the rest of the country, the government is implementing numerous programmes that affect the pastoral communities of the Tibetan Plateau. Resettlement of pastoralists away from their traditional grazing lands is the most significant intervention into their way of life. According to the design of resettlement programmes, small village settlements are viewed as ideal for establishing new lifestyles for Tibetan pastoralists who should afford increased comfort and income possibilities through better connections to infrastructure. The research results from my case study area of Zeku County, Qinghai Province, demonstrate that implementation of resettlement away from grasslands often does not reflect official planning, and that there has been a general lack of consultation with Tibetan pastoralists throughout the entire process. Due to insufficient state support for pastoralists to obtain alternative skills and livelihoods, the many resettlement villages now appearing across the eastern Tibetan plateau remain populated by unemployed pastoralists facing an uncertain future.

Keywords Resettlement • Sedentarization • Development • Governmental programmes • Grassland degradation • Tibet

12.1 Introduction

Following the introduction of the *Opening the West* (Chin: 西部大开发 *Xibu da kaifa*) development strategy, launched by Jiang Zemin in 1999 (Heath 2005), the central government shifted its focus from the coastal areas in the east of China to

---

J. Ptackova (✉)
Central Asian Seminar, Institute for Asian and African Studies, Humboldt University, Unter den Linden 6, D - 10099 Berlin, Germany
e-mail: jptackova@googlemail.com

H. Kreutzmann (ed.), *Pastoral practices in High Asia*, Advances in Asian Human-Environmental Research, DOI 10.1007/978-94-007-3846-1_12,

the western provinces and autonomous regions. The state maintains that social and political stability in the country as a whole depends upon narrowing the growing socio-economic gap between the East, that has been developed with the aid of huge state investment since the 1980s (Goodman 1989), and the West, where many regions feature at the lowest positions in statistics for national GDP (QD 2009). Chinese propaganda often uses the word 'backward' to describe the western regions, which are overwhelmingly rural. On the eastern Tibetan Plateau, which nowadays extends over most of Qinghai Province and parts of Sichuan, Gansu and Yunnan Provinces, 'backward' refers to the traditional way of life of Tibetan pastoral communities and emphasizes their lack of modern facilities used in daily life and production processes. Many Tibetan pastoralists themselves also currently adopt this term when describing their situation. Whilst admitting their 'backwardness' in comparison with other types of Chinese citizens, they also agree with state plans to develop and modernize their lands by way of the *Opening the West* development strategy.

The first decade-long phase of *Opening the West* concerned the development of infrastructure and extension of urban areas (Paul and Cheng 2011). Thus, nowadays roads connecting all administrative centres and existing towns and cities across the eastern Tibetan Plateau have been rebuilt and enlarged, and new urban settlements have been established. The second phase of the development strategy uses this new infrastructural network to reach rural inhabitants for implementation of programmes that directly modify daily life practices of local communities and which will have a significant impact on their future development. In this chapter, I will outline such development programmes that directly concern the pastoralists of the eastern Tibetan Plateau. My primary case study area is in the Zeku County (Chin: 泽库县 *Zeku xian*, Tib: *rTsekhog*) of Huangnan Prefecture, Qinghai Province. Zeku County is a purely pastoral area and one of the poorest places in Qinghai Province. The 2007 population was 57,923 people in 11,165 households, of which 56,000 people were recorded as engaged in animal husbandry (ADGM I). With an average altitude over 3,500 m, grassland vegetation in Zeku is not as rich as in neighbouring pastoral counties situated at lower altitudes, such as Henan County, and suffers more from degradation (ADGM VII).

Current state programmes that influence the lives of Tibetan pastoralists most directly are those featuring large-scale resettlement and sedentarization measures. The major programmes of this nature on the eastern Plateau are named *Turning Pastureland into Grassland* (Chin: 退牧还草 *Tuimu huancao*), *Ecological Resettlement* (Chin: 生态移民 *Shengtai yimin*) and *Nomadic Settlement* (Chin: 游牧民定居 *Youmumin dingju*). Each programme has a clear policy outline and detailed implementation plan for affected areas. Nevertheless, the execution of these programmes in practice often differs significantly from their theory and guidelines (Table 12.1). Each local government modulates guidelines according to the current situation in any affected area under their jurisdiction, in order to benefit either the local communities concerned or the implementing officials themselves. In this chapter, I will present examples from the implementation of the three programmes just mentioned as they occurred in my case study area of Zeku County (Fig. 12.1).

Table 12.1 Overview: implementation of resettlement in theory and practice

Theory	Practice
Socio-economic improvement:	
New house	Partial or total abandonment of grassland and reduction of livestock
Increase of household income, e.g. through available facilities on site to grow vegetables or establish business	Increase of cash need, difficulties to establish new source of income (no skills to grow vegetables or get engaged in business)
Access to science and technology, widening of skills of the pastoralists through vocational training offers	Lack of state funds to cover vocational training costs, no assistance to adapt to the new environment
Access to infrastructure and facilities, including main water supplies, electricity, roads, schools, medical and veterinary services and television broadcasting	Insufficient funding causes shortfalls to the formally scheduled subsidies, the subsidy is too low to cover expenditures of a household
Regular payment of subsidy for the resettlement period	
Environmental protection:	
Recovery of grassland vegetation, protection of high-altitude animals and natural resources, water retention in the catchment areas of the three rivers, reduction of soil erosion and prevention of desertification	Fundamental change of grassland ecosystem
Total reduction of livestock, restriction of land use rights	Pastoralist household splitting in order to keep grassland and get a new house
Return to grassland forbidden during the resettlement period	Reselling of resettlement houses and returning to grassland
	Political benefits:
	Exploitation of natural resources
	Close political control and better integration of ethnic minorities

Source: Own compilation

My data are drawn primarily from Chinese state documents and from interviews conducted from 2007 to 2009 with programme officials in implementing bureaus and Tibetan pastoralists directly affected by the programmes.

12.2 State Programmes Featuring Resettlement and Sedentarization

The Chinese state is promoting its regional development policies in the west of the country as a way to improve household income levels, usually measured in cash terms. In Tibetan populated areas, pastoralist households have typically had low levels of cash income and are therefore considered to be poor, an appraisal which

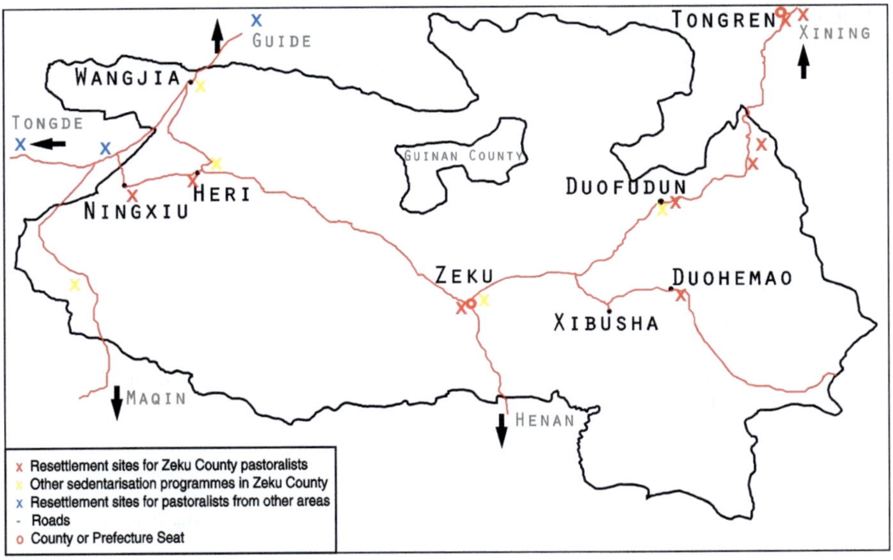

Fig. 12.1 Sketch map of Zeku County resettlement sites (Design: Jarmila Ptackova)

disregards their investments in livestock. In addition to local economies in China's west, the state has also expressed concern about the level and quality of vegetation cover and the spreading of surface erosion on the grasslands. Such environmental factors of the high plateau environment have a downstream impact on the main Chinese river system, the Yellow River, the Mekong River and the Yangtze River, all of which have their sources in Qinghai Province and supply most of lowland China with fresh water. Increasing erosion leads to sedimentation downstream, which in turn causes flooding in low-lying regions. This encouraged the state to include environmental protection of the Qinghai-Tibet Plateau within its *Opening the West* policy (Yeh 2005). In 2003, a huge preservation zone named *Three Rivers' Headwaters National Nature Reserve* (Chin: 三江源国家级自然保护区 *Sanjiangyuan guojia ji ziran baohu qu*) (= Sanjiangyuan) was established to protect the upper reaches of the Yellow, Mekong and Yangtze Rivers (Foggin 2005). With an area of 152,300 km^2 (ADGM III), the zone covers the entire southern part of Qinghai Province. Within its boundaries, the policies of relocation of pastoralists away from their traditional grassland areas and reduction of livestock numbers have been stressed more than in any other Tibetan Plateau areas.

Currently, resettlement and sedentarization are promoted by the state as measures for both socio-economic improvement and environmental protection. According to the state's discourse, the potential ecological benefits from implementation of resettlement programmes and adoption of grassland resting and rotational grazing systems will ease pressure on grassland habitats, stimulate recovery of grassland vegetation and aid protection of high-altitude animals and natural resources. Longer-term ecological benefits would be better water retention in the catchment areas of the three rivers, reduction of soil erosion and prevention of desertification. Second, the state claims a wide range of potential benefits for local pastoralism and pastoralists from its resettlement and sedentarization programmes. Achieving balanced numbers of grazing animals and livestock reduction will benefit carrying capacity of grasslands and should mitigate the lack of grass for animals in winter. In terms of socio-economic improvement, these programmes should benefit pastoralists by offering training to raise their level of skills. They should help to increase the income of households, reduce mortality of livestock, improve selling rates of animal products and accelerate the fattening of lambs in order to be able to sell them within the first year. Livestock turnover could be hastened and animal husbandry improved due to use of animal sheds which could double as greenhouses to plant vegetables during the summer period. Adopting the resettlement and sedentarization programmes will also result in new access to infrastructure and facilities, including mains water supplies, electricity, roads, schools, medical and veterinary services and television broadcasting to each new village. It should enable access to science and technology and help pastoralists to absorb, extend and apply knowledge, such as education about prevention and cure of animal diseases, the use of greenhouse-based horticulture and so on (ADGM II).

Large-scale resettlement of Tibetan pastoralists commenced in eastern areas of the Plateau during 2003, with the implementation of the *Turning Pastureland into Grassland* programme. This programme has a primarily ecological agenda and concentrates on the restoration of degraded areas of grassland. In places with high levels of degradation, the pastureland is completely excluded from herding and a grazing ban is instituted. The grazing ban entails a year-round grazing prohibition scheduled for a period of 10 years. Affected households are being resettled, and the number of their livestock holdings significantly reduced. During 2004, additionally, the *Ecological Resettlement* programme was introduced in the Sanjiangyuan protection zone. This programme focuses on poverty alleviation and improvement of the socio-economic situation of pastoral households. In order to help pastoral households from regions with severe grassland degradation, and to allow for grassland ecosystems to recover, this programme resettles households from affected regions into newly constructed settlements (Ptackova 2010).

As participants of both *Turning Pastureland into Grassland* and *Ecological Resettlement*, local pastoralists should receive a new house in a resettlement community near to an already existing administrative centre or at least along main roads, and they should obtain a government subsidy which is officially scheduled for a period of 10 years. Following the resettlement period mentioned in the official contracts between the state and the pastoralist participants, participating households should have the possibility to return back to a pastoralist way of life. Apart from the official agenda of both of these programmes, their actual content is very similar, and in practice, it is often difficult to distinguish between them. In fact, some officials claim that it is the same programme which was split administratively between the Department of Agriculture and Animal Husbandry which administers *Turning Pastureland into Grassland* and the Development and Reform Committee which administers *Ecological Resettlement*, in order to obtain double subsidies from the Central government which again enables the resettlement of twice as many pastoralists.

The aim of the *Nomadic Settlement* programme is relocation of pastoralists from their grasslands as well. Nevertheless, it has a slightly different agenda to the two aforementioned programmes. *Nomadic Settlement* was launched in 2009 as a reaction to widespread riots throughout Tibetan areas during 2008. By providing Tibetan pastoralists with new houses in settlements established along main roads, the government intends to demonstrate to the Tibetan population that it cares for them and wants to help them to attain a more comfortable life. On the other hand, a police station situated in each of these new villages enables the government a far closer degree of control over Tibetan pastoralists, who otherwise live scattered over vast areas of grasslands (Ptackova 2011).

A significant difference to the other programmes, and a most important point for local Tibetan pastoralists themselves, is that currently the *Nomadic Settlement* programme does not require any abandonment of the grassland or reduction of livestock on the part of participating pastoralists. Depending upon local implementation office, pastoralists can either obtain a fixed amount of money as a government grant towards building a new house (the case, e.g. in Maqin and Hongyuan Counties) or the pastoralists pay very low prices to purchase a new house constructed by the government (e.g. in Zeku County). The pastoralists usually appreciate such governmental assistance, especially if it comes in additional to maintenance of their pastureland and herds and does not require any restrictions upon their lifestyle as pastoralists.

All of the programmes I have just described that include resettlement or sedentarization of Tibetan pastoralists do sound reasonable and also beneficial for their participants, at least on the level of state policy documents that describe their agendas and implementation schemes. Nevertheless, the reality of these programmes on the ground often appears to be very different. The main reason for this discrepancy between state policy and the reality of implementation is to be found in the Chinese administrative system and the almost absolute authority of local implementation officials within the areas under their jurisdiction. Therefore, actual implementation and its circumstances vary from place to place. In the following section, I will describe some examples of the local execution of resettlement and sedentarization policy with reference mainly to Zeku County, Qinghai Province.

12.3 Local Implementation of the Turning Pastureland into Grassland and Ecological Resettlement Programmes

According to the government of Qinghai Province, over 500,000 Tibetan pastoralists should be relocated from grassland areas by the end of 2014 (Du 2009). In Zeku County, the first resettlement phase was scheduled for the years 2003–2006. Initially, this resettlement has been implemented as part of the *Turning Pastureland into Grassland* programme. The *Ecological Resettlement* programme was then introduced into Zeku County during 2005. These programmes form the main aspects of the overall Sanjiangyuan zone resettlement policy. In 2007, *Ecological Resettlement* was made programme number one in Zeku County. From each bureau on the township and county levels, one member was selected to participate in project administration as part of the new Sanjiangyuan office (ADGM X). This office is responsible for selecting future resettlement sites, planning new villages and supervising construction works and the overall resettlement process (ADGM XI). During the first period of 2003–2006, 1,093 households from throughout the entire County were scheduled to be resettled to nine different resettlement sites. Seven of these sites (Zeku County seat, Longzang village and Duolong village in Duofudun Township, Duofudun Township administration centre, Duohemao Township administration centre, Ningxiu Township administration centre and Heri Township administration centre) were located within the county itself, close to the township and county seats. Two resettlement locations, Laka site and the CP school site, were located in neighbouring Tongren County near or within the prefectural seat (ADGM XII) (see Fig. 12.1).

By 2006, only 400 households of the planned total were resettled in only two of the designed new villages, being those which were partly finished in Heri and Ningxiu Townships. The large-scale resettlement wave did not start in Zeku County until the year 2007. According to the Sanjiangyuan *Ecological Resettlement* plan, during 2007, 765 households with 3,620 people should have been resettled in Zeku County (ADGM VI). Similar plans were made throughout the whole of the Sanjiangyuan zone. Each resettlement construction site was marked by an information board presenting information about the agenda of the resettlement programme, the exact number of households and people that should move in there, site plans with sizes of houses and courtyards and a list of additional facilities. The new houses were constructed either in the form of bungalows, usually with small courtyards, double-storied houses with shop facilities on the ground floors, or blocks of flats in the case of the Communist party school resettlement site in Tongren town. The information boards also informed about the dates of start and completion of all construction works on each site. During 2007 in Zeku County, commencement of construction work was scheduled for the start of May, and the resettlement sites should have been completed already by the beginning of October of the same year. Nevertheless, none of the sites was completed by this date. Official data from the Chinese government state that by 2007 in Zeku County, 3,991 pastoralists already participated in or were assigned for the *Ecological Resettlement* programme and 500 pastoralists already participated in or were assigned for *Turning Pastureland into Grassland*. Throughout the whole Sanjiangyuan area, by 2007, the total number of participants of both of these programmes is listed as 69,283, which means almost 15% of the entire pastoralist population of the Sanjiangyuan zone (ADGM I).

In 2008, the information boards informing the public about each construction site were removed, and it became difficult to distinguish between the resettlement sites and their background programmes. The construction of new houses continued far in excess of the previously scheduled numbers. In 2009, additional localities were chosen to situate new resettlement villages to be raised within the framework of the new *Nomadic Settlement* programme. In Zeku County, the *Nomadic Settlement* programme should fulfil the resettlement and settlement efforts. Within a further 3-year period, the government plans to involve all remaining pastoral households still located on the grasslands within this latest programme.

12.4 Selection Process for Participation in Resettlement

Officially, any public participation in resettlement and sedentarization programmes is voluntarily. However, during the planning period, the government sets targets for the number of households that must be relocated in each county or township. The exact numbers of households which are going to be resettled as mentioned in the programme implementation documents are the result of state examination of local grassland carrying capacity, and thus represent the number of households that exceeds the local capacity. The task of the local implementation officials is then to persuade this specific number of households to participate. Similar to the implementation process for other state policies in China, the designated county and township government officials visit the communities[1] under their jurisdiction and inform the community leaders and heads of each household about the programme during a meeting. Alternatively, the community leaders are invited to the seat of township government, where the county and township officials educate them about the new policy and subordinated programmes, and give them instructions on how to implement this in practice. The community leaders then inform the heads of each household in their community at a meeting, or they visit each household individually. Whilst informing the households, the mediators stress the benefits of the planned programme, such as a new comfortable house achieved under very advantageous conditions, the easy access to roads, transportation and schools, available government subsidy and so on. They also inform the people that an unwillingness to participate could result in restrictions in the form of no further governmental assistance in future or similar measures.

The benefits of resettlement mentioned in these sessions are certainly real, but without knowing the entire agenda of the overall policy and the commitments entailed by participation, such as partial or total abandonment of grassland and reduction of livestock, the pastoralists themselves cannot imagine the thoroughgoing impact that relocation programmes will have upon their daily lives. Motivated by the promised benefits, numerous households apply to participate. The first targeted households, and also those who most readily agree to relocate, are the poorest households with few or no livestock. Additionally, richer households with enough livestock to secure a comfortable living on the grassland are also interested in the prospect of a cheap new house near an urban area. Nevertheless, such households would only agree to participate if the new houses came in addition to, and not instead of, their pastoralist lifestyle. The skilful presentation of the new state programmes usually causes a high enough interest to cover the set quota or target of households designated for resettlement. In Zeku County so far, there have often been even more households applying for resettlement than the number of available new houses. In such cases, the community leader decides by lot, for example, by pulling names of applicant households from a hat. If not enough households from one community should be interested, the officials can appoint more households from another community instead, or they simply decide which families from the reluctant community must move to the resettlement.

After the households in a community are selected for resettlement, the heads of each household proceed to the township government to sign the resettlement contract. Over 50% of my pastoralist informants from Zeku County who were appointed to resettle into the Duofudun town, Zeku County seat and Tongren Laka resettlement sites are completely illiterate and had to sign a contract they could not understand. Those who do know how to read only found out during the signing procedure itself that their participation entails that they must voluntarily reallocate from 50% to 100% of their pastureland to the government. Nobody dared to change their mind at this point about their participation in the resettlement programme as they feared the possible negative consequences.

According to the original government plans, on average, each participating household should pay 18,000 RMB (from the total of 55,300 RMB) as their own contribution to new house construction. Nevertheless, pastoralist households in Zeku County are comparatively poor and those households which actively participate in resettlement programmes represent the poorest without livestock. For this reason, it was decided that in Zeku County the share of the resettlement construction costs normally contributed by pastoralists will only be set at 3,000 per household (ADGM XII). This of course caused financial shortages in government construction planning and was one of the reasons for implementation delays. According to my pastoralist informants, each household involved in the resettlement programme initially had to pay 6,000 RMB for the new house. Later the government repaid them 3,000 RMB. This latter amount, however, matches the annual government subsidy granted to the pastoralists by the state as set out in the programme policy documents. By the end of 2009, Zeku pastoralists with resettlement contracts dating from 2007 had received only two annual subsidies of 3,000 RMB per household for the 2007–2008 period for those who resettled within the county itself and 3,500 RMB per household for those who moved to resettlement sites within neighbouring Tongren County. Additionally, the government granted them 500 RMB each winter for fuel costs. In the official programme documents, this annual government subsidy is scheduled as a regular payment to participating pastoralists for the duration of 10 years (ADGM XII). However, the pastoralists themselves who participate in the programme in Zeku County possess no written document stating the amount of the payment or the duration of entitlement. Every year now they have to wait to see whether any further support will be granted or not.

12.5 The Resettlement

After moving to a resettlement site, pastoralist households find themselves in a completely new environment. Giving up the land and livestock which formerly provided much of their subsistence needs, their daily costs for basic needs rise enormously. The pastoralists suddenly realize that everything must be purchased for cash. Their modest government subsidy cannot cover daily expenditures for a whole household, and the pastoralists must seek additional sources of income. Part of the state's implementation guidelines includes plans for future economic supplements in resettlement areas. Depending upon the specific resettlement locality, some sites were designated for farming, for example, the Laka site near Tongren town. It is intended that others secure their income through harvesting caterpillar fungus (*Ophiocordyceps sinensis*, Tib. *yartsa gunbu*) and trade, as was scheduled, for example, for the Communist party school resettlement in Tongren town. Other resettlements near cultural sites or spectacular landscapes of interest to tourists, like the Longzang village near Maixiu forest, are meant to somehow profit from tourism and related service industries (ADGM X). Unfortunately, at least by the end of the year 2009, these plans all remained mostly on paper and were not converted into practice. The implementation reports by local governments often concede that the available funds are not sufficient to cover necessary costs for the vocational training of resettled pastoralists, nor for establishing the required capital investments to enable them to start new businesses. The insufficient funding even causes shortfalls to the formally scheduled subsidies. Therefore, not enough help is offered to the resettled pastoralists to obtain new skills and qualifications to be able to follow the new economic concept presented in the state's programme documents. If there is any vocational training given at all, then it is only a 1-month course, which is far from sufficient to impart enough knowledge about the new domain and enough self confidence in it to start new business ventures.

Obtaining new qualifications and creating new income bases is a key task in order to enable pastoralists to adapt to the new environment of the resettlement sites and to urban social life in general. Without a secured existence for their households, resettled pastoralists will never find satisfaction in the new villages. At the majority of resettlement sites for Zeku pastoralists, I observed situationally the types of problems just described. For example, in the resettlements of Duofudun, Zeku County seat, Longzang, Ningxiu and Tongren, the pastoralists complain that there is no possibility to find a new job. First, there are no jobs on offer, and second, they lack any qualification apart from skills in pastoralism.

In several resettlement sites, the pastoralists have obtained houses with courtyards which could be used to grow vegetables, whilst in others, shops were included to enable business opportunities. In reality, the pastoralists say they do not know how to grow vegetables, and the high local altitudes do not allow the plants to gain their usual size. The new shops are in most cases managed by local Tibetans from nearby townships and not by the households from the resettlement sites themselves. For resettled pastoralists to be able to secure their livelihoods, there remains the option of using their own savings as investment capital or of seeking work at state construction sites in the region. The construction work brings only a small salary home, and the work is physically demanding. This leads most resettled pastoralists to revert to the recently lucrative alternative of harvesting medicinal plants, especially caterpillar fungus, which is sold for very high prices on the Chinese market (Winkler 2010).

Facing difficulties in adapting to their new environment and in securing enough income to support the whole family, many pastoralists attempt to resell their new houses and return to the grassland. Theoretically, this should not be possible within the framework of the state's programmes, but the pastoralists find their own ways to negotiate the resettlement policy and gain the most benefit out of it for themselves.

According to the agenda of the original resettlement programmes, all pastoralists who resettle must sell all their livestock. Nevertheless, in many areas, such as Zeku, Henan or Maqin Counties, we can observe pastoralist households who possess a new resettlement house whilst keeping their herds at the same time. One reason for this is that, in order to extract the maximum official benefits, numerous households officially split into two, appointing the elderly resident members of the 'grandparents' generation as a separate household unit. This way, one original household is able to send the grandparents and children to inhabit the new house in a resettlement site and to enjoy all the benefits of a comfortable house with a road and school access plus government subsidy but also keep their herd on the grasslands as a source of food and income.

In some areas, for example, in Duofudun or Heri Townships in Zeku County, the government did not require the pastoralist households to sell all their livestock and give up the pastureland contracted to them. In Duofudun, those pastoralists designated for resettlement in 2007 had to sign a contract, which assigned from half to all of their pastures to the government. Nevertheless, the government has not enforced its right to the land, and so the majority of pastoralist households intended for resettlement still remained at their grassland locations during 2010 and hope that the government will not enforce its claim to the land. In Heri Township, the pastoralists who participated in resettlement claim that they did not even have to sign any contract and that they are allowed to keep their land and livestock in addition to their new state houses. These households still have some possibility to abandon the resettlement site and return to the grassland in case they dislike life in the more urban locations. In case a household should reconsider its participation in the resettlement process and decides to return to their original pastureland, it usually tries to resell the house in the resettlement site. Especially in resettlements sites established in county or prefecture seats, such as those at the Communist party school in Tongren, or even those nearby to towns such as the Zeku County seat resettlement site, pastoralists can earn at least double the amount they invested in building their resettlement apartment or house by reselling it. Both reselling of such houses and returning to the pastureland is technically illegal under the state's programmes, but ultimately it depends on how such breaches are dealt with by the implementing office of the local government.

But those pastoralist households who had no livestock, or who sold it all before moving into the resettlement sites, and whose pasture is now being controlled by the government, have none of these options and must stay in the resettlement, an example being the resettlement of Maduo pastoralists in Tongde County on the Zeku County border.

During my field study, it was not the case in all resettlement sites that pastoralists complained about the difficulties of adaptation or finding new sources of income. There are some locations where the resettlement programme appears to be successful. For example, in the Heri Township resettlement site, all my pastoralist informants claim that the relocation enables them a more comfortable life and a better income. The reason for this is the local tradition of stone carving in this particular pastoralist area. In almost every resettled household, there is someone who masters this handcraft. The availability of more free time and better connection to the market allowed by the resettlement enables these people to produce more and better sell their stone products. Nevertheless, these particular pastoralist households at Heri still maintain their pastures and livestock on the side. They all agree upon the fact that without the additional economic possibility of stone carving, life in the resettlement would be a harder life than that as a pastoralist on the grassland.

In Zeku County, the government actually promotes the Heri site as an example of a resettlement success (ADGM IX), although it was not primarily the government which was responsible for this apparent positive result.

12.6 Further Sedentarization Efforts of the *Nomadic Settlement* Programme

It could be said that, in general, the pastoralists of eastern Tibetan areas are not against sedentarization programmes and what they entail. However, they only want to accept the new housing provided by the programmes in addition to their pastureland and livestock and not instead of it. Therefore, the ideal solution for both the state's intention to relocate part of the pastoralist population from the grassland to accessible urban locations and the wish of the pastoralists themselves to possess a new house on top of their existing life as pastoralists seems to be offered by the *Nomadic Settlement* programme. This allows the pastoralists to maintain their existing pastureland and livestock whilst additionally enabling them to obtain a new house in an urban area.

However, the *Nomadic Settlement* programme does not end merely with providing pastoralists with a new house. Under the label 'Development of Tibetan Areas', the state wants to demonstrate to the pastoralists that, following the period of riots during 2008, it cares for them and their welfare and comfort. Supplying pastoralist households with material goods and houses on the one hand, the government pursues its unpublicized aim of tighter social control over Tibetan pastoralist communities on the other hand since they now should reside in permanent houses in highly accessible locations where social life can be easily monitored. The *Nomadic*

Photo 12.1 Longzang resettlement village in Ningxiu, Zeku County (Photograph © Jarmila Ptackova, July 2008)

Settlement programme targets all remaining pastoralist households, not engaged into any resettlement programme yet (ADGM VIII). From the beginning of the implementation of the *Nomadic Settlement* programme in 2009, the Zeku County government planned to resettle approximately 30% of the remaining pastoralist population from the grassland each successive year. Under this programme, each participating household must contribute 5,000 RMB for their new house, whilst the remainder of the ca. 40,000 RMB costs is covered by the government. The government itself takes responsibility for the entire construction work involved and selects the actual settlement locations. Similar to the resettlement sites of the *Ecological Resettlement* programme, the new houses built for *Nomadic Settlement* in Zeku County form well-laid out housing schemes (Photo 12.1) at the periphery of county seats or townships, or they are situated directly along major roads.

So far, the participants in the *Nomadic Settlements* programme can maintain their livestock and land and must move only part of their household into the new houses. During 2009 in Zeku County, housing projects for *Nomadic Settlements* were already under construction. Nevertheless, none of these new urbanizations had been finished by the end of the year. Therefore, local pastoralists did not start inhabiting these new houses during my fieldwork period, and it was not possible to examine their situations after joining the *Nomadic Settlement* programme. In other pastoralist areas, such as Maqin County in Guoluo Prefecture or Hongyuan County in Aba Prefecture in Sichuan, the construction work proceeded much faster, and participating pastoralist households could already enjoy their new houses during 2009. In these counties, the implementation methods of the programme vary from those applied in Zeku County. In Maqin County, the pastoralists first build a house that complied with the parameters set down by the government, and then they can apply for a financial subsidy of 40,000 RMB. The pastoralists could erect the new house either at their regular winter pasture site or decide in favour of a house in an existing settlement near the county or prefectural seat. In Hongyuan County, the households first applied for and were granted 20,000 RMB to build new houses in a settlement location selected by the government, or they could allow the government to build the new house for them (Ptackova 2011).

Regardless of how promising the recent implementation of *Nomadic Settlement* appears to be, as it reflects the current wishes of the pastoralists to continue their lifestyle as pastoralists in grassland areas with an additional chance to obtain a living base in an urban area. Beyond this, the Chinese state is already developing much grander plans that will affect the entire pastoralist population and its future right across the Tibetan Plateau. It is the state's current view that the whole 'backward' system of Tibetan-style pastoralism should be modernized (ADGM IV). This means, for example, focusing on the rapid fattening of marketable animals in sheds (ADGM V), or introduction of new livestock breeds, along with a restraint upon the practice of traditional grassland pastoralism. In the long term, all such measures would ultimately lead to more sedentary lives for Tibetan pastoralists.

12.7 Conclusions

Due to various complex factors, grassland conditions across the eastern Tibetan Plateau—including all of Qinghai Province—are changing, and in some areas, vegetation cover is significantly degrading (Harris 2010). In several pastoral areas, where grassland degradation is relatively advanced, animal husbandry alone cannot now cover the increasing cash demand of Tibetan pastoralist households. Representatives from such pastoralist communities, for example, from Ningxiu Township in Zeku County, but also from Golok, Tianzhu and so on, went to the responsible local government officials to ask for help to secure their living standards. In such cases, local governments have only few possibilities to respond to such requests. They merely control funds granted by the Central government, most of which are already appointed for specific programmes and are not flexible enough to deal with each request by local people on an individual basis. Only if there is a programme already designed by the state at the central level to be implemented in pastoral areas can the local government offer people chances to participate in these and thus gain assistance. This is also the case with resettlement. The government offers the pastoralists participation in resettlement programmes as a realistic solution for their difficult situations, as a way out of poverty, or as a possibility to establish an additional income base if the grassland is not good enough to support their livestock. Nevertheless, even if the pastoralists do not specifically request any help, the task of local government is also to ensure enough participation by the Tibetan pastoralist communities they administer in order to fulfil the quotas or targets set by the Central and Province level governments.

The fundamental problem is that none of the current resettlement and sedentarization programmes were primarily designed to suit the pastoralists and their needs. In the first instance, the implemented policy matches developmental and political aims of the state to modernize the western part of the country. Whilst this is often phrased by the state as a discourse of 'bringing the poor west up the same standard of the affluent east', its goals are clearly to better exploit the abundant and diverse resources available in the west of the country and maximize its economic potential, whilst also further integrating the region's various (and often politically disruptive) ethnic minority populations into the mainstream of Han Chinese population. In some grassland areas, the government requires back the use rights over the land from the Tibetan pastoralists assigned for the resettlement programmes. Officially, the full government management authority over the grassland is needed to monitor the grassland conditions and assure a restoration of degraded vegetation. However, we can also find increasing number of cases, where the grassland areas controlled by government are used for mining activity (Lustgarten 2008), which certainly is in conflict with ecological preservation. The inhabitants of grassland areas, the Tibetan pastoralists, must give way to the implementation of government policy. Although they are those most profoundly affected, the Tibetan pastoralists were not consulted during any stage of the design and preparation of all current resettlement programmes. Apart from insufficient consultation during the planning and preparation phase of programmes, the most important deficit in the implementation phase is lack of assistance during the adaptation period in the new sedentary, village-style environments at resettlement sites. After the resettlement and abandonment of pastureland and livestock, the cash demand of a household increases enormously. Having difficulties to supply the family with basic alimentary products, the pastoralists experience the life in a resettlement as a greater hardship then the tough life in the grasslands. Moreover, at least in Zeku County at the time of my research, the majority of the pastoralists remained unaware that resettlement life might become a permanent state for them, and they still count on a future return to their original pastureland after a certain period of time mentioned in the resettlement contract has passed or after the government approves the grassland conditions and allows them to go back. Therefore, due to such perceptions, most resettled pastoralists in areas like Zeku County do not make much effort to establish new income bases and only try to survive in resettlement sites with the help of their own savings, their government subsidies or by trading medicinal plants. Regardless of whether they sold all their livestock and gave up their pastures before relocating or not, they still consider themselves as pastoralists or nomads (Tibetan: *'Brog-pa*).

If the government wants to ensure a successful and possibly permanent relocation of Tibetan pastoralists into new urban settlements, it must provide vocational training in new skills and abilities to these people in a sufficient way during their transformation period.

Furthermore, the lack of detailed information given to pastoralists about the resettlement and settlement conditions during the application period and the hasty implementation of these programmes both hinder the possibility of a successful adaptation. The pastoralists have neither enough information nor time given to them to consider the benefits and disadvantages of a resettlement step. In a very short period of time, they are requested to relocate and start a completely new life in a totally new environment. A successful adaptation under such conditions is very difficult, if not impossible. Many households could remain long-term dependents upon governmental subsidies, and thus resettlement communities might become a far worse social problem for the government then the current, mainly self-sufficient Tibetan pastoral communities living on their traditional grasslands.

It is also not clear whether the removal of pastoralists from the grassland environment through resettlement and leaving grasslands fallow and ungrazed by livestock for a longer period of time, as required, for example, in the *Turning Pastureland into Grassland* programme, will actually have the planned positive influence upon the recovery of grassland vegetation. The majority of households which actively participate in resettlement programmes are the poorest ones, with no or only few livestock. Therefore, concerning ecological impact and the reduction of overgrazing, removing such households alone cannot have the desired influence upon ecosystem recovery.

Moreover, although the majority of my pastoralist informants from Zeku County claim that the grassland condition in their community grazing areas has worsened in comparison with the situation prior to the 1980s, only 10% of them believe that resettlement implemented by the government might benefit the improvement of grassland vegetation. On the contrary, the same informants claim that long periods without livestock grazing on pastures would actually harm the ecosystem and modify it permanently.

It is clear that achieving ecological balance in grassland areas across the eastern Tibetan Plateau system whilst also raising the living standards of Tibetan pastoralists closer to those enjoyed by the Chinese population in the east of the country will not be easy tasks to achieve. Therefore, the Chinese state might better now slow down the implementation speed of its development policies in order to really match the complex local needs of those most affected, the Tibetan pastoralists, and reconsider exactly how much impact this implementation is contributing towards attaining their stated ecological goals.

Acknowledgements I would like to express my thanks to Hermann Kreutzmann for inviting me to participate on this issue and to Toni Huber for his encouragement and support to write this chapter. Further I would like to thank to sKalbzang Tshebrtan for his assistance during the fieldwork. I am also grateful for the chance to be a member of the RETPEC project funded by the European Commission, which enabled a study of recent strategy of development and environmental policy implementation in comparative rangeland areas of Eastern Tibet.

Note

1. Community or village (Chin: 村*cun*) refers here to the smallest administrative unit in the Chinese state administrative system. It describes an area inhabited and used by people who administratively belong to one community. The next higher administrative levels are township (Chin: 乡 *xiang*), county (Chin: 县 xian), prefecture (Chin: 洲 *zhou*), province (Chin: 省 *sheng*) or autonomous region (Chin: 自治区*zizhiqu*) and central government (Chin: 中央 *zhongyang*).

References

(i) Primary Chinese Sources: Administrative Documents Concerning Grassland Management (ADGM)

I. Chen G (ed) (2007) Sanjiangyuan ziran baohu qu shengtai baohu yi jianshe 三江源自然保护区生态保护与建设 (Ecological preservation and constructions in the Nature preservation zone of the three river sources). Qinghai peoples publishing house, Xining

II. Guanyu dui Zeku xian "sanjiangyuan" ziran baohu qu shengtai yimin xiangmu jianshe qingkuang de diaoyan 关于对泽库县三江原自然保护区生态移民项目建设情况的调研 (Investigation of the Ecological resettlement project construction situation in the Zeku County Sanjiangyuan nature reserve). 10.7.2007. Standing committee of Zeku County National people's congress.

III. Guojia ji ziran baohu qu minglü 国家级自然保护区名录 (The name records of state level natural preservation areas). 2005. www.gov.cn/test/2005–07/28/content_17872.htm

IV. Jiakuai zangqu you mumin dingju gongcheng jianshe 加快藏区游牧民定居工程 建设 (Speeding up the construction of Nomadic settlement project in Tibetan areas). 24.04.2009. Qinghai Daily. http://xz.people.com.cn/

V. Jiaqiang shengtai baohu yu jianshe cujin ren yu ziran hexie fazhan加强生态保护与建设 促进人与自然和谐发展 (Strengten ecological protection and construction works. Promote harmonious development for people and nature). 2007. Development and reform department of Tongren County

VI. Huangnan zhou sanjiangyuan shengtai yimin gongzuo jingyan yu silu黄南洲 三江源 生态移民工作经验与思路 (Experiences and ideas concerning the Sanjiangyuan Ecological resettlement work in Huangnan Prefecture). 31.7.2007. Huangnan Prefecture Sanjiangyuan office

VII. Qinghai sheng caodi gongzuo ziliao xuanbian 1995 青海省草地工作资料选编 1995 (Selected material of the grassland work in Qinghai Province in 1995), July 1996. Qinghai department of animal husbandry, Xining

VIII. Qinghai sheng zangqu you mumin dingju gongcheng 2009 nian shishi fangan 青海省藏区游牧民定居工程2009年实施方(Implementation plan for the *Nomadic settlement* project in Tibetan areas of Qinghai province in 2009). 2009. Qinghai department of agriculture and animal husbandry, Xining

IX. Zeku huanying nin泽库欢迎您 (Welcome to Zeku). 2007. Zeku people's government

X. Zeku xian sanjinagyuan shengtai yimin huibao cailiao 泽库县三江生态移民汇报材料 (Report data of the Zeku Sanjiangyuan ecological resettlement). 26.7.2007. Zeku County government

XI. Zeku xian sanjiangyuan ziran baohu qu yidi banqian banqian xiangmu shishi qingkuang huibao 泽库县三江原自然保护区易地搬迁搬迁项目实施情况会报 (Situation report about the implementation of the relocation project in the Zeku County Sanjiangyuan nature reserve). 4.9.2007. Zeku County Sanjinagyuan office

XII. Zeku xian sanjiangyuan ziran baohu qu 2003–2006 nian yidi banqian banqian xiangmu shishi qingkuang 泽库县三江原自然保护区 2003–2006年易地搬迁搬迁项目实施情况会报 (Situation report about the implementation of the relocation project in the Zeku County Sanjiangyuan nature reserve). 4.9.2009. Zeku County Sanjiangyuan office

(ii) Secondary Sources

Du F (2009) (draft version). From eco-refugees to eco-migrants: a research in Maduo County in the source area of the Yellow River

Foggin M (2005) Promoting biodiversity conservation and community development in the Sanjiangyuan region. Plateau Perspectives. www.plateauperspectives.org

Goodman D (ed) (1989) China's regional development. Routledge, London

Harris R (2010) Rangeland degradation on the Qinghai-Tibetan plateau: a review of the evidence of its magnitude and causes. J Arid Environ 74:1–12

Heath J (2005) Tibet and China in the 21st century. Saqi, London

Lustgarten A (2008) China's grosser Zug. Die Eroberung Tibets durch die höchste Eisenbahn der Welt. Scherz, Frankfurt

Paul C, Cheng L (2011) China verstehen. Sankt Ulrich, Augsburg

Ptackova J (2010) The sedentarization process in Tibetan pastoralist areas of Qinghai, China. Mongolo-Tibetica Pragensia 10:155–180

Ptackova J (2011) Sedentarisation of Tibetan pastoralists in China: implementation of the nomadic settlement project in the Tibetan Amdo area; Qinghai and Sichuan provinces. Pastor Res Policy Prac 1:4. http://www.pastoralismjournal.com/content/1/1/4

Qinghai daily (QD). 24 Apr 2009. http://xz.people.com.cn/GB/139197/139211/141080/9192660.htm

Winkler D (2010) Caterpillar Fungus (Ophiocordyceps sinensis) production and sustainability on the Tibetan plateau and in the Himalayas. Chin J Grassl 32:96–108

Yeh E (2005) Green governmentality and pastoralism in western China: 'converting pastures to grasslands'. Nomad Peoples 9:9–29

Chapter 13
'Everybody Likes Houses. Even Birds Are Coming!'

Housing Tibetan Pastoralists in Golok: Policies and Everyday Realities

Emilia Róża Sułek

Abstract This chapter discusses the trend visible amongst Tibetan pastoralists of Golok Tibetan Autonomous Prefecture in Qinghai Province, China, to invest increasing amounts of money in building houses upon their winter grazing lands. It reveals the beginnings of this phenomenon and brings data on the newest state policies aimed at encouraging the pastoralist population to construct houses. Analysis of successive waves of house construction is accompanied by a discussion of the reasons pastoralists themselves give for building new houses. The chapter analyses the roles which the houses play in the lives of their owners and discusses whether or not the pastoralists perceive the living in houses as conflicting with their self-image.

Keywords Pastoralists • Sedentarization • Resettlement • Tibet

13.1 Introduction

On my first visit to the Golok highlands during 2007, my hosts invited me: 'Come to stay with us in summer! We'll be staying in black tents, that's how the pastoralists do it!' When I returned and visited the same friends' summer camp and lived in their black tent, they said: 'Come to stay with us in winter! We'll be back in the house. It'll be warmer and more comfortable'. It was a rainy summer and tents did not seem to be the most admired form of dwelling under the constant showers and upon soil soaked with rainwater. But was that only about the bad weather?

E.R. Sułek (✉)
Central Asian Seminar, Institute of Asian and African Studies,
Humboldt University, Unter den Linden 6, D - 10099 Berlin, Germany
e-mail: emisulek@yahoo.com

H. Kreutzmann (ed.), *Pastoral practices in High Asia*, Advances in Asian Human-Environmental Research, DOI 10.1007/978-94-007-3846-1_13,

13.1.1 Aims of the Study

Visitors to the pastoral region of Golok (*mGo log*)[1] Tibetan Autonomous Prefecture (hereafter TAP) in Qinghai Province, China, can observe rows of houses sitting next to each other in the pastoralists' winter quarters. Garages for vehicles are often part of the architecture, and block walls enclose the yards. Such images immediately beg various questions for the uninformed observer. Are these houses built due to state intervention? Are all Tibetan pastoralists participating in state-run house construction projects? This study deals with the voluntary settlement by Tibetan pastoralists. It asks what functions do the houses play in the contemporary pastoral society and whether or not the pastoralists, who spend much of the year living in a house, perceive this fact as a danger to their self-image. The study argues that in spite of the Chinese state's involvement in changing the housing realities on the Tibetan plateau, Tibetan pastoralists may be and are interested in constructing houses for themselves and that growing number of houses on the high Tibetan grasslands is not only a result of the state pressure on sedentarization but a function of growing affluence of the pastoral society.

This study is based on long-term anthropological research conducted in Golok TAP between 2007 and 2010. The research focused on social and economic consequences of the emergence of the trade in caterpillar fungus (*Ophiocordyceps sinensis*) for pastoralists living in Domkhok (*sDom khog*) Township of Machen (*rMa chen*) County in Golok. Domkhok[2] remains the study's main focus, although additional information from other townships is added to gain a comparative perspective. Focused on a pastoral community inhabiting one township, this study will not be representative of the whole of pastoral Tibet. However, neither the area studied nor Golok as a whole exists in a space niche independent from others, and whilst details of administrative solutions describe this township only, broader tendencies observed are less locally specific and run across larger portions of the Tibetan plateau.

13.1.2 The Study Area of Domkhok

Domkhok is located within a short distance of the Amnye Machen (*A myes rma chen*) mountain range (the highest peak counts 6,282 m above sea level). The average altitude in Domkhok is 3,800 m, and mean annual temperatures oscillate around 0.6°C, with the strong diurnal variations and quick changes of weather characteristic of much of the Tibet highlands (MQ 2005: 90). Snowfalls are not uncommon even in summer, and on hot July days when the sun drops below the horizon, the air becomes chilly in an instant. Winters are cold and dry, summers short and humid. This humidity has additionally increased since the area became targeted by artificial rain-making programmes aimed at improving environmental conditions, albeit considered as a nuisance by local populations.

Domkhok is a pastoral area with extensive livestock production as the main source of subsistence for its residents, although this pattern has begun to change recently due to the caterpillar fungus trade. Pastoralists in Domkhok have mainly relied upon yak and sheep breeding. They keep small numbers of horses, and in the 'old society' (prior to 1958) they also kept yak-cow crossbreeds called *dzomo* (*mdzo mo*). This animal is rarely seen in Golok nowadays. Formerly, yaks played a more significant role due in part to their additional function as pack animals for seasonal movement to different grazing camps—a function now replaced by modern motorized vehicles. Sheep, on the other hand, are frequently referred to as a better 'cash provider'. This reflects attempts during the people's commune period (1960s–1980s) to strengthen the position of sheep within the local pastoral economy as a more productive unit than larger stock (Manderscheid 2002: 279). Pastoralists in Golok, however, are not 'sheep producers' in the same sense that pastoralists in the Tibet Autonomous Region (hereafter TAR) are—they do not milk their ewes, for example. Yaks have once again recently come to dominate local herd composition, especially during the last decade, when sheep populations underwent a dramatic decline (Sułek 2010a).

Domkhok is settled almost purely by Tibetans who officially constitute 99% of its residents (MSY 2009: 11–12). The pastoralists in Domkhok do not belong to the well-known 'Golok tribes' mentioned frequently in the literature on eastern Tibetan areas. They identify themselves instead as belonging to the Wranakh *tsowa* (*sBra nag tsho ba*) in a subgrouping called Metsang *dewa* (*dMe tshang sde ba*).[3] Wranakh is a name referring to a wider group of Tibetan pastoralists inhabiting an area stretching from the Amnye Machen range right up to the north of Qinghai Province (Sułek 2010b). This group is believed to form one of three main Tibetan linguistic units in northeast Tibet (Roerich 1958: 6). The Metsang pastoralists claim to have arrived in their present territories from Rebkong (*Reb gong*; Tongren County, Huangnan Prefecture, Qinghai) and became allied with the Golok tribes as their *philde* (*phyi sde*) or 'outer tribe'.[4]

13.1.3 Resettlement in Domkhok

Tibetan pastoralists in Golok have been subjected to various kinds of state resettlement programmes linked to environmental improvement policies. Some initial studies reveal the variation of such resettlement programmes (Ptackova 2010; Sodnamkyid 2008). Regarding their scope—the size of territories and numbers of people affected—officially published data can contribute some insight. Golok TAP's official website states that only one of the programmes, 'Turning Pastureland into Grassland' (Chin. *Tuimu huancao*), has been implemented in 26 of the 44 Golok townships (Chin. *xiang*) and towns (Chin. *zhen*) allowed nearly 20,920 km² of grassland to be retired from grazing (GTH 2010). Already this example reveals that resettlement projects, although large in scale, have not impacted every part of the prefecture with a similar intensity. Presently, a programme of state subsidies for house building is being implemented in Golok, and this can be seen as another step in the series of state interventions into the living conditions of Tibetan pastoralists. This programme, however, does not require or imply removal of pastoralists from their original living place, and hence cannot be counted as a resettlement initiative. Details on the house-building subsidies programme are discussed below.

13.2 Houses on the Grasslands: First Encounters

During the collectivization period (at least post-1961), pastoralists in Domkhok lived in their own tents. The state project of modernizing livestock production in the people's communes included attempts to introduce new animal breeds and winter fodder production. Constructing animal shelters was another dimension of this programme. These animal shelters, as some persons recall today, gave them their first chance to experience the benefits of having something more than a tent roof over one's head. Some informants admitted that they moved into the abandoned shelter walls in the early period of de-collectivization. Others say that they used to do this even earlier. But, as a matter of record, at least in Domkhok, the picture of pastoralists living in their black tents next to the commune-built animal shelters prevailed, and the first houses intended for *people* were built only after the communes were disbanded. It was de-collectivization which set the sedentarization process in motion.

It is generally recognized that the state took an active role in directing Tibetan pastoralists towards house construction in the years following the closing down of the communes (Clarke 1988; Manderscheid 2002). Some pastoralists in Domkhok recalled that the 'government encouraged people to build houses', but anything like coercion to move into a house they could not confirm. Many persons did not even remember anything like a milder incentive programme on the part of local administration to increase people's interest in houses back in that period. It could, of course, be the case that more subtle measures were taken to introduce the idea of building houses into the local society, as had occurred in Xueshan (Tibetan name Gangri, Gangs ri)—a township neighbouring Domkhok. There, as a local leader reported, the process started with several 'pioneers' who paved the way for the houses. Later on, the less enthusiastic members of the community joined too:

> There were families who refused to build a house, but some families started their own ones, and after having the experience of how much better it is to stay in a house, other families started too. So it began with two to three families, then people learned that houses are warmer, and others joined in. For example, with solar panels, it was the same. The government said that it would be good to buy them, but no family wanted to invest. People said: 'We only want butter lamps and candles'. So the party members were forced to buy them first. Later, people said: 'That's really good', and all of them bought [panels].

Whilst Xueshan Township pastoralists started building their houses in the beginning of 1990s (as the township leader recalled), the same process in Domkhok commenced already during the 1980s. However, an important difference between two townships was state involvement in financing the house construction. Xueshan informants called the house building a 'government programme' in which township residents and local institutional bodies were said to have shared its costs as follows: 20% was paid from the state budget, 30% by the Herders Association or *Drok u khang* (*'brog u khang*),[5] and the remaining part was contributed by each household involved. If a household did not have enough of its own funds, it could easily get a state loan for house building—the informant stressed.[6]

In Domkhok, the situation looked different. The majority of informants whose first houses were built in the 1980s reported having financed them with their own funds. Persons who worked during that period in the township administration all confirmed that building houses and fencing pastures in Domkhok was paid from the pastoralists' own pockets.[7] Some possibilities for receiving financial aid for house construction also existed in Domkhok, but none of the informants recalled anybody 'poor enough' to be given such aid. Only one person (out of fifty reported cases) confessed that his family's first house (built in 1987) was co-sponsored by the state. His costs for the building materials and hiring workers totalled 6,000 yuan, of which 20% was paid from the 'government' budget—the informant declared.

Receiving financial aid was not something my informants easily admitted to during research interviews, and some caution might be needed to interpret their reports. Receipt of money from the state—in the form of state aid or as retirement funds which some ex-officials received—was repeatedly left undeclared or was revealed only after initial hesitation. Given this reluctance to disclose the state input into the family budgets, it may well be that state financial input into house construction could have been more significant than reported. But the informants' denials of the state's agency during the early phase of house building might also be an indication that they not only accepted this development but took it over as their own project and—by now—probably identified with it.

13.3 The Present Situation in Domkhok

During my field research, almost every household in Domkhok already had several houses located alongside each other at the family's winter quarters. Apart from newly married couples, no household in Domkhok was limited to owning *just one* house, with two, three or sometimes even four houses being the rule. Households with already established houses during the 1980s and early 1990s are essential for understanding of the dynamics of the house-building phenomenon: these households were *there* when the house building started and witnessed subsequent developments in the housing situation.

Houses in Domkhok were constructed in several construction periods or waves. The first construction period started in the mid-1980s (the first houses so far documented in my research were built in 1986) and lasted until the early 1990. During the mid-1990s, the second construction period began and lasted until the beginning of 2000s. The most recent construction phase commenced in 2009 and continues

Photo 13.1 View of a pastoral household's living quarters, Domkhok Township, Machen County, Golok TAP (Photograph © Emilia Sułek August 2009)

until the time of writing. All households studied built their houses in the following sequence: each had its first house built between 1986 and 1991, then a second house built between 1994 and 2001, and post-2009 they have each built or were building a third new house. All these households have in their domestic compounds all types of buildings built between the 1980s and the present day. Pastoralist families do not raze their old houses when constructing new ones on the same spot, they merely add it alongside. Lack of space for construction works is not an issue for inhabitants of the expansive Tibetan grasslands, neither is there any concept of reusing building materials from a previous house to build a new one.[8] That makes the pastoralists' settlements appear like colonies of houses springing up next to each other, sometimes in a row, where the newer houses 'mushroom' off the older ones (see Photo 13.1).

The houses built in these three construction waves differ in certain respects. The paramount difference can be observed in the construction materials themselves. The walls have changed from being built of earth in the earliest houses to concrete blocks during the 1990s and early 2000s and now to bricks during the most recent construction period. The size of houses also changed—whilst older houses were only approximately 50 m^2, newer ones average around 80 m^2. Another change is window size and number, which in early houses were small and few, whilst in later ones the larger window panels can cover a large part of the entire front wall. Gabled roofs have taken the place of older shed or flat roofs, and ceramic tiles have replaced corrugated metal and pounded earth roofs—in the oldest houses, the roofs rested on wooden poles and were isolated with layers of dry bushes, straw, rubber and plastic. Rammed earthen floors have been exchanged for cement floors covered with PVC. Finally, newer houses have been aesthetically transformed due to their new façade decoration. Whilst the rammed earth walls of the earliest houses were left in their 'pure' state, later walls received a coating of ceramic tiles, and Buddhist symbolic motifs on them give the house a specific 'Tibetan' character.

What has not changed significantly is the shape of houses nor the organization of the space inside. The pastoralists' houses in Golok remain rectangular in their ground plan and are never higher than a single storey. Organization of the house's inner space has remained determined by the fact that the windows face only one direction, whilst the back of the house sits against a hillside or mountain slope. In these respects, pastoralists' houses replicate the manner in which their tents are erected, with their entrance facing the bottom of the valley, often to the south, and their rear walls facing the rising slopes behind them. Thus, such houses cannot be expanded in their depth, and all rooms have to be arranged in one long row. The actual size of the fully habitable space in the house is not calculated by the number of rooms, but by the number of stoves a house contains. As houses have become larger, additional stoves have been added into the new rooms.

How the houses relate to the household's biography two examples show. Mr. Churtod, from the first household, is 44 years old (born 1967). He lives together with his wife Yibo (same age) and their two elder children aged between 20 and 22 (b. 1989 and 1990). The youngest daughter (b. 1992) studies (since 2001) in a boarding school in Dawu. Churtod's mother (b. 1947) lives together year-round with his family, whilst his father (b. 1944) moved (in 2004) to town to look after the youngest daughter. Before moving into their first house, Churtod and Yibo lived in a black tent in the same spot where their house stands today. Their first earth house was built in 1991 by workers from Rebkong. Churtod recalled that building it involved an expense of some 6,000 yuan, of which all was paid by his parents. Their second concrete house was constructed in 2001 for approximately 60,000 yuan. Their newest brick house was built in 2009, and the family spent 110,000 yuan for its construction (of which 39,000 yuan was reimbursed—more on that below). In terms of living spaces which the three houses offered, the oldest one had just one room and one stove, the second and third each have three rooms and two stoves. Churtod owns a small house located on the family's autumn grazing land, too. The proportion between the number of household members in Churtod's family and the houses owned is shown in Fig. 13.1.

That the number of houses owned does not have to correspond with the growing number of people inhabiting them is revealed in the second example. Mr. Dudu is 44 years old (born 1967) and his wife Tseko is 43 (born 1968). They have three children aged between 12 and 20 (born 1999, 1995 and 1991). Their first house, which Dudu and Tseko moved to, was built in 1987. It was paid for by Tseko's parents and constructed by Hui builders from Hualong Hui Autonomous County (Haidong Prefecture, Qinghai). The second house, of concrete blocks, was built in 1998. Dudu and Tseko recall that it had cost approximately 50,000 yuan and was built by Han workers from Huangzhong County (under jurisdiction of Xining City, Qinghai). Their most recent house was built in 2009 by Tibetans from Xunhua Salar

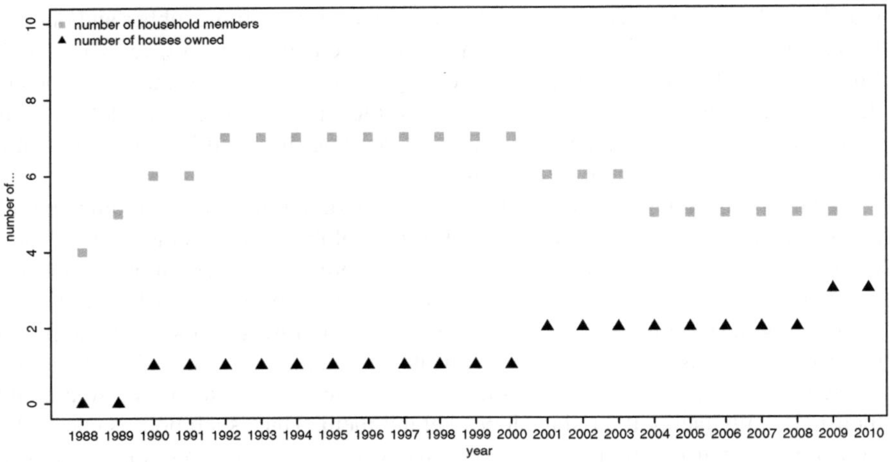

Fig. 13.1 Houses and the development of the household, Mr. Churtod and Mrs. Yibo, Domkhok Township, Machen County, Golok TAP, 2010 (Source: Fieldwork by the author)

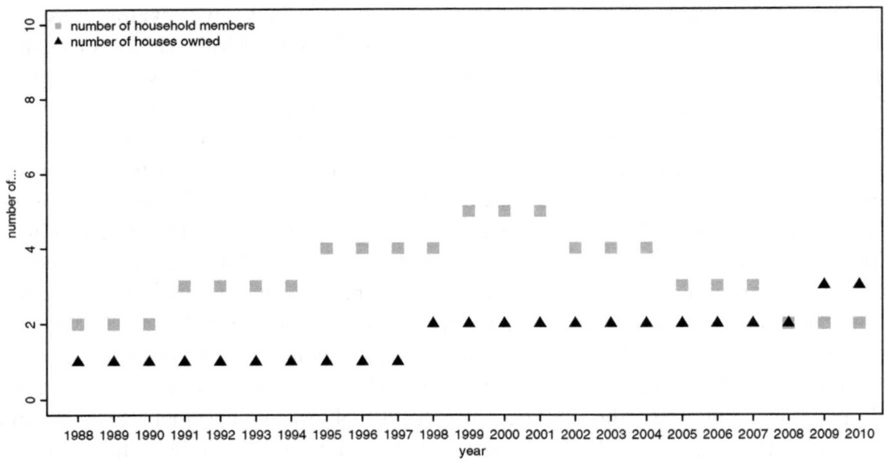

Fig. 13.2 Houses and the development of the household, Mr. Dudu and Mrs. Tseko, Domkhok Township, Machen County, Golok TAP, 2010 (Source: Fieldwork by the author)

Autonomous County (Haidong Prefecture, Qinghai) at an expense of 110,000 yuan (of which 40,000 yuan was reimbursed by a state programme). Dudu thought about buying a house in Dawu town, but this required a further 140,000–150,000 yuan investment. In Dudu's household, all children attend (since 2002, 2005 and 2008) schools in town. Thus, the three houses Dudu has serve only two persons at the moment. The changes in Dudu and Tseko's household size and houses owned are shown in Fig. 13.2.

Of all the houses owned by the households in these examples, only those raised in 2009 and after were said to have been subsidized by the state. This was done within the framework of a programme called *You mumin dingju* or Nomadic Settlement.

13.3.1 Nomadic Settlement Programme

My research coincided with the first implementation of the Nomadic Settlement programme in Tibetan populated areas of Qinghai. This programme implies that pastoral households could receive a subsidy to build a new house.

Formally speaking, the Nomadic Settlement programme is aimed at households that either do not have a permanent house or who have a house which is 'unsteady' and 'in a danger of collapse'—originally defined as a house made of earth and wood or one that 'has not been repaired for a long time' (Ptackova 2011: 4). If this aspect of the programme was followed as a guiding principle, then in Domkhok there could only ever be very limited implementation possible. 'Unsteady' houses exist in Domkhok, too, but they are usually only one amongst several different houses which a family owns. Moreover, it is impossible to find anyone in Domkhok who lacks a house. Whilst in present-day Golok newly married couples are the only ones who might live all year round in a tent for the first 2 or 3 years of establishing a household, such cases were never encountered in Domkhok. Thus, in the Domkhok study area, the housing transitions supported by Nomadic Settlement actually took place between existing houses and newer ones. What then is the qualitative difference between these newer houses subsidized by the state and those which the pastoralists built by and for themselves in the years preceding the Nomadic Settlement programme?

In order to receive subsidies, any planned new dwellings must fit the criteria of the Nomadic Settlement programme. First, a subsidized house has to be built of bricks and never of concrete block (it can theoretically be built of stones, too, but this is not practised). Following the 2010 earthquake in Jyekundo (*sKye dgu mdo*), Yushu (*Yul shul*) Tibetan Autonomous Prefecture Qinghai, concrete has been labelled as a potentially dangerous construction material. Although the Nomadic Settlement programme started in Golok 1 year earlier, statements about the need to increase building security in case of seismic movements were used by my informants in 2010 to justify why the programme is being implemented. Moreover, a new house needs to be at least 80 m^2—whilst larger houses are accepted.[9] New houses must thus offer a certain size of comfortable living space for the inhabitants. In addition, the house itself should be enclosed within a high concrete wall, with an iron gate leading to the yard. This is a novelty since older houses merely sat on the family's land and the land itself was fenced, but the house usually was not. If houses were fenced—because a family lived close to the road or had a preference for extra security—the fence was nothing more than a *cagra* (*lcags ra*) or metal wire used to fence pastures.[10] Another novelty is the toilet: after 2009, characteristic toilet buildings sprang from the ground at some distance from new houses. Finally, animal sheds (roofed buildings and not just corrals) are mentioned in the programme outlines (Ptackova 2011: 4) and are indeed appearing in Domkhok, but these are conceived of as a separate enterprise and are not built together with the house.

To build a brick house of the minimum size required by the programme is not cheap. The cost of such a house—based upon the investment of persons who had built or were building them in 2009–2010—ranged from 110,000 to 150,000 yuan. The household initially needs to raise this sum by itself and pay for both the materials and labour from its own budget. Only when the house is complete and passes a quality control check will a percentage of the total cost be reimbursed. In Domkhok, this reimbursement has been paid up to 40,000 yuan, of which 36,000 yuan comes from the Qinghai Provincial budget, with the remainder covered by prefectural funds.[11] Whether or not the house follows the building guidelines and is eligible for a subsidy is decided during inspection tours by a group of officials from both the Grassland and the Financial Bureaus, plus members of other administrative bodies at the prefecture and township level. Each October, when construction works stop before the winter sets on, the 14-member team tours the township to evaluate the quality of the newly erected buildings.

In 2010, a township official showed me a series of photographs on his laptop taken during such an inspection tour from 2009—a comprehensive series of images picturing houses, some of them still empty and some filled with the first pieces of furniture. It appeared that the programme had been overwhelmingly embraced by the local population. My visits to Domkhok in 2007–2008 before the programme was commenced and after that also suggested wide participation in the programme.[12] During the first year of implementation, 40 families in Domkhok built new houses. A year later, 125 families were in the process of house construction. In 2011, 61 households were expected to start new construction work on their land. By 2012, as a township official informed me, every family in Domkhok was expected to have a new house: 'That's a government plan', he said.

If every household receives 40,000 yuan of subsidy, it is easy to calculate how much—in only one township—the programme will cost the state. Summing up all households that already took part in the programme, or are taking part in 2011, a total of 9,040,000 yuan is reached. This money does not reach the hands of the rural residents before the house is accepted as qualifying. Thus, it is still the local people who cover the major part of the programme's costs. How do they manage to raise such money?

Reports from other Tibetan areas where this and other housing programmes have been implemented revealed that households took out bank loans to pay for house construction—loans which not everybody was likely to pay back in due time (Yeh 2011: 312). The state's ambition to provide Tibetan populations with new houses was thus viewed as damaging fragile household budgets. In Domkhok—according to the information from one of the banks in Dawu—local residents apply for loans for house construction, although not on a large scale. According to the bank employee, the pastoralists tend to use banks to store their savings rather than to take out loans. The loans, if taken, oscillate between 50,000 and 60,000 yuan at 10–11% interest rates and are taken out over a 6-month to 2-year period. Loans for house construction constitute around a half of all loans issued—the remainder are issued for the purpose of buying cars and fencing the pasturelands.

The second source of income used to finance house construction is trade in caterpillar fungus. Money from gathering the fungus and from leasing pasturelands to other persons for gathering has become the basis of the pastoral household economy in present-day Golok (Sułek 2009). Caterpillar fungus income allows smooth and unbroken functioning of household economies but also allows for considerable savings, which can later be used as capital for large investments. In 2010, half a kilogram of dry fungus sold for between 30,000 and 70,000 yuan, and one household easily managed to collect more than this amount per annum. Fees which the gatherers paid for using the pastureland started (in 2010) at 5,000 and reached up to 20,000 yuan per person per season—they are decided according to current fungus prices and quality of the grassland leased. A single household can lease the land to at least several gatherers. In extreme cases, these numbers may reach into the hundreds. Knowing the scale of caterpillar fungus-related income, estimations of the savings one household can make after 4–5 weeks of the gathering season do not surprise. My informants, when asked about their savings from one season, declared 30,000–50,000 yuan for their own households and up to 500,000–600,000 yuan for other households they know. Whilst the latter sum might be overstated, the former sum seems too modest. In any case, the caterpillar fungus money was singled out by both the pastoralists and the township officials as the main financial 'fuel' for house-building investments. It fuels the house building in two ways: first, by paying for building materials and construction workers, and second, by repaying bank loans taken out for house building. The relation between caterpillar fungus income and investment made in house building is visible in the timing of the pastoralists' repayments of their loans. As the bank employee reported, about 80% of Domkhok pastoralists repay loans in July and August, that is, soon after the fungus-gathering season is over.

13.3.2 If Not Nomadic Settlement Then What?

The case study data presented above demonstrates that Domkhok residents readily embraced housing when they had the opportunity to do so. Yet, commonsense or entrenched understandings of what 'pastoralism' means might cause outside observers to view houses and pastoralists as somehow antithetical. This issue refers directly to the Tibetan concept of *drokpa* (*'brog pa*)—a term which pastoralists on the Tibetan plateau use to describe themselves.

Drokpa refers to inhabitants of highland grassland zones beyond the upper reaches of farming areas and where herding and animal husbandry are of necessity the only sustainable and primary means of subsistence. Being a *drokpa*, in the Domkhok informants' understanding, indicates not only sharing a place and a way of life (in terms of economy), but entails participating in a certain 'pastoral culture' or *droksul* (*'brog srol*) and speaking the 'pastoral language' or *drokked* (*'brog skad*). What is not an immediate constituent of this category is mobility and—connected with it—dwelling style (Gruschke 2008: 14). None of the informants indicated tents (at least as a year-round basis for dwelling) as an important criterion for being a *drokpa*. Had there ever been a latent assertion that life in tents is an essential part of the cultural 'luggage' the term *drokpa* carries, this assertion must have been reformed when houses and not tents belong to the pastoralists' everyday experience. Houses, on the other hand, did not have to be totally alien to Tibetan *drokpas*. This was pointed out by Cynthia Beall and Melvyn Goldstein, who observed that for their Western Tibetan informants 'living in a house instead of a tent was a matter of comfort, not basic identity', and that already in the 'old society' having a house was perceived by pastoralists as a status symbol (1990: 64).

When asked how they view their own tent dwelling of the past, Golok informants showed no willingness to think, even hypothetically, about staying in a tent for the entire year. Many called houses an improvement in their lives, and one that could even be extended, if possible, to life in their distant and higher summer pastures as well. In answer to the question 'Do you enjoy the tent?', a man of 60+ years replied, 'No. In the house it is better. But I can't build a house here [in the summer pasture]. In the house it is warmer and everybody has his/her own bed'. A man in his 40s responded, 'We have to live here in tents, because the roads are so bad. If we always stayed here in summer we could also build a house here'. A woman in her 40s replied, 'Everybody likes houses. Even birds are coming!' and pointed at a sparrow hopping through her room, and the comment of a 32-year-old man was simply 'I don't miss anything about the tent when I'm in the house'.

Not all pastoralists who were interviewed were unanimous in their view of houses as neutral for the *drokpa* identity. But *all* of them nevertheless live in houses. Even if harbouring nostalgic sentiments about their parents' or grandparents' tent-dwelling past, my informants saw houses as a cherished opportunity offered by life in the present times and a function of increased security and levels of affluence.[13] It can be questioned how is this view influenced by the state promoted vision of modernity and of the Tibetan society in the past as simple reverse of what the Chinese state strives for in terms of material development. There is however little doubt that the battle between sentiment and practicality has been won by practicality.

A question may arise as to what would be happening in Domkhok in the housing situation had not the programme of Nomadic Settlement been announced. Pastoralists interviewed believed that—whether the state provided money for house building or not—people would build new houses anyway. It was simply because 'the pastoralists were crazy about houses', as many persons stated. The programme of house construction subsidies offered what informants perceived as 'free money' to fulfil aims they would normally desire. Next to gold, local pastoralists considered houses the best way of investing one's savings. When asked what steps they were planning in their lives for the near future, informants indicated 'building' or 'repairing' the house as one of the first responses.

13.3.3 Houses as Good to Have. Not Necessarily to Live in

A house is a material object of a particular kind. Being exposed, on a lasting basis, to the general public gaze, the house constitutes a form of property which 'expresses or betrays, in a more decisive way than many other goods, the social being of its owners, [or] the extent of their "means"' (Bourdieu 2005: 19). It is the owners' business card, a testimony of their pecuniary repute (Veblen 1987[1899]: 54) But not only that: the house reveals its owners' 'taste, the classification system they deploy in their acts of appropriation and which, in assuming objective form in visible goods, provides a purchase for the symbolic appropriations of others, who are thereby enabled to situate the owners in social space by situating them within the space of tastes' (Bourdieu 2005: 19). The house counts not only within the space of material or economic differences but within the space of tastes and cultural loyalties and can inform observers of the owners' advancement on the path towards 'being modern'. By having a house of this or that kind, one not only satisfies his/her immediate survival needs and even more emotional needs but also communicates to the neighbours: we are like this, we can afford that, we are advanced to such-and-such a degree in terms of adhering to the trends of the day.

Tibetan houses in the pastoral regions, specifically those located in the grazing areas rather than township villages or county towns, lack certain traits which a house as an object on the housing market has. Pastoral houses are not objects of buying, selling or renting, for example. When pastoralists move to town—voluntarily and not within any resettlement programme—they keep their grasslands and houses and hire workers from lower altitude areas who manage their livestock at the family's outpost on the grazing lands. Thus, pastoral houses do not have the same trajectory which other enduring commodities can have: they are not subject of trade transactions.[14] Transferred between generations, they remain within their owners' family. If 'commodity status' occurred episodically in an object's life history, then Tibetan pastoral houses would be commodities in their dormant phase; one day they may be commodified and their value converted into cash upon sale, but so far their economic biography (Kopytoff 2003: 68)—of sales, resales, growing and falling values—is only rather brief.

As already observed, pastoralists do not dismantle their old house to replace them with new ones. On the contrary, old houses are gradually renovated, with new roofs or tile-coating façades being added. The fact that all houses—whether 20 years old or new—sit together on a single living space strengthens the impression of pastoralists' now having collections of houses. Not all of these houses are in actual use. It was common in the households I visited that they lived in their older houses, whilst the new houses were kept for special occasions. Nicely furnished and richly equipped, they were opened up and heated only when any special guests were expected. But on a normal day the door might stay closed, and the owners might follow the old habits of staying where they used to stay before.

This specialness of the newest house—which one has, but does not really use—is apparent from the pastoralists' own statements about use patterns. Mrs. Tsedron (she has three houses: the first one was built '20–30 years ago', the second in 2003 and the third in 2010) admitted spending most of her time in the first and second house, and about the newest one she exclaimed: 'We don't even go inside!' Mr. Tsela (he has four houses: the first one built in 1985, the second from the mid-1990s and two newer ones: from 2007 and 2010) declared spending most of his time in the third house. The most recent, fourth house, he used only for the New Year and similar important religious and community events. Mr. Dudu (his household was discussed above) admitted that in the oldest house his family used to 'keep things', in the second one they cooked and watched television and in the newest one they sometimes slept (cooking, watching television and sleeping were used in the survey as three activities which define use of domestic space by the informants). Although the house use patterns cannot be too easily subsumed under one heading, the impression remains that some of the houses the pastoralists own are redundant from the point of view of their everyday needs.

Houses, in a certain sense, resemble the position of livestock in the pastoralists' lives. Tibetan pastoralists were noted as being reluctant to specialize in commercialized meat production (Levine 1999). This lack of enthusiasm for producing animals for the market has its reasons not in 'irrational and non-economic love of their livestock' but in values ascribed to livestock: more than a mere means of subsistence, livestock is 'a form of wealth, social capital and source of prestige or esteem connected with specific cultural values' (Khazanov 1998: 14) and functions as an 'investment commodity' (Salzman and Galaty 1990: 26). Houses now hold a similar position: they provide their owners with much more than the space under their roofs. But it is not only the house itself, or the number of houses which are owned as a token of one's pecuniary status, but also what is contained within the house.

13.3.4 House as Container

The housing market in Golok is probably as alive as never before, but it is not particularly diversified in terms of the possibilities it offers. This is especially visible in the example of houses built within the Nomadic Settlement programme: differing in minor details, these houses have to adhere to the prescribed blueprint. If houses are to fulfil the function of demonstrating one's financial standing, the house's outer shell cannot serve this goal due its uniformity. This goal has to be accomplished by furnishing the house's inside, and so the family's standing is shown by what the house-shell contains.

The function of the house as a 'container' is highlighted by the pastoralists' own reflections on the costs of building and of furnishing the house. Providing furniture and all equipment considered indispensable for a comfortable living turns out to be another financial challenge. 'Compared to a house, what is inside is sometimes even more expensive', one pastoralist remarked. Other explanations made this point even more explicit: the pastoralists need houses to 'put things inside', as another informant stated. What then do the pastoralists keep in their homes?

The newer, brick or concrete houses are divided in two rooms. The bigger room is the more representative, arranged with piety and without regard to cost. The smaller room provides a more informal space. In the main room, there is everything one could wish to have at home: a stove, elegant furniture sets and consumer electronic equipment. Along the front and side wall, low wooden seats or sofas run covered with Tibetan-style carpets. A row of long tables stands in front of them.

Photo 13.2 Inside the house. Note the metal safe and double bed. Domkhok Township, Machen County, Golok TAP (Photograph © Emilia Sułek Tibetan New Year, February 2008)

The tables and seats are all made in 'Tibetan style', same as is the main piece of furniture on the back wall: a cabinet, with shelves and drawers and with an extra shelf to hold a television set (carpenters in Dawu claimed in 2010 that 'anything without a television shelf did not sell these days'). The cabinets, measuring up to 7 m in length and over 2 m in height, are all of carved wood painted with dragons, wind horses, other creatures of Tibetan and Chinese imagery and Buddhist religious symbols (Photo 13.2). A metal safe sits in the corner of the room securing valuables, documents and cash. Next to the door, a 'domestic corner' is arranged with cupboards for cooking pots and other kitchen utensils. Finally, a refrigerator and other home appliances are there, too—if the owners do not keep them in another house.

How much money does a household need in order to provide the house with this much in the way of chattels and furnishings? And what do pastoralists regard as necessary to have in the home? Table 13.1 shows a selection of important elements in every house's inventory, including only goods placed in the largest room.

This table does not provide a complete list of all that is necessary to furnish the house. This list is also changing and gets regularly upgraded; the 'emulation effect' requires that levels of consumption change together with the perceived standards of living well (Olson 1998: 191). But even this short list from one room gives a good indication of how strong momentum for consumption is in relation to house building. If the house is a container to fill, then those who supply the necessary 'filling' must enjoy a rise in their turnover. Additionally, since older houses continue to be used, there is little or no possible recycling or second use of chattels and furnishings; every new house constructed entails the purchase of further new chattels and furnishings.

Table 13.1 Selection of home chattels and furnishings. Domkhok Township, Machen County, Golok TAP

Item	Price in yuan (and size)
Tibetan-style cabinet	5,000 (2.3 m × 6 m) to 9,500 (2.3 m × 7 m)
Tibetan-style table (minimum three necessary)	500 (60 cm × 1.5 m)
Tibetan-style bench (minimum three necessary)	500 (2 m × 80 cm) to 2,000 (4 m × 1 m)
Metal stove	750–900 (60 × 80 cm)
Television set	2,000–8,990
VCD player	300
Total	11,050–27,190

Source: Field enquiries the author in 2010

Tibetan pastoral areas in China are nowadays an expanding market for furniture. This market is increasingly competitive, and shop owners interviewed during my research complained that compared to their earnings from a decade ago, now they are able to sell more products but earn less per item sold. The pastoralists used to buy less in the past, but also the supply was much smaller; 10 years ago, two shops in Dawu sold furniture, and today there are almost 20. The 'traffic' in the shops speeds up in summer or autumn, when new house builders complete their work. At the busiest point in the year—one trader reported—'some people come and buy just everything [they need at once] for the house, for nine or eleven thousand yuan at a time, so sometimes our entire furniture stock is sold out in one day'. Although the pastoralists 'sometimes don't know how to put the furniture together' (as another trader put it), they are ready to 'spend more money than officials do'. Whilst previously the pastoralists 'didn't care about fashion, just bought the cheapest things', today some customers even skip the prefecture markets and go shopping to centres such as Ngawa (*rNga ba*; town in Ngaba Tibetan and Qiang Autonomous Prefecture in Sichuan Province) or Xining City where the quality is said to be better, the prices lower and the choice broader.

The market both fuels fashion and satisfies tastes. From what local traders report, it seems that the main impetus for changes in furniture trends amongst pastoralist buyers comes from beyond Golok. The pastoralists copycat their neighbours, the traders said. The fashion for Tibetan-style furniture in Golok has, for example, only a short history. Whilst in the rest of Qinghai Province (as carpenters in Dawu reported) the market for Tibetan furniture opened some 20 years ago, in Golok, this trend for buying 'things Tibetan' developed only within the past decade. Beforehand, pastoralists purchased 'Chinese' furniture[15]; the fact that older houses are not demolished provides a good opportunity to observe the former trends and styles of pastoralist furnishing endeavours.

The majority of households covered in this study already had electricity connected to their houses. Power generators are a solution for those who still await electrification or during power cuts; during summer of 2009, when the power cuts lasted for weeks and towns sunk into darkness, in the pastoralists' houses people still watched television. Television sets feature in every house, independently of the quality of its access to electricity. Although more conservative or poorer customers stick to 'traditional' old-fashioned models, large television sets with flat 32-in. screens sell too. For sales in just one appliance shop within one summer month, with an estimated 80–90% of customers being pastoralists, a shop owner reported sales of 30–40 television sets, 70–80 VCD players, 40–50 refrigerators and around 50 washing machines.

The need to furnish and equip new pastoralists' houses has created opportunities of the expansion of markets into areas of rural Tibet where relatively few consumer goods used to be sold in the past. Recently, pastoralists in Golok have acquired unprecedented economic power due to caterpillar fungus revenues. Some of this income is saved, and some spent—on consumer goods answering those more immediate and those more status-serving needs. Houses, in this context, can be viewed in two ways. First, they are a drain on the family budget: having a house to furnish makes pastoralists actively engage in the consumer culture. By directing the pastoralists onto the path of house building, the state creates large numbers of new consumers ready to participate in markets for goods that they previously could not buy, mainly due to a trivial lack of domestic storage space. Second, houses offer storage for growing numbers of family belongings. By so doing, they create a new opportunity to manifest a household's material standing—a house with all that it stores becomes like a trophy. When a society becomes more affluent, the need for new status manifestations becomes more imperative, and here houses now have an important role to play.

13.4 Final Considerations

Houses are a relatively recent, imported element into Tibetan pastoral regions such as Golok, but the last 20 years of developments reveal how rapid and evolving their local 'career' has been. The great success of the house in pastoral regions has been enabled by several factors. The Chinese state's pro-house and pro-permanent settlement policies are only one of them, albeit certainly a powerful one. A second factor which must not be overlooked is what can be referred to as the tangible realities of life on the Tibetan grasslands. The average altitudes and temperatures in Domkhok certainly also condition local debates and choices about tents versus houses as suitable domestic dwellings. The third factor is the rising affluence of the pastoral society, its intense relationship with the commodity market and a need for new means of manifesting one's economic standing.

Houses are 'new arrivals' into Golok's cultural space, but for a novelty like this the lack of academic interest can surprise. One reason why pastoral houses are academically neglected may be that, by their linking to the resettlement programmes, the houses are evilized as a topic. They are associated, by some outside observers, with the state exercised pressure, with the politics of de-culturation and cultural uprooting. The sedentarization policies are even called an 'ethnocide' or attack on the existence of an ethnographic group such as Tibetan pastoralists: the houses become a symbol of a dead-end road the pastoralists are believed to go (Tenzin 2010). This maybe does not influence the scholars directly, but blinds them for certain themes which seem less appropriate than others. In this sense, the array of topics the scholars choose is in itself a topic worth investigation and an indicator of intellectual and political trends prevailing in the academic milieus.

Proper understanding of Chinese state projects which aim at or result in increased sedentarization in pastoral regions of Tibet is undeniably important. However, underscoring only the state's interventions into pastoral contexts can be distorting. It risks underestimating the agency of local actors and can generate a perception of pastoralists as mere recipients of state policies. Although certainly affected by them, the pastoralists are more than that. Concentrating solely on the state actions also carries a danger of taking from the people away the responsibility for the moves they make—an effect which can serve both the informants' and the scholars' agenda. On a more academic level, this bias makes one miss a chance to learn how innovations in pastoral contexts like this spread and how do local populations envision and proceed with changing their own society.

Domkhok Township is suitable to show the inhabitants' agency in shaping their new life because the state's interference into their living practices has been limited, compared to areas covered with the resettlement programmes. In these latter cases, studying the pastoralists' attitudes towards houses can be more difficult, as state policies, pastoralists' life preferences, expectations and sometimes brute realities form a body difficult to disentangle. The resettling of the highland populations of ecologically endangered zones has not always been carried out on a free-choice basis, and its consequences were certainly unclear to many. The resettled pastoralists, however, do not exhaust the pastoral population of the Tibetan plateau. Those who are not affected by the resettlement deserve attention, too, and their relationship with their houses, understandings and choices, can—in a bigger perspective—inform about how pastoralists would wish to shape their lives and neighbourhoods in the future.

Acknowledgements I owe my thanks to Hermann Kreutzmann and Toni Huber for their encouragement to contribute to this volume and generous help with giving this paper its final look. I also thank Sodnamkyid and Aku Suoba for their help with gathering material, Emily Yeh for sharing her book manuscript, Melvyn Goldstein, Andreas Gruschke and Daniel Winkler for keeping an academic 'hotline' with me throughout my work, my colleagues Diana Altner and Jarmila Ptackova for sharing their field experiences, Gabriel Lafitte for his positive feedback, Huadan Zhaxi for his linguistic consultations, Roman Frąckowski for finding even the most inaccessible literature in the United States libraries, Jens Albrecht and Mariusz Grzęda for help with the illustrations, and my students at Humboldt University in Berlin for discussions which shaped the structure of this chapter. As always, I am very indebted to the Rosa und Antoni Stiftung—without its support I would not be able to accomplish my work.

Notes

1. Tibetan words are given in approximate phonetic rendering, reflecting the dialect of the study area, and followed by their transliteration according to the Wylie system. Chinese words are preceded by 'Chin'.

2. In 2008, Domkhok had 1812 residents dispersed over nearly 780 km^2 of rugged terrain; MSY (2009: 11).
3. *Tsowa* refers to groups inhabiting large territories and claiming origins from a distant common ancestor. *Tsowa* should, theoretically speaking, split into a number of *dewa*, due to historical and demographic developments. Both terms are translated as 'tribe'. For more, see Levine (n.d.).
4. For more information about 'outer' and 'inner' tribes, see Gelek (1998: 51).
5. The Herders Association (Chin. *mumin weiyuanhui*) is an administrative body in the township. It was created from the former production brigades and corresponds to 'village' (Chin. *cun*), albeit existing in the pastoral context. At the time which is being discussed here, the Herders Association drew its budget from one main source: caterpillar fungus-gathering fees, which collectors coming to the township had to pay.
6. For more information about Xueshan in the mid 1990s, cf. Goldstein (1996).
7. *Dadui* and *xiaodui* are Chinese translations for Tibetan *richen* (*ru chen*) and *richung* (*ru chung*), being administrative divisions under the township level. The terms translated as 'production brigade' and 'production team' come from the people's commune period and are still used by local inhabitants.
8. This is different from what Diana Altner observed in the TAR, where villagers at Yamdrok Lake (*Yar 'brog mtsho*) razed their old houses to build new ones on the same sites (Altner 2009).
9. The same was reported from Yushu TAP (Dzartod County/*rDza stod*); Daniel Winkler, email communication, 30 May 2011.
10. Fencing can certainly be seen as an intervention of a political nature, although my informants expressed enthusiasm about it. For a discussion of different opinions about fencing, cf. Bauer 2005. One may wonder if fencing houses with concrete walls will stir similar debates and how the new walls might impact the shape of community life. Since they are only presently appearing in Golok, one has a unique chance to compare attitudes towards them today with those in several years time.
11. This is more that what is paid in similar programmes in the TAR, where values between 15,000 yuan and a little over 24,000 yuan were noted (Yeh 2011: 305; Goldstein et al. 2010: 67). Pastoralists in Kakhok (*rKa khog*) County, Sichuan Province, received grants of 20,000 yuan (Ptackova 2011: 7). In Cigdril (*gCig sgril*) County, Golok TAP, these grants amounted to only 6,000 yuan in the beginning of the programme's implementation and were raised subsequently.
12. It was reported from other parts of Golok and from the TAR that more households wanted to receive subsidies than the existing funds allowed (Goldstein et al. 2010: 63). The programme could thus be limited to newly wed couples, or the households were chosen by drawing lots.
13. This connection between higher security levels and voluntary sedentarization has been made for the Bedouin of Israel; Medzini (1998: 60).
14. In contrast to houses sitting in the pastoralists' winter quarters, those in county towns, townships, or resettlement villages are subject to complicated trade exchanges.
15. 'Chinese (*rgya*) furniture', similar to 'Chinese clothes', refers to mass-produced goods of rather 'Western' style and in fact lacking any 'Chineseness' which one might anticipate from a product carrying such a name. Any modern clothing or furniture, lacking Tibetan features, is thus 'Chinese' in the local context.

References

Altner D (2009) Die Verkleinerung der Yakhautboote. Fischerkulturen in Zentral- und Südtibet im sozioökonomischen Wandel des modernen China. Harrassowitz, Wiesbaden

Bauer K (2005) Development and the enclosure movement in pastoral Tibet since the 1980s. Nomadic Peoples 9(1–2):53–81

Bourdieu P (2005) The social structures of the economy. Polity, Cambridge
Clarke GE (1988) China's reforms of Tibet, and their effects on Pastoralism, Kailash. J Himalayan Stud 14(1–2):63–132
Gelek (1998) The Washu Serthar. A nomadic community of eastern Tibet. In: Clarke G (ed) Development, society and environment in Tibet. Verlag der Österreichischen Akademie der Wissenschaften, Vienna, pp 47–58
Goldstein MC (1996) Nomads of Golok, Qinghai. A report. http://www.case.edu/affil/tibet/tibetan-Nomads/books.htm. Accessed 27 Sept 2010
Goldstein MC, Beall CM (1990) Nomads of western Tibet. The survival of a way of life. University of California Press, Berkeley
Goldstein MC, Childs G, Wangdui P (2010) Beijing's "people first" development initiative for the Tibet autonomous region's rural sector. A case study from the Shigatse area. China J 63:58–75
Gruschke A (2008) Nomads without Pastures? Globalization, regionalization, and livelihood security of Nomads and Former Nomads in Northern Khams. J Int Asso Tibetan Stud 4. http://www.thlib.org/collections/texts/jiats/#jiats=/04/gruschke/. Accessed 27 Sept 2010
GTH (2010) Guoluo zhou tuimu huancao gongcheng jingshi chengxiao xianzhe (Guoluo Prefecture's Tuimu huancao construction program is a remarkable success). http://www.guoluo.gov.cn. Accessed 2 Sept 2010
Khazanov AM (1998) Pastoralists in the contemporary World. The problem of survival. In: Ginat J, Khazanov AM (eds) Changing nomads in a changing world. Sussex Academic Press, Brighton, pp 7–23
Kopytoff I (2003) The cultural biography of things. Commoditization as a process. In: Appadurai A (ed) The social life of things. Cambridge University Press, Cambridge, pp 64–91
Levine N (n.d.) Reconstructing tradition: persistence and change in Golog social structure. http://www.case.edu/affil/tibet/booksAndPapers/Segpaper.htm?nw_view=1307345682&. Accessed 6 June 2011
Levine N (1999) Cattle and the cash economy: responses to change among Tibetan nomadic pastoralists in Sichuan, China. Hum Organ 58:161–72
Manderscheid A (2002) Revival of a nomadic lifestyle: a survival strategy for Dzam Thang's pastoralists. In: Huber T (ed) Amdo Tibetans in transition: society and culture in the post-Mao era. Brill, Leiden, pp 271–289
Medzini A (1998) Bedouin settlement policy in Israel, 1964–1996. In: Ginat J, Khazanov AM (eds) Changing Nomads in a changing world. Sussex Academic Press, Brighton, pp 58–67
MQ (2005) Maqin xian zhi (Annals of Maqin County). Qinghai Renmin Chubanshe, Xining
MSY (2009) Guomin jingji he shehui fazhan tongji ziliao huibian 2008 (Statistical yearbook on people's economic and social development 2008). Maqin xian tongjiju, Tawu
Olson P (1998) My dam is bigger than yours. Emulation in global capitalism. In: Brown D (ed) Thorsten Veblen in the twenty-first century. A commemoration of the "theory of the leisure class" (1989–1999). Edward Elgar Publishing, Cheltenham, pp 189–207
Ptackova J (2010) The sedentarization process in Tibetan nomadic areas of Qinghai, China. Mongolo-Tibetica Pragensia 10:155–180
Ptackova J (2011) Sedentarization of Tibetan nomads in China: implementation of the nomadic settlement project in the Tibetan Amdo area; Qinghai and Sichuan provinces. Pastor Res Policy Prac 1(4). http://www.pastoralismjournal.com/content/pdf/2041-7136-1-4.pdf. Accessed 30 May 2011
Roerich G (1958) Le Parler de'l Amdo. Étude d'un Dialecte Archaïque du Tibet. Instituto Italiano per il Medio ed Estremo Oriente, Roma
Salzman PC, Galaty JG (1990) Nomads in a changing world: issues and problems. In: Salzman PC, Galaty JG (eds) Nomads in a changing world. Instituto Universitario Orientale, Naples, pp 3–21
Sodnamkyid (2008) The implementation of a resettlement development project: socio-economic changes accompanying the transition from Nomadic to Town Life in Sogrima Town, Western China. M.A. Thesis in International Studies, Miriam College, Quenzon City, Philippines

Sułek ER (2009) In the land of checkpoints. Yartsa gunbu business in Golok 2007, a preliminary report from the field. In: Dotson B, Gurung KN, Halkias G, Myatt T (eds) Contemporary visions in Tibetan studies. Proceedings of the first international seminar of young Tibetologists. Serindia, Chicago, pp 15–44

Sułek ER (2010a) Disappearing Sheep. The unexpected consequences of the emergence of the Caterpillar Fungus economy in Golok, Qinghai, China. Himalaya. J Assoc Nepal Himalayan Stud 30(1–2):7–22

Sułek ER (2010b) The Wranakhs. A story of discovering and forgetting the Tibetan tribes of Amnye Machen. In: Bareja-Starzyńska A, Majkowski F, Rogala J, Tulisow J (eds) Altaica et Tibetica. A volume in honour of Prof. S. Godziński on his 70th Birthday. Rocznik Orientalistyczny (Warsaw) 63(1):237–252

Tenzin D (2010) The last Nomad: can Tibetan nomads survive China's ethnocide? Presentation at a conference: pure Earths? The environmental challenges facing Tibet in the twenty-first century, 15 Apr 2011. School of Oriental and African Studies, London

Veblen T (1987[1899]) The theory of the leisure class. Penguin, New York

Yeh ET (2011) Development as gift: state incorporation and landscape transformation in Tibet. Unpublished manuscript

Chapter 14
Change and Continuity in a Nomadic Pastoralism Community in the Tibet Autonomous Region, 1959–2009

Melvyn C. Goldstein

Abstract This chapter examines the process of change and adaptation that a group of Tibetan nomadic pastoralists have experienced from the traditional (pre-socialist) period to the present. The data are based on anthropological fieldwork over a 25-year period conducted in Phala, a nomadic community located about 500 km west of Lhasa in the Tibet Autonomous Region.

Keywords Nomads • Pastures • Change • Stocking rates • Privatization • Fencing

14.1 Introduction

Tibetan nomadic pastoralists have resided on the Qinghai-Tibet Plateau (QTP) at altitudes too high for farming for centuries, if not millennia. All share certain basic features—they raise combinations of four kinds of livestock (yak, sheep, goats, and horses) that graze all year on natural vegetation—about 4 months on fresh vegetation and 8 months on senescent vegetation.[1] They all also move their livestock seasonally at least several times a year, harvesting a wide variety of products from their animals, some of which they consume directly and some they trade with neighbouring farmers or, nowadays, sell to outsider traders and nearby towns. However, beyond such basic aspects of the nomadic pastoral adaptation, it is difficult to generalize about socio-political-historical organization and environmental conditions, either at present or in the traditional (pre-socialist) era because the QTP contains significant diversity. Ecologically, the western part of the QTP is higher and drier, and nomads living there predominately raise sheep and goats, whereas in the eastern portion, yaks were the more important animal raised.

M.C. Goldstein (✉)
Department of Anthropology, Center for Research on Tibet, Case Western Reserve University, Cleveland, OH 44106, USA
e-mail: mcg2@case.edu

H. Kreutzmann (ed.), *Pastoral practices in High Asia*, Advances in Asian Human-Environmental Research, DOI 10.1007/978-94-007-3846-1_14,

Politically, Tibetan nomads have existed under very different political systems. Hugh Richardson, the well-known British diplomat/historian, differentiated the Tibetan world into 'political' Tibet, the polity ruled by the Dalai Lamas, and 'ethnographic' Tibet, the other ethnic Tibetan areas in the east (Amdo and Kham) that were outside that state. In 'political' Tibet, the Tibetan government ruled continuously from the earliest times down to 1951, whereas in 'ethnographic' Tibet, the Dalai Lama's government exercised jurisdiction only in certain places and at irregular intervals with local lay or monastic chiefs being in control of districts of varying size (Richardson 1984, 1–2). Another diplomat/historian conveyed the basic political differences on the QTP as follows:

> At the beginning of the present [20th] century, before the British expedition to Lhasa in 1904 and the subsequent Chinese forward movement in Kam, that portion of High Asia inhabited by Tibetan-speaking peoples, and labeled *Tibet* on European maps, consisted of three separate entities, firstly, the Lama Kingdom of Tibet with its provinces and dependencies, secondly, the semi-independent Native States of Kham under Chinese protection, and thirdly, the Kokonor [Amdo] Territory under the control of the Chinese Amban residing at Sining and Kansu. (Teichman 1922, 7–8)

These historical differences were carried over into the PRC after its creation in 1949. The Dalai Lama's Tibet was transformed into the Tibet Autonomous Region, and the ethnic Tibetan areas in the east (Kham and Amdo) were incorporated into different administrative units—the provinces of Qinghai, Gansu, Sichuan and Yunnan. Consequently, for many organizational, administrative, linguistic and even sub-cultural aspects, today's nomadic pastoralists derive from very different backgrounds and administrations. Compounding this is the fact that over the past two decades, the pastoral policies of the current Chinese government have been implemented differently in the TAR and in eastern areas like Qinghai.

Beijing's pastoral policies are based on the belief that serious pasture degradation is underway on the QTP due to livestock overstocking which it considers is primarily the result of the irrational management system of Tibetan nomads whose focus, the government claims, is solely on each household raising as many animals as possible without consideration of the carrying capacity and future of the grasslands. The government and Chinese scientists see this as a classic example of the 'Tragedy of the Commons'. For example, the English language *China Daily* newspaper (27 January 1987) reported that 15% of China's grasslands had deteriorated by the mid-1970s and that this had increased to 30% by the mid-1980s. Similarly, a Chinese scientific expedition to the QTP reported a 113% increase in herd size during the 23-year period from 1959 to 1981, that is to say, a 3.3% increase per annum with a doubling time of 21 years. The same source explained the reasons for this as:

> To date, the animal structure in Tibet is … irrational … The increased rate of total animals had been the main criterion for measuring the development of animal husbandry and no attention has been paid to [the yields of] animal products. (Chen et al. 1984, 51)

This negative portrayal of traditional pastoralism has become the dominant government view in China and has resulted in the government taking a proactive role in dealing with Tibetan pastoralists, especially those in Qinghai Province where environmental degradation in the grassland area that contains the headwaters of the

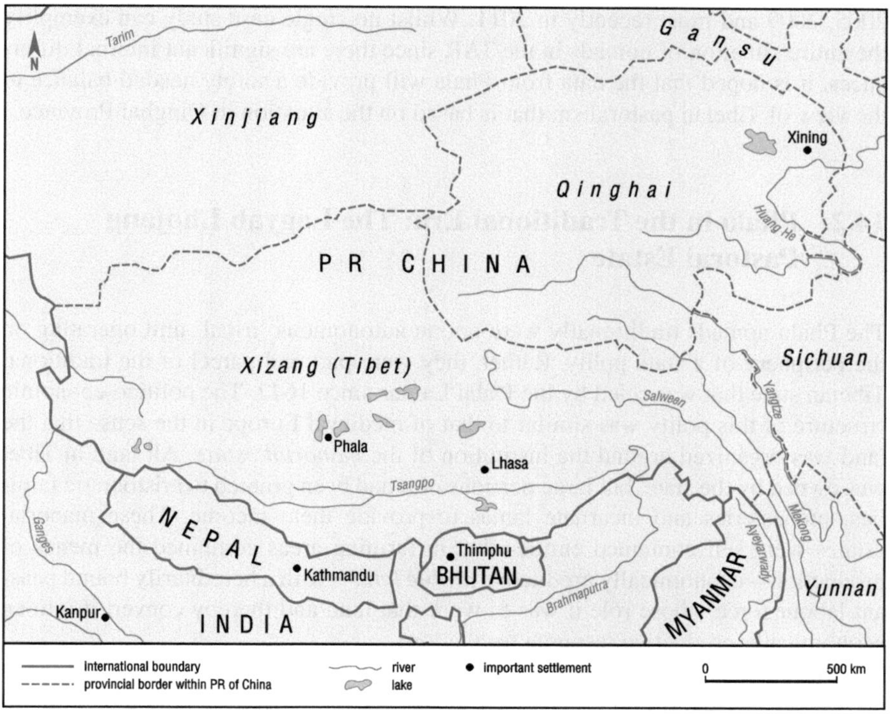

Fig. 14.1 Location of the fieldwork region in Phala (Source: Based on design by Goldstein)

Yangtse and Yellow Rivers (which impact millions of people in the lowlands of China and Southeast Asia) brought grassland conservation in the QTP to the forefront of Beijing's attention. The Chinese government has responded by developing policies that sought to radically reorganize the traditional pastoral management system and transform the nomadic pastoralists into something akin to small family ranchers who would control their own pastures and adopt modern 'scientific' strategies of animal husbandry and grassland management. To this end, starting in Qinghai, it has instituted a series of major policy interventions including the privatization and fencing of pastures on a household basis, sedentarization, the resettlement of herders to towns and programmes that set aside pasture areas for varying periods of time.

The situation of Tibetan pastoralists in Qinghai, however, differs in significant ways from that found in the TAR where very little research has been done and very little is known. This chapter, therefore, seeks to expand our understanding of Tibetan pastoralism on the QTP by means of a case study of a nomadic pastoralist area in the TAR called Phala (tib. *bar la*) that is located about 500 km west of Lhasa (Fig. 14.1). Anthropological research was started there by the author and his colleague Professor Cynthia M. Beall in 1986 and has continued for the past 25 years, including stints of fieldwork in 1986, 1987–1988, 1990, 1993, 1997, 1998, 2000,

2005, 2009 and most recently in 2011. Whilst no single case study can exemplify the entire situation of nomads in the TAR since there are significant internal differences, it is hoped that the data from Phala will provide a sorely needed balance to the view of Tibetan pastoralism that is based on the situation in Qinghai Province.

14.2 Phala in the Traditional Era: The Lagyab Lhojang Pastoral Estate

The Phala nomads traditionally were not an autonomous 'tribal' unit operating on the periphery of a state polity. Rather, they were part and parcel of the traditional Tibetan state that was ruled by the Dalai Lamas since 1642. The politico-economic structure of this polity was similar to that of medieval Europe in the sense that the land was organized around the institution of the *manorial estate*. All land in Tibet was owned by the state, but large portions of it had been granted to aristocratic families, monasteries and incarnate lamas to provide them income. These manorial estates were self-contained entities that in farming areas combined the means of production—economically productive *arable land*—with a hereditarily bound peasant labour force whose role it was to work that land and thereby convert it into an economically productive resource for the lord.

Tibet also contained what we can think of as purely pastoral estates. Like farming estates, these pastoral estates were controlled by lords, who were either aristocrats, monasteries, incarnate lamas or the government itself, and combined the means of production—economically productive *pasture land*—with attached nomadic pastoralists whose role was to raise livestock and pay taxes of animal products like butter to their lord, that is, to transform the grassland resource into economic profit for the lord.

Phala was part of a large pastoral estate called Lagyab Lhojang (tib. *la rgyab lho byang*) whose lord was one of Tibet's most powerful religious incarnations, the line of Panchen Lamas.[2] All the pastureland used by the Phala nomads, therefore, belonged to the Panchen Lama who administered it through a staff of managers and local officials.[3]

Lagyab Lhojang encompassed an area of about 2,500 km^2, all of which was situated between 4,700 m and about 5,500 m. Internally, it was divided into ten nomad groups called *tsho* (tib. *tsho*), one of which was Phala. The entire area was headed by a chief called the Garpön who was chosen from the nomads by the lord. Each *tsho* also had a headman chosen from the local nomads. Within each *tsho*, the most important social and economic unit was the extended household since it was the unit in which primary management and production decisions were made.

A core element of Lagyab Lhojang's pastoral management system was a sophisticated triennial system of pasture reallocation that insured stable income for the lord whilst preventing the degradation of his valuable resource through overgrazing.

Traditionally, Lagyab Lhojang's ten *tsho* were divided into hundreds of pasture areas of various sizes, each of which had explicitly demarked borders that were identified by geographical markers such as streams and hillocks and formally recorded in a register book. Each named pasture was considered suitable to support a fixed number of livestock calculated on the basis of a measurement unit called a *marke* [tib. *mar khal*].

In the 1950s, a one *marke* unit of pasture in Phala had a carrying capacity of 13 yaks or yak equivalents (each yak being equivalent to seven goats or six sheep). Thus, a one *marke* unit of pasture had a carrying capacity of 13 yaks or 91 goats or 78 sheep (or some combination). This rating system was monitored by the lord who allocated pasture units to his nomad subject households on the basis of the number of animals each household held. For example, a household with 26 yaks (or yak equivalents) would be given exclusive usufruct rights over specific pastures having a total rating of two *marke*. These *marke* were defined by vegetative productivity not areal size, in other words, a larger pasture area with poor vegetation could be rated as having fewer *marke* than a smaller one with higher quality grass.

This system of pasture allocation, however, was not permanent. Instead, pastures were reallocated by the lord every 3 years based on a livestock census. Households whose livestock had increased over the previous 3 years received additional pastures according to the new number of yak equivalents they held at the time of the triennial census, and those whose herd had decreased over the prior 3 years lost pastures. Following this reallocation, each household again had exclusive usufruct rights over its new set of pastures for the next 3 years.

The long-term viability of this flexible pasture allocation system rested on an indigenous assumption that uncontrollable variables such as disease and climatic disasters such as blizzards operated unevenly, for example, wiping out many animals in one locality but not others. Data on livestock in Phala from de-collectivization in 1981 to 1988 reveals such fluctuations, showing decreases in herd size for 5 years, followed by a substantial increase in the sixth year (Table 14.1).

This same pattern also occurred at the household level. For example, whilst some households in Phala suffered 100% neonatal mortality of sheep and goats in the spring of 1988, their neighbours lost none or just a few percent. And, in the early summer of 1986, one area just west of Phala lost about 30% of its sheep and goats due to a snow storm, although Phala was unaffected. Such fluctuations can be seen clearly in Table 14.2.

Thus, in any given year, some households and sub-areas within *Lagyab Lhojang* would have expanding herds, whilst others' herds would be shrinking. Any single household, therefore, theoretically could experience sustained net growth in herd size over 3 years, whilst a neighbouring household might be experiencing decreases in livestock numbers and underutilization of their allocated pastures. Phala's triennial pasture management system accommodated this fundamental reality by reallocating pastures every 3 years to balance the gains and losses in livestock numbers in specific areas/households. This maximized productivity by rewarding nomad households when they were successful with more pasture whilst minimizing overgrazing of the pastures by *not allowing* stocking rates to increase beyond the carrying capacity of each pasture.[4]

Table 14.1 Number of livestock in Phala, 1981–1988

Year	Yaks	Sheep	Goats	Total	Change from 1981	
					No.	%
1981	1,211	6,838	2,738	10,787		
1983	1,164	5,441	2,929	9,534	−1,253	−12
1984	995	4,548	2,930	8,473	−2,314	−22
1985	909	4,369	2,963	8,241	−2,546	−24
1986	898	4,276	2,950	8,124	−2,663	−25
1987	1,024	5,425	3,886	10,335	−452	−4

Source: Data derived from handwritten records found at the *xiang* headquarters and head counts conducted during the course of the authors' research (Goldstein et al. 1990)

Table 14.2 Change in total number of livestock for households in two contiguous home-based encampments (*Dzuk*), 1981–1986 and 1986–1987

Household	Number of livestock 1981	Number of livestock 1986	% Change from 1981–1986	Number of livestock 1987	Change from 1986–1987 (number of livestock (%))
DZUK A					
Household 1	361	321	−11%	153	−168 (−48%)
Household 2	306	159	−48%	213	+72 (+45%)
Household 3	296	262	−12%	376	+114 (+44%)
Total	**963**	**742**	**−23%**	**760**	**+18 (+2%)**
DZUK B					
Household 1	356	680	+91%	634	−46 (−7%)
Household 2	501	782	+56%	845	+63 (+8%)
Household 3	204	152	−26%	136	−16 (−11%)
Household 4	245	96	−61%	80	−16 (−17%)
Household 5	308	334	+8%	345	+11 (+3%)
Household 6	40	60	+50%	60	+0 (+0%)
Total	**1,654**	**2,104**	**+27%**	**2,100**	**−4 (0.2%)**

Source: Data collected by Goldstein et al. (1990)

For this system to work effectively for hundreds of years, three kinds of reallocations were utilized:

1. Shifting pasture areas every 3 years amongst households within a single nomad *tsho* such as Phala
2. Transferring one or more pasture areas from the control of one of the ten *tsho* to another
3. Moving entire households and their herds from one *tsho* where herd size had increased significantly to another *tsho* where herd size had decreased

14.3 The Socialist Era

14.3.1 Phase One: Democratic Reforms and Collectivization (1959–1980)

The People's Republic of China incorporated political Tibet into its new state in 1951, but did not immediately end the traditional manorial estate system or dissolve the Dalai Lama's government.[5] The estate system, in fact, continued until the flight of the Dalai Lama to India in 1959 at which point Beijing ended the old socio-political system and began to implement 'democratic reforms', that is, changes starting the transforming of traditional society into socialist society. Consequently, in 1959, the nomads of Phala now found themselves subordinate to a new 'lord' (political entity) that held a totally different ideology about social, political and economic organization.

The initial reforms that began in 1959 involved the creation of a new class hierarchy in which the poor were valourized and placed at the apex of the social hierarchy, whilst the rich were decried as exploiters and relegated to its bottom. At this time, the government confiscated the large herd of the most powerful local nomad chief, the Garpön, redistributing his livestock and possessions to the poorest nomads. Other nomad households, however, continued to manage their production and consumption, and even those households labelled as 'rich nomads' (with the exception of the Garpön) were allowed to retain ownership over their animals and to manage their own herds. Day-to-day pastoral production, therefore, did not change although pasture allocation did.

The year 1959 was scheduled to be a triennial census year, but the new government did not want to continue a system from the manorial estate era, so it chose not to do the census/reapportionment. However, it also did not want to implement communes at this time since it felt that Tibetan herders were not ready to handle these. Consequently, they initially did nothing and told every household to keep their previously allocated pastures regardless of any increases or decreases in their herd size over the past 3 years. The system of triennial pasture reallocations, therefore, ended in 1959, but households continued to control their own pastures.

This situation did not last long, and 2 years later, in 1961, a programme called 'mutual aid' teams was instituted. It created small *Pasture Groups* consisting of several poor- and middle-class nomad households (5–15) who shared pastures and cooperated in herding and production. These were not collectives, however, since each household retained ownership over its own animals as well as ownership over all the products its own animals produced. This was seen by the state as a modest first step towards the CCP's ultimate goal of replacing household production with communal production.

A decade later, in 1969, the government initiated real socialist reforms in Phala by reorganizing the nomads into full communes. At this point, every household had to transfer its livestock and implements (e.g. churns, saddles) to the new commune entity which also assumed control over all pastureland. All aspects of pastoral production, for example, when to milk and where to herd, were now decided by the commune leadership, not by nomad households. Each nomad, therefore, became in essence a worker for the commune (or in its own representation, an owner-worker in the commune). A complex system of 'work points' was utilized in which each task was rated from 1 to 10 points, and each worker earned points based on the type and duration of work done. Payments of food and other needed goods were provided by the commune based mainly on these work points and a basic ration amount.[6]

Consequently, with the exception of a few goats that each household was allowed to keep privately for its own food needs (the equivalent of the private vegetable garden in farming communes), nomad households now functioned only as units of consumption. They owned no animals and made no decisions about production tasks or the movement of livestock to different pastures. Each household member worked separately at tasks and locations determined by the commune's leaders. Nevertheless, the basic pastoral work tasks such as herding, milking, churning and shearing were done the same as before, as was the pattern of moving herds, but now its organization was managed by the commune leadership not the household.

The pastoral commune in Phala (and others throughout the TAR) remained in operation for roughly 11 years (1969–1980). During that time, no attempt was made to diminish the geographic scope of pastoralism, for example, by converting large pasture areas into farmland, although in Phala, an experiment with growing barley in one small area was tried without success.

14.3.2 Phase Two: De-collectivization and Market Economics

Despite the government's belief that nomadic and agricultural collectives were a more efficient and modern form of production, they in fact were an economic disaster that caused a sharp decline in the standard of living throughout the TAR (and the rest of China). Consequently, with the rise to power of Deng Xiaoping in 1979–1980, the Chinese Communist Party reversed its economic ideology/policy and ended communes. China now embraced a programme of major modernization by adopting the essence of the capitalist market economic system where working for profit was an accepted goal. Modern scientific and business methods became the valued means not just for industries and enterprises but also for individual farming and herding households. Deng Xiaoping's call to all in China to strive to modernize production, increase productivity and 'get rich' was enthusiastically conveyed even in remote areas like Phala where nomadic pastoralist households were urged to produce more effectively for China's new market economic system. At the same time, communes were dissolved and replaced by what was called the household 'responsibility' system in which households again became the basic unit of production. In Phala, this was accomplished by dividing the commune's animals equally in 1981 on a per capita basis—every nomad alive on the day of division regardless of

Table 14.3 Economic differentiation by household, 1981–1986

	1981	1986
Number of households with <30 animals	0%	19%
Number of households with 30–60 animals	100%	62%
Number of households with >70 animals	0%	19%

Source: Data collected by author

age received an equal share of 39 animals (4.5 yaks, 27 sheep and 7.5 goats).[7] Individuals and households now again owned their own animals and were *responsible* for all aspects of production and marketing, just as they had been in the traditional, pre-socialist society.

How to allocate pastures in the post-commune era, however, was problematic. Some Phala herders wanted the government to reinstate the traditional triennial reallocation system, whilst others wanted to continue the collective's common-pasture system in which all households in Phala would be free to use any of Phala's pastures. Government officials generally thought the marke system was too difficult for them to administer, and so decided to implement a middle course by dividing Phala's pastures into ten mutually exclusive pasture-sharing units called *dzashog* (tib. *rtswa shog*). Each of these pasture-sharing groups was comprised of between 5 and 15 households who shared a delimited set of pastures which they used exclusively in accordance with local rules of use. Despite this sharing of pastures, each household in the pasture-sharing group remained economically autonomous, and there was no requirement that these households cooperate in herding or marketing. Each household made its own milking, shearing and marketing decisions. However, the state continued to exercise ownership over all pastureland, so the nomads only received *usufruct* rights to these pastures.[8]

Under this new system, households initially were free to increase their initial herd size and most worked energetically to do so. As shown in Table 14.3, within 5 years, significant disparities in wealth had emerged. It should be noted that Phala had excess pastureland at this time, that is, they had pastures that they did not use.[9]

14.3.3 Phase Three: Stocking Limits and Privatization

Beijing's overall concerns with overgrazing and pasture degradation reached Phala in 1987 when a 20% reduction in livestock was imposed, followed in 1988 by formal stocking limits setting the final 1987 livestock total as the stocking limit. After this, the county annually gave each of its nomad townships (ch. *xiang*) a document stating the number of livestock it had to eliminate ('kill/eat or sell' (tib. *söjö* [*gsod spyod*]). The *xiang* then passed this down to each nomad community (tib. *trongtso* [*grong tsho*]) and household. Generally, the annual *söjö* figure (the% reduction) was based on the number of newborns that survived that year; that is, however many newborns survived, that percent would be reduced. In normal years, this was about 30% of the number of animals present in the fall.[10] When I asked officials why they were imposing limits when there was still excess pastureland in Phala, a county-level Tibetan official explained, ‘The nomads have to be educated to understand that just rearing more and more animals is not the answer’.

This system continued for the next decade at which time the government introduced further changes regarding privatization of pastures.

In 1996, the government sought to implement the more radical Qinghai programme of changes in Phala and the TAR. As mentioned earlier, the new programme called for the replacement of the system of shared pastures with a system of privatized pastures (especially winter pastures) on a household basis. The goal was for each household to control and fence off its own pastures so that each household could pursue modern and scientific animal husbandry practices, for example, raising fewer, higher quality animals. The government's rationale was that if each nomad household controlled its own pastureland, it would be motivated to invest time and resources to improve the quality of the vegetation and animals. Nomads, therefore, would in the end become transformed into something akin to autonomous family ranchers.

Although these changes had been implemented in many areas in Qinghai, in Phala and most areas in the TAR, privatization and fencing of individual pastures was strongly opposed by herders as well as their local officials who believed this would cause serious problems and be difficult to administer. They argued that the more flexible system of pasture-use groups should be continued and were successful in persuading the government to allow this. Nevertheless, a kind of privatization was implemented that we can think of as *Virtual Privatization.*

14.3.3.1 Virtual Privatization

The pasture privatization implemented in Phala differed fundamentally from *real* privatization experienced in Qinghai Province. In both systems, each nomad household was allocated its own share of the pastureland, but with virtual privatization, this share was *never specified at ground level.* Nomad household knew that they had a certain number of shares of pastureland, but not where those pastures were located so they could not exercise exclusive usufruct rights over them. This system of virtual privatization had two main steps.

First, the carrying capacity of Phala was determined by the number of livestock present at the end of 1996. For example, if Phala hypothetically had 4,000 sheep at the end of 1996, that was established a priori as the carrying capacity of Phala. At this time, Phala was still using the old *marke* unit, so if hypothetically there were 100 *marke* of pastureland in Phala in 1996, the carrying capacity of each *marke* would be 40 sheep (4,000/100).

The second step was to allocate pastures to each household. Although animals consume grass, pastures were not allocated to households on the basis of the number of the animals they had at the end of 1996. Instead, shares of the pastureland were allocated to each household based on both *the number of people in the household as well as the number of animals each household had.*

Initially, the government recommended dividing pastures using a ratio of 70% based on the number of people and 30% based on the number of livestock. One half of Phala decided instead to use a 65/35% ratio and the other half, a 60/40% ratio. In the latter arrangement, 60% of the pastures were divided on the basis of the number of people in the area and only 40% based on the number of livestock. This worked as follows:

Given the hypothetical size of 100 *marke* of pastureland, 60% of this pastureland (60 *marke*) was divided based on the number of people. If the total hypothetical population was 60 nomads, each nomad therefore would have received 1 *marke* of pasture as his/her share of the overall pasture (based on people). The remaining 40% of the pastureland (40 *marke*) was divided on the basis of the number of livestock. As there were 4,000 sheep in the hypothetical example, each sheep would have received a share of 0.01 *marke* (40/4000) for the animal share. These two together determined the final share of each household.

Consequently, if we take a hypothetical household called 'A' that had 5 people and 400 sheep, it would have received five *marke* based on the number of people in the household (each person was entitled to a one share) and four *marke* of pastureland based on the number of animals it possessed since each animal was entitled to 0.01 *marke* of pastureland, that is, four *marke* (400×0.01). Household A's share of the pastureland, therefore, was a total of nine *marke* of pastureland. Since each *marke* could hold 40 sheep, that meant that household A's share of Phala's pastures would allow them to keep 360 animals. However, because they actually owned 400 animals, their pasture allocation was *less* than the number of animals they then actually had.

If we take another household, 'B', and say hypothetically that it had no livestock but four members, it would have received a share of pastureland based on the number of people in the household, that is, four *marke* of pastures, one for each member. Since each *marke's* carrying capacity would have been 40 sheep, it was entitled to graze 160 sheep, even though the household actually had none.

This system of *virtual* privatization of pastures, therefore, gave each household a fixed share of the overall pastureland without ever specifying where each household's pastures were located. Households only knew the number of pasture units they held, not where these pastures actually were located. The nomads, therefore, continued to herd and share pastures through pasture-use sharing groups as before.

Virtual privatization clearly did not fulfil the government's aim of giving individual households control over their own pastures in order to motivate them to fence off their pastures and modernize animal husbandry, but it had other functions.

First, it allowed officials in localities like Phala to report to higher ups that they had privatized their pastures as the state had called for—without actually having to force the unwilling nomads to do so.

Second, it created a system of poverty alleviation. By implementing virtual privatization on the basis of people as well as animals, poor households with few or no animals received a significant share of the pastureland which they could then lease to richer households who had more animals than their pasture share warranted. To return to the hypothetical example cited above, household A had 40 too many animals, so it could lease pastures from household B for a fee and thus be able to keep these 'excess' animals. Consequently, in Phala and most of the TAR, the 1997 privatization was actually a mechanism for poverty alleviation rather than a means for transforming pastoral production and management.

14.3.3.2 The 2005–2006 Re-privatization Initiative

Although the 1997 pasture division was announced as permanent, a decade later in 2005/2006, the government revisited the pasture privatization law allowing each area to adjust the previous pasture allocations to accommodate herd increases and decreases based on the number of animals at the end of 2005. This time, the government again advocated real privatization of individual pastures and again proposed that the ratio of people to animals to be increased in favour of people to 70/30 to further help the poor. At the same time, use of the old society term '*marke*' was replaced with the standard Chinese areal measure of called mu. However, this was done by simply creating a conversion rate—there was still no empirical research on the real carrying capacity of Phala, which continued to have no pasture degradation and excess pastures.

As was the case in 1996, the nomads and their officials oppose real privatization and were able again to prevail, the government allowing them to continue the pasture-sharing system and the system of virtual privatization, although the people-to-animal ratio was increased from 60/40% to 65/35%.

Since 2006, the government has continued to intervene in Phala. In 2009–2010, for example, a new township official agreed that the area had excess pastures, and so removed the stocking limits for 2 years to allow the area's total number of livestock to better fit its real carrying capacity.

However, in 2011, a new government initiative was again in the works to implement the basic Qinghai approach. Discussion are now going on about real privatization, and now for the first time, there were also serious discussion about starting a programme to set aside a significant portion of Phala's pastures for 5 years to allow 'regeneration'. This reduction in pastures would be balanced by drastically reducing overall herd size, with the government paying compensation to each nomad household for the number of livestock it eliminated. Since there are excess pastures in Phala, it is unclear whether it will actually be implemented or whether the nomads will again be permitted to use all of their pastures by pasture-sharing groups.

Despite these repeated interventions by the government and the imposition of stocking rules that precluded the nomads from keeping livestock numbers that reflected the real carrying capacity of the area's pastureland, the Phala nomads have done well economically. By 2009, almost 50% of the households had motorcycles, many had cell phones and five had trucks or tractor-pulled carts. Roads to this remote area have been improved markedly, and in 2009 the government paid for 224 new houses (at a cost of 5.5 million RMB) in the township that Phala is part of along with community centres for each nomad village. Most households already had houses, so the additional houses made it possible for the rest of the nomads to acquire houses at their main campsites. However, having houses did not mean sedentarization or that nomadic pastoralism has ended. The nomads in Phala like houses which they consider more comfortable in winter than tents, but all Phala households

Photo 14.1 The camp of a wealthy Phala household that is balancing modernization, in the sense of owning a motorcycle, tractor and truck, and traditional nomad culture, in the sense of continuing to live in a tent in summer and fall despite having a nice house (Photograph © Melvyn Goldstein 2005)

Table 14.4 Changes in price of nomad products in Phala, 1986–2005

Product	Percent change in price from 1986 to 2005 (%)
Sheep	+650
Yak	+655
Goat cashmere	+669
Yak skins	+1,200
Sheep skins	+208
Goat skins	+246
Sheep wool	+132
Yak wool	+100

Source: Fieldwork by author

still move with their animals to fall pasture sites in September where they remain living in tents for 4 months (Photo 14.1). They also moved sub-sets of animals to more distant pastures at different times throughout the year, setting up satellite camps with tents.

The economic gains experienced in Phala were not the result of nomads going out of their area as migrant labourers to earn cash income, nor was it the result of significant increases in the number of livestock—the number of animals per capita was virtually the same in 2005 as in 1986. Rather, it was the result of major increases in the value of the nomad's animal products, as can be seen in the following table (Table 14.4).[11]

14.4 Conclusions

The Phala data raise serious questions about the validity of the Chinese government's view that a radical restructuring of Tibetan pastoralism is necessary to preserve the integrity of the QTP's grasslands due to serious overstocking and pasture degradation. Clearly this was not the case in Phala where there are excess pastures. This is not to say that pasture degradation is not a problem in other parts of the TAR, but that is something that must be determined empirically not simply asserted. Just as it would be incorrect to generalize from Phala to the entire TAR or the whole QTP, it is also incorrect to generalize from some areas in Qinghai to the entire TAR. Policies for pastoral development and change must be grounded in careful scientific investigation of local conditions and local requirements.

Similarly, the Phala data also raise serious questions about the government's assertion that traditional Tibetan pastoral management systems are irrational and destructive. Traditionally, Phala and other nomads in Tibet clearly had a sophisticated system of pasture reallocation based on carrying capacities. The decisions not to use this in 1959 and 1981 were made by the government, not by the nomads.

The Phala data also show that whilst the government's pastoral policy seeks to privatize the grasslands on an individual household basis, it has allowed nomads such as those in Phala to continue their system of shared pastures. Consequently, 30 years after de-collectivization in 1981, nomadic pastoralism is continuing in Phala. Motorcycles may be replacing horses and tractors and trucks replacing yaks for transportation, but the traditional activities of raising livestock, harvesting their products and trading these for items made elsewhere is ongoing.

However, government pressure to implement real privatization and fencing is continuing, so it is unclear whether Phala will be able to withstand this pressure in the coming years. Consequently, whilst protecting China's unique QTP is a matter of national and world concern, the Phala data suggest that the government needs to adopt a more nuanced pastoral policy that takes into consideration real local conditions and supports more traditional forms of nomadic pastoralism in areas like Phala where there is no grassland degradation and where the nomads oppose privatization. It would indeed be ironic—and tragic—if, after surviving the destructive madness of the Cultural Revolution and successfully revitalizing their society after de-collectivization, the way of life of nomads like Phala is undermined by inappropriate notions of conservation and development that are based on faulty evidence, negative stereotypes and untested assumptions.

Notes

1. Many nomad groups also harvested small amounts of wild vegetation (hay) from set-aside pastures, which they fed as supplements to weak pregnant and lactating animals at birthing time in Spring.

2. The Panchen Lama appears to have obtained this area in the eighteenth century.
3. The ninth in this lineage died in 1933 and the tenth in 1989.
4. This triennial re-allocation system appears to be widespread in at least Western Tibet, for example, nomads in Sangsang, Saga, Drongpa and Porong also utilized it, and there is some evidence that it was also in operation in Nagtsang in the western part of today's Nagchuka Prefecture and in Tshochen in Ngari Prefecture.
5. For the history of this period, see Goldstein (1989, 2007).
6. The actual method of classification and payment in the communal era is too complex to be discussed here.
7. A household of five therefore received five shares of the commune's livestock or 185 animals (25 yaks, 125 sheep and 35 goats).
8. There was a great deal of ambiguity over exactly how many years the pastures were allocated for, some saying 30 years and some saying 50 years, but all believed that it was for a long time.
9. Excess pastureland refers to the presence of pasture areas that the nomads did not use for grazing during the year. See Cincotta et al. (1991) for a discussion of our data on grazing intensity that were collected from exclosures in Phala. Also see Goldstein et al. (1990); Goldstein and Beall (1989, 1990); Goldstein (1994).
10. There was, however, some flexibility since increases in the herd size of households within a pasture-use group were tacitly permitted so long as this was balanced by losses in other households, in other words, so long as the overall total size of the pasture-use group stayed the same.
11. Since 1986, only one nomad has left the area to work as a migrant labourer. Farmers in the TAR, however, face very different problems since the value of barley and wheat has hardly increased since 1981. They have responded by sending household members out as migrant labourers to earn cash income. For a discussion of this, see Goldstein et al. (2008).

References

Chen H, Ni Zh, Sun Sh et al (1984) Agricultural geography of Xizang (Tibet). Science Press, Beijing (Chinese)

Cincotta RP, van Soest PJ, Robertson JB, Goldstein MC, Beall CM (1991) Foraging ecology of livestock on the Tibetan Changtang: a comparison of three adjacent grazing areas. Arctic Alpine Res 23(2):149–161

Goldstein MC (1989) A history of modern Tibet, 1913–1951: the Demise of the Lamaist State. University of California Press, Berkeley

Goldstein MC (1991) Change and continuity in nomadic pastoralism on the Western Tibetan plateau. Nomad Peoples 28:105–123

Goldstein MC (1994) Change, conflict and continuity among a community of nomadic pastoralists in western Tibet, 1950–90. In: Barnett R, Akiner S (eds) Resistance and reform in Tibet. C. Hurst & Co, London, pp 76–111

Goldstein MC (2007) A history of modern Tibet, vol 2, 1951–1955: the calm before the storm. University of California Press, Berkeley

Goldstein MC, Beall CM (1989) The impact of China's cultural and economic reform policy on nomadic pastoralists in western Tibet. Asian Surv 29(6):619–641

Goldstein MC, Beall CM (1990) Nomads of western Tibet: the survival of a way of life. University of California Press, Berkeley

Goldstein MC, Beall CM, Cincotta RP (1990) Traditional nomadic pastoralism and ecological conservation on Tibet's northern plateau. Nat Geogr Res 6(2):139–156

Goldstein MC, Childs G, Wangdui P (2008) "Going for income" in village Tibet: a longitudinal analysis of change and adaptation, 1997–2007. Asian Surv 48(3):513–534
Richardson HE (1984) Tibet and its history. Shambala, Boulder/London
Teichman E (1922) Travels of a Consular Officer in eastern Tibet – together with a history of the relations between China, Tibet and India. Cambridge University Press, Cambridge

Chapter 15
Tibetan Pastoralists in Transition. Political Change and State Interventions in Nomad Societies

Andreas Gruschke

Abstract Past and present Chinese policies towards Tibetan pastoralists developed in the context of the Communist Party's ideological agenda from which implemented project measures can be deduced. This context needs to be understood in order to assess what kind of practical implications of pastoral policies were and are aimed at. Examples from case studies in Yushu, southern Qinghai, will demonstrate what kind of transformational processes underlie changes both in the pastoralist society and in the policies. This paper will argue that policies are imposed with regard to both the difficult livelihood situation of the people and new efforts for ecological conservation. However, the policy's objectives and the results of its implementation often diverge very strongly. A preliminary analysis will seek to explain this.

Keywords Tibetan pastoralism • Political change • State interventions • Settlement and migration • Rangeland availability • Declining significance of animal husbandry

15.1 Introduction

When collectivisation started in the People's Republic of China (PRC), mobile pastoralism in Tibet was thought to perish. The economic liberalisation starting in 1980, however, apparently brought about a 're-nomadisation' (Gruschke 2008, 3). Pastureland all over the Tibetan Plateau was 'reconquered' by the typical black tents of nomadic[1] households and their herds, and prospects for the market orientation of the pastoral groups were supported by government policies. Obvious features of

A. Gruschke (✉)
Institute of Oriental Studies, University of Leipzig, Schillerstr. 6, D-04109 Leipzig, Germany
e-mail: gruschke@uni-leipzig.de

H. Kreutzmann (ed.), *Pastoral practices in High Asia*, Advances in Asian Human-Environmental Research, DOI 10.1007/978-94-007-3846-1_15,

better livelihoods – motorbikes, mobiles, etc. – were interpreted as signs of a sound mobile pastoralism and animal husbandry in Tibet. The PRC's leadership is the Chinese Communist Party (CCP), and yet outside observers neglect the ideological background as a constituent of its policies. Rather, mere speculations about alleged motives disseminate. Thus, new government schemes aiming at severe changes in rangeland management, the transformation of property rights, settling nomads and policies to encourage households to give up animal husbandry and resettle in urban settlements are often interpreted as the mere ambition of a socialist government to get a vagrant people under its definite control. To understand the policies, probe its shortcomings and assess where it might have strengths, the underlying ideas need to be considered. For this reason, the Sino-Marxist[2] background, political changes since the PRC's establishment and their consequences for policies in Tibetan pastoral areas are addressed here. This ideological background then is contrasted with a number of practical experiences and local testimonies on the policies' impacts, hinting at the complex nature of sociopolitical processes and the difficulty in finding solutions to problems, when only few factors are understood.

15.2 Methodology and Research Region

The objectives of Chinese policies and their implementation in Tibetan pasture areas are studied according to their blueprint representation in official and semi-official (grey) literature.[3] Their comparison to actual circumstances on the ground is based on data collected in 2004–2010 during field trips to pastoralist areas in the eastern part of the Tibetan plateau. Major research methods include participative observation, non-formal, semi-structured, biographical and standardised household interviews as well as some focus group discussions. The interviews cover a big spectrum of household types and single protagonists in their various roles and positions in pastoral and non-pastoral society, economy and administration. The main focus of research lays in the Tibetan autonomous prefecture (TAP) Yushu, a region extending in the east-central part of the Tibetan plateau and the south of Qinghai Province. The administrative seat of the prefecture is the town of Gyêgu (Tib. *Jyekundo*), widely known as Yushu, that was destroyed by an earthquake in April 2010. Although recent processes of urbanisation are effective (Gertel et al. 2009), the nomadic population of this region still outnumbers farmers and agro-pastoralists. Different processes of modern reconfiguration are now following the recent re-nomadisation. Interview sites included 41 of 46 townships in all the six counties of Yushu TAP and are complemented by interviews and field observations in other Tibetan areas.

15.3 Development Processes Since the Communist Takeover

Since the mid-twentieth century, Tibet underwent processes of transformation to an extent hitherto unknown. With mobile pastoralism representing the major livelihood in the vast expanse of the Tibetan highland, nomads were seriously affected.

The takeover of the CCP in China in 1949 entailed socio-economic changes initiated from outside the pastoralist society. Those changes, however, did not occur all of a sudden. In 1942, Mao Zedong had advocated that 'the economic activities of the army, Party and government should harmonise with the economic activities of the people. Anything which damages the people's interests or causes them dissatisfaction is not allowed'.[4] Initially, the CCP's actions indeed interfered less with the nomadic way of life than with other groups of China's population. When summing up the living circumstances of Tibetan pastoralists during the first years of CCP rule, Ekvall (1972, 285) noted:

> They have imposed less coercive restriction on the high-pasturage ones [= drokpas] than on any other segment of the population and appear concerned mainly with leaving the herdsmen undisturbed, provided only that the latter continue to produce. They bemoan the fact that production is less than it should be, and they have some things to say about increasing the birthrate by elimination of venereal disease. With great pride, they report increase in livestock and in the production of wool, meat, hides, butter, and animal fats; and they are obviously experimenting with projects and devices calculated to reduce nomadizing while maintaining full-scale pastoralism.

Ekvall's observations were made from a remote position and his conclusions drawn from secondary sources, but his assessment of the situation in the early 1950s corresponds with statements made in biographical interviews. A short while later, the hasty collectivisation during the so-called 'Democratic Reforms' in 1958–1959 and the ideologisation of daily life in the Cultural Revolution (CR) in 1966–1976 revolutionised the social and economic sphere (Goldstein 1994). The nomads were then deprived of their responsibility for, and the free disposal of, herds and pastures as all animals were transferred into collective property and management. Essential innovations in the livestock economy and production were the introduction of veterinarian measures and state-organised processing and marketing of animal products. Obvious endeavours in the pastureland were large tracts being enclosed with walls made from sods and clay. Apparently, those did not disturb seasonal migration cycles since the enclosures were delimiting community borders that followed traditional clan borderlines. This was still the case when collectivisation measures started in the mid-1950s and people's communes were formed after 1958. The nomads' way of life, however, underwent serious shifts, and their social organisation was revolutionised. The economy experienced the total reorganisation of work distribution and management responsibilities, but pastures and herds were still organised according to schemes of mobile pastoralism. Biographical interviews with elder nomads in Yushu between 2005 and 2007 support this view. Goldstein had similar findings amongst the Phala nomads in western Tibet.[5] Production brigades used seasonal camps even during the CR, and animals were herded on seasonal pastures as it was done before. This is confirmed in an interview with 75-year-old (2006) Phüntsog Nyima, a nomad of Xialaxiu who sold his animals and settled in Yushu county town in 2000:

> Yes, in the old society we had different camps for winter and summer, but now after [re-] distribution no more, but before [i.e., during the CR] we also had different winter and summer camps. We want different camps for winter and summer pastures, but now it is impossible, we don't have enough pastureland [any more].

With Mao's death in 1976, the Cultural Revolution ended. China initiated a new phase of development with crucial effects on pastoral areas. The CCP under Deng Xiaoping gradually realised the economic construction programme known as 'Reform and Opening' (*gaige kaifang*) by instigating the 'Four Modernisations' (*sige xiandaihua*): of industry, agriculture, national defence and science and technology. This alteration of the political agenda is frequently taken as a pragmatic course of the Chinese leadership and as an indication that the long-term objective of communism was abandoned. This is not the case, but a shift of perspective within Sino-Marxist dialectics as von Senger (1982, 2008, 103–125) has made clear. The major concept of argumentation in dialectical materialism is the defining of the principal contradiction. Since 1937, it represented a central idea in Mao's line of reasoning, and remained so unaffectedly under Deng Xiaoping and his successors.

The Sino-Marxist conception denies the epistemological significance of philosophy but perceives it as a means for changing the world. To incite changes, the CCP defines a principal contradiction that needs to be solved by the Chinese people. In the eyes of the party leadership, redefinitions of the principal contradiction delimit different stages of the country's social development.[6] According to that definition, it is decided which of the upcoming tasks have the highest priority (von Senger 2008, 99–111). Until the end of the CR, class struggle was seen as major task but was rated secondary when, in 1978, the party defined that 'at the present stage, the principal contradiction in Chinese society is one between the ever-growing material and cultural needs of the people and the low level of production'.[7] This resolution acknowledged that the leadership under Mao had introduced the commune system too early. It was now explained that China is a country in the 'initial stadium of socialism and will, for a longer period of time, remain in this stage'. The redefinition determines the primary task of 'deliberating the productive forces and gradually realizing the country's socialist modernisation' to fight poverty and backwardness in China.[8]

The party's confidence in positive developments in Tibet led to the Dalai Lama being invited to dispatch a so-called 'fact-finding delegation' in 1979. The latter's critical report about poor life conditions of Tibetans incited CCP chairman Hu Yaobang to lead his own delegation to Tibet and verify the observations. The drastic reform measures he suggested for Tibet were implemented right away (Goldstein 1994, 96–100). In order to improve the living standard of Tibetan farmers and herdsmen, the Beijing government enacted their exemption from dues[9]:

> The major reform … dissolved the communes and restored the household as the basic unit of production. For the nomads in Pala, this resulted in all the commune's animals being divided equally among its 57 households with all infants and senior citizens receiving the same share. The nomads owned these animals and were free to utilise them as they wished. Pastures were also divided at this time, but were allocated to small groups of several households (called *dzug*) rather than to individuals. These *dzug* then held exclusive usufruct rights over them: that is, the families in a *dzug* had exclusive right to use these pastures. (Goldstein 1994, 99–100)

This grave alteration of the leadership's attitude vis-à-vis the people's communes led, in the early 1980s, to the ultimate dissolution of the collective system and to the introduction of the 'household responsibility system'. The distribution of the communes' animals to the pastoral households led to the revival of mobile pastoral systems that existed before collectivisation. Notably in naturally favourable regions of eastern Tibet, pastoralists achieved certain affluence by marketing their highly valued animal products to urban markets. Counties like Zhidoi and Qumarlêb in Yushu TAP were then considered wealthy since due to their low population density, they had large herds grazing on low-stocked rangeland resources. In 1992, herders there disposed of 66 and 75 sheep units (SU) per capita,[10] whilst in comparatively densely populated Nangqên, it was not even half that figure (Fig. 15.1). The increasing productivity that followed the privatisation of livestock and the better market integration initially seemed to be the key for raising the living standard in rural areas.

In 1987, the CCP enacted a new comprehensive three-step strategy for China's economic construction aiming at (1) doubling the 1980 value of the gross national product and solving problems in nourishing and clothing the population, (2) quadrupling the 1980 GNP by the end of the twentieth century and (3) completing the country's modernisation by the mid-twenty-first century.

By then, the average per capita GNP is intended to equate the level of a developed country of medium standard and the Chinese citizens to lead a relatively comfortable life (von Senger 2008, 119–120). To attain these national goals, regional development plans had to be designed to suit the respective local conditions. In pastoral areas, the improvement of infrastructure (highways, services) was envisioned to enable a better market integration of the pastoral economy and consequently enhance public welfare.

After marketing was conferred to the private sector, however, marketing chains through state enterprises built up in the collective period disintegrated. Only households in relative proximity to bigger cities – like Xining, Lanzhou and Lhasa – were able to profit from the urban demand for livestock products. The densely populated pastoral areas in south-eastern Yushu experienced a growing demand for such products within their own sphere. As the human population grew faster than livestock numbers, the per capita availability of animals decreased, resulting in a shift back to a subsistence-oriented animal husbandry amongst the majority of rural households. The conditions were further aggravated by three subsequent snow hazards in the 1990s (Gruschke 2009, 90–91). Still, the average stocking rates in all counties evened out at a level a little higher than the 1950s – with livestock availability considerably below subsistence level (25 SU) for most parts of the region (cf. Fig. 15.1).

The situation seems further exacerbated by pasture degradation. In a persuasive paper, though, Harris (2010) has drawn the extent and alleged causes of rangeland degradation into question. The various hypotheses to explain range degradation may be invalidated as *general* explanations for the phenomenon and further challenged due to the lack of supporting primary data. Existing Chinese data are certainly far from being thoroughly analysed. Observations in the field and interviews made in Yushu, however, support the view that combination of several factors results in serious pasture degradation.[11] Its magnitude may be difficult to calculate and

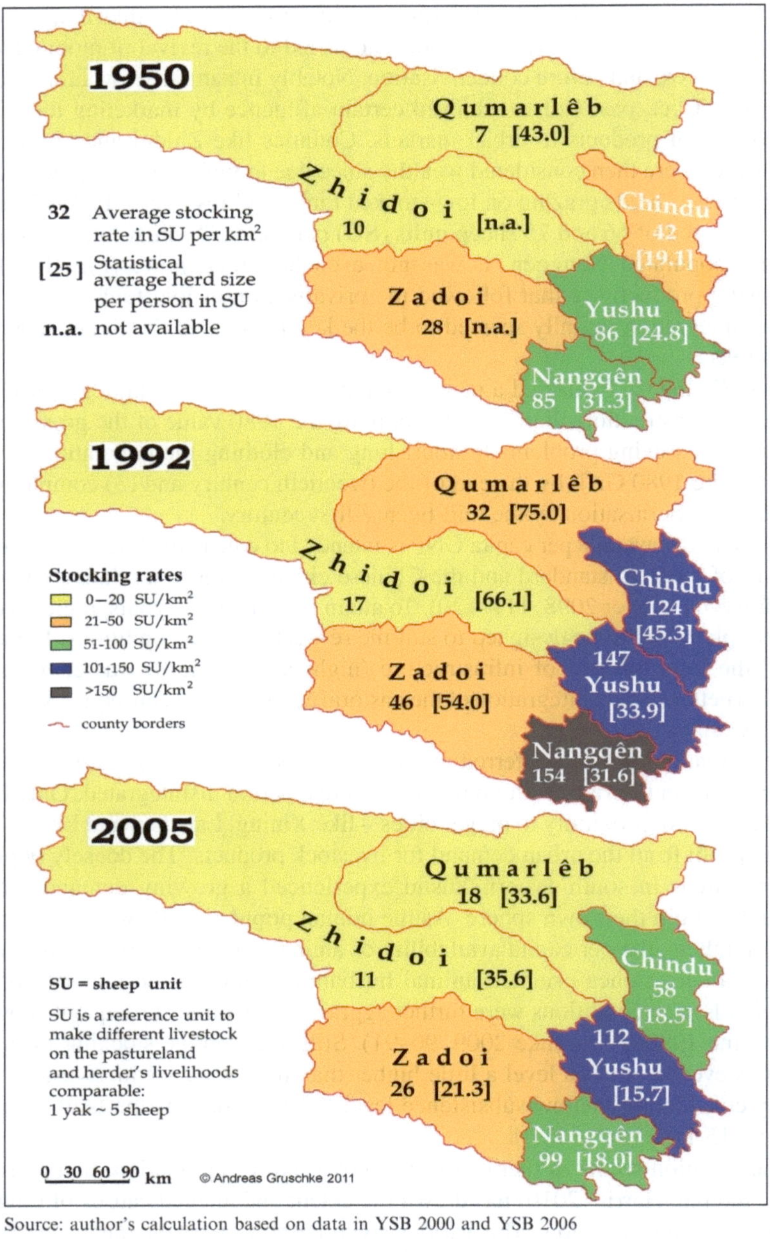

Source: author's calculation based on data in YSB 2000 and YSB 2006

Fig. 15.1 Development of stocking rates and per capita herd size in Yushu (1950–2005) (Source: Author's calculation based on data in YSB 2000 and YSB 2006)

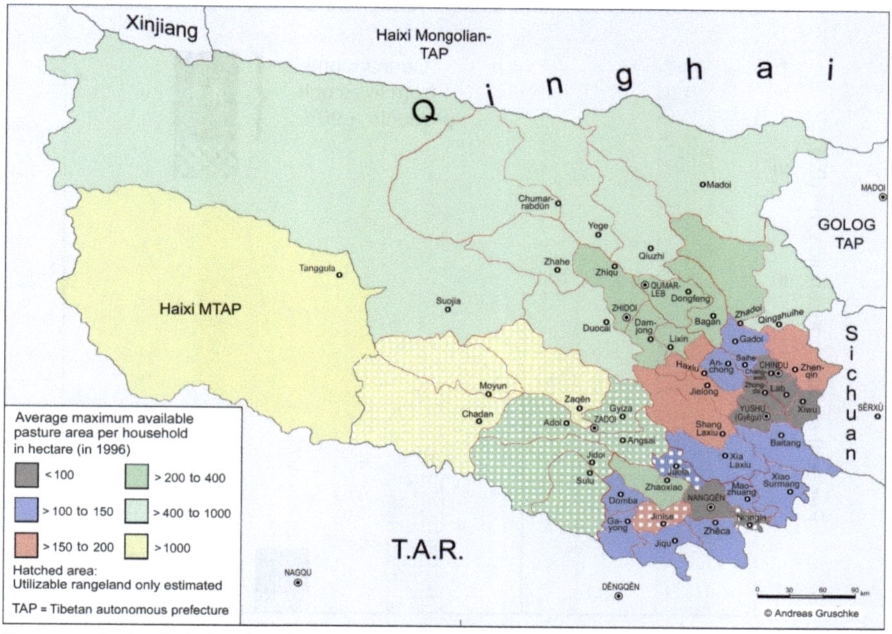

Sources: author's calculations based on data in YSB 2000 and YZZ 2005

Fig. 15.2 Uneven rangeland distribution in Yushu (Sources: Author's calculations based on data in YSB 2000 and YZZ 2005)

express in numbers, but the extent is perceptible and felt by pastoralists. Almost 70% of the 296 Yushu households we surveyed in 2006/2007 referred to degradation as a major concern, slightly less than their complaint about the general lack of pastures (74.3%).

Unlike in historic times, the region does not offer anymore unused space to make new grassland accessible elsewhere. The availability of pastureland is far from being balanced (Fig. 15.2). The most extensive pasture area is available where the population density is low. New households of the young generation had no chance to develop bigger herds and a sound basis for animal husbandry since all utilisable pastures were already distributed. Resettlements to more distant pastures would entail conflicts with herders who historically claimed and at present have an entitlement to that land. Remaining possibilities are the intensification of animal husbandry production, finding new modes of herd management – with pasture availability and lack of means for investment being the main limiting factors for enhancing productivity – or to extend income portfolios through non-pastoral activities.

Still, rural areas in Yushu appear to harbour a sound mobile pastoralism. When analysed in detail, however, the 2006–2007 sample survey of Yushu households reveals that less than half of them own enough animals to live above the subsistence level (25 SU per person). Between 12% and 20% of rural households did not even

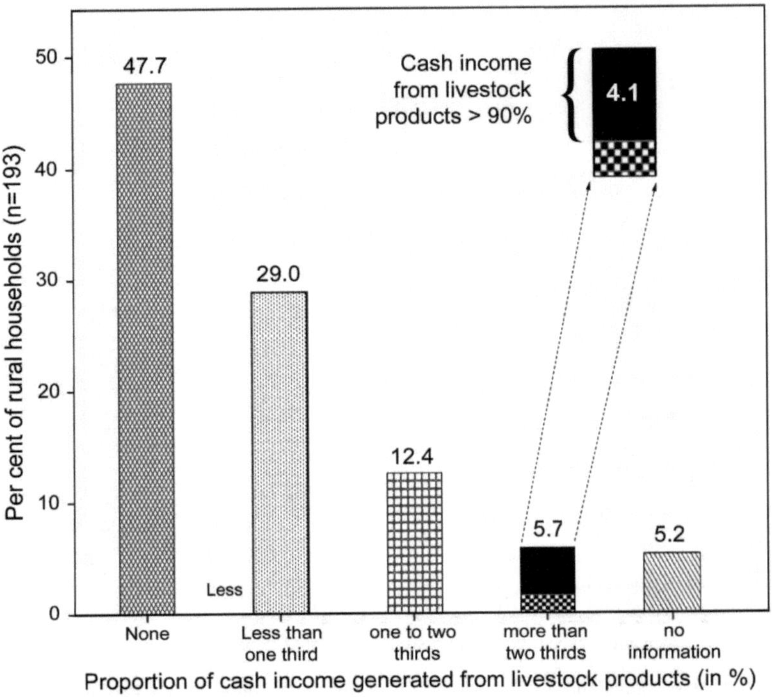

Source: Gruschke (2009)

Fig. 15.3 Decreasing importance of livestock for income generation of Yushu pastoralists (Source: Gruschke 2009)

own livestock. Almost half of the livestock owners interviewed (47.7%) could not earn any cash from animal products, most of them not even producing enough for their own sustenance. Another 29% earn less than a third of their income from pastoral economy. Only every sixth to eighth pastoral household of the Yushu survey could establish animal husbandry as main source of living (cf. Fig. 15.3; Gruschke 2009, 396–397).[12]

Under such circumstances, rural outmigration is a foreseeable consequence, and this is what gradually started in the 1990s. The extent of migration is yet smaller than could be expected since many nomads in the eastern part of the plateau have access to a unique resource: caterpillar fungus. This tonic is collected in specific parts of the grassland and helps them making up for the deficits of the pastoralist economy (Winkler 2008). It enables both rural people to stay in their home villages even if they have too few animals, and to subsist in townships and urban areas, where their lack of education and vocational training make it rather difficult to find employment. Rural migrants to towns run a high risk of being marginalised given that they compete on urban labour markets with Han, Hui and Salar immigrants. Many herders have thus become aware of the importance of education, which is corroborated by the increasing number that moves to towns because they want their children to attend better schools.

This meets the intentions of the state. By upgrading and extending the infrastructure in pastoral areas, political measures intervene more strongly into the organisation of their livestock-based economy. This applied especially for the period from 1996 to 2004 when the 'Comprehensive Set of the Four Constructions' (*sipeitao jianshe*) was implemented in the grassland areas. This project's goals were the construction of fixed settlements on winter pastures, the installation of fenced pasture areas, the construction of schools and health stations and forming long-term leasehold contracts for pastureland, which endow household with territorially bound usufruct rights for a period of 50 years.

These measures aim at minimising risks for pastoralist households by enclosing reserve pastures for the most difficult time in spring, creating possibilities for additional fodder supply and equipment of winter quarters. The consequences of the agenda for mobile pastoralism and the household economy are, however, reflected in controversial discussions.[13] Whilst the project intended to lessen the vulnerability of pastoralists, some reports indicate that it may have brought about more negative effects: restricting pastoral mobility through fences, reducing the income by limiting the number of animals, causing overgrazing near winter housing and the indebtedness of households as they have to bear their share of the costs. Such problems cannot be generalised, however, for they are far from thoroughly researched. Again, there is a lack of quantitative data, and qualitative data procure ambiguous results at times. Some statistical figures, on the other hand, overstate certain developments. The so-called enclosure movement, for instance, is widely seen as fencing privately used pastures contracted to the households. In Yushu, the fenced area fluctuated considerably during the 1980s, but was already increased before the above-mentioned 'Four-sets' project started off (Table 15.1).

The project thus amplified developments that were already underway. This is all the more true with the construction of winter houses. Before collectivisation, few nomads possessed houses in winter camps, as can be learned from biographies like 70-year-old Lobsang's[14]:

> When I was a servant [in his youth], we stayed in the tent in summer and lived in a house in winter. Nowadays, we stay in the tent from April to August and live in the house from September to next April. I would like to tell you foreigners that if you stay in the tent in winter, you'll be frozen without any doubt.[15]

In the 1980s and 1990s, more and more households started to build simple houses. With the 'Four-sets' agenda, the state supported this development, now called 'construction of fixed settlements'. Meanwhile, almost every Tibetan pastoralist household possesses a winter house, and yet can mostly continue herding on seasonal pastures. But as Lobsang recounts, other things have also become relevant for them:

> Currently, a new road is important for us [since] to fetch water is a difficult task .., and some families have a shortage of fences for grassland. Fences are important for us; if we have [some], the grass will grow well and herding animals is easier for us.

Table 15.1 Enclosure development in Yushu TAP

Year	Fenced area in ha	Time period	Increase by
1983	33,370		
1984	12,610	5 years	200%
1985	10,170		
1986	51,790		
1988	15,340		
1989	37,730	5 years	158%
1990	58,310		
1994	97,490	5 years	257%
1995	105,990		
1996	119,280		
1997	152,540		
1998	197,520		
1999	348,130	6 years	178%
2005	966,000		

Source: YSB (2000, 2006)

Fencing is not meant to interrupt the system of mobile pastoralism, even though figures seem to reproduce a different trend. Cadres use statistics to demonstrate the progress of their project implementation, but the growth rates, remarkable at a first glance, need to be compared to the general situation. In two decades the increase rate of enclosures was very high in the beginning of the 'Four-sets' (257%), but slowed down to the average before the project's start (Table 15.1). When state subsidies for fences ended in 2005, the total fenced area in Yushu was at 16.9% of its utilisable rangeland (162,800 km^2). Thus, it does not represent a movement to split up the rangeland into pieces of privately used land, but reserve pastures for harvesting fodder during the difficult spring period.

The purely socio-economic target of the Chinese long-term development strategy unfolded serious consequences. According to the Sino-Marxist inside view, the single focus on the principal contradiction disregarded 'secondary contradictions' too long. Their resolution having been neglected, existing problems increased and new 'contradictions' added on top. Thus, for instance, the disparities between Chinese coastal provinces and the 'hinterland' increased, the wealth gap between poor and rich widened and ecological problems long ignored became conspicuous. Ultimately, the 'secondary contradiction between economic development and environmental conservation' had to be addressed as it contributes to the aggravation of the 'contradiction between the modernisation of the country and the protection of its cultural peculiarities (notably of minority groups)' (von Senger 2008, 146).

In order to realise the nationwide success of modernisation efforts until 2049, the strategic goal the CCP pursues is to reduce regional disparities and to overcome the development gap between the Han majority and ethnic minority regions, thus contributing to the maintenance of social peace and national unity.[16] The means the CCP chose to overcome those disparities is the strategic 'Big Opening of the West' (*xibu da kaifa*) agenda. At the turn of the millennium this state programme started off with enormous investments in order to accelerate the economic development of poorer areas in the western half of China. The Tibetan plateau and its vast rangelands were embedded in a national framework aiming at integrating all areas that had been neglected in the hitherto existing development processes. The implementation of the projects related and investments handled needed to be adapted by regional and local administrations. This was done within political structures gradually reformed since the late 1980s. New competences to people's congresses at different levels were defined and responsibilities to regional and local administrations handed over (Paul and Cheng 2011, 107–118). This entailed political challenges, like decentralisation that initially was not procured intentionally, but yet hazarded:

> Without accompanying decentralisation of political power and the conferring of substantial degree of regional autonomy in the control and use of local resources, ethnic minorities may perceive the central State's projects as attempts at internal colonisation, leading to their outright opposition to the whole regional development strategy itself, thus exacerbating the already simmering ethnoregional tensions. (Yeoh 2008, 23–24)

One of the major consequences of decentralisation is that the outcome of project implementation is quite inhomogeneous. Major infrastructural projects – like the building of new highways and solar energy plants – were completed at different pace from province to province. Other projects like the construction of settlements, electrification and rural water supply varied extremely depending on priorities given by regional administrations – not to speak of financial 'losses' by defrauded funds.

Paying attention to the contradiction between economic development and environmental conservation, the state decided on measures to protect the fragile ecology on the Tibetan highland. A major result was the establishment of the 'Three Rivers' Headwaters (*Sanjiangyuan*) National Nature Reserve' where poorly managed mining, logging, hunting and grazing were contained, in order to protect rangelands and wildlife and promote sustainable economic development (Foggin 2005, 5–6). In designated 'core zones', herding had to be stopped, whilst pastoralist activities in all other areas were controlled. This was a reaction to the growing number of concerns coming from academics as well as from locals about ecological problems and the impoverishment of the population. The policy 'Pastures to [unused] Grassland' (*tuimu huancao*) went against overgrazing and classified the rangeland into zones where herding was to be abandoned for a number of years or seasonally.

The realisation of this policy began in 2002. It has definite consequences for the livelihood of nomads, even more so as it is accompanied by measures for 'Poverty Alleviation by Resettlement' (*yimin fupin*). These measures were unambiguously reflected in the 11th Five-year Plan (2006–2010) that is subordinate to the above-mentioned three-step strategy. The latter aims at optimising the structure of industries and expects to increase the efficiency of resources' utilisation, strengthen sustainable development efforts, bring the system of the market economy to perfection and raise the people's standard of living.[17] In autumn 2007, Party Secretary Hu Jintao emphasised the significance of a 'scientific development conception' in the sense of a comprehensive, well-coordinated and sustainable development to attain a 'harmonious society'. The Five-year Plan was supplemented by the 'National Plan for Environmental Protection' intending investments in environmental conservation on the level of 1.35% of the GNP (von Senger 2008, 120, 146–147).

One of the best-known interventions in this context is the so-called 'Eco-migration' (*shengtai yimin*), a programme that was started in ecological problem zones. The emphasis on environmental conservation entailed the resettlement of herders in towns to take pressure from rangelands – first implemented in Inner Mongolia and later in Tibetan pastoral areas. In Qinghai alone, the investment for the 'Eco-migration' scheme amounted to 7.5 billion Yuan, covering the construction of houses, related development of infrastructure, financial support for the households concerned, promotion of vocational training, etc. (Gruschke 2009, 356). This programme is different from the 'Four-sets' settlement policy, although the housing built make it look the same. Within the resettlement scheme, households receive terminable subsidies in order to become residents of county towns and townships, sell their animals and waive the land use rights of their pastures temporarily (5–10 years). There is, however, the expectation that many will continue to stay in urban settlements in case they find employment. Still, according to their agreement with the government, they are entitled to dispose of their contracted pastureland again once the 5–10 years are over.

On the other hand, the Yushu survey (Gruschke 2009) revealed that a considerable number of migrants from rural areas already moved to county towns and Gyêgu. In two of the sample villages, Yarcer and Zhêca of Nangqên County, between one third and half of the registered households had already taken residence in the county town during the past 15 years. Pastoralists who have little prospects to live from animal husbandry as snow hazards made them lose their livestock or their herd is too small, are often willing to accept government resettlement schemes. As such, the 'Eco-migration' fails in its objectives since such families do not help to reduce pressure on the grassland (whilst they help to lessen the demand for resources in their village): it rather turns out to be a 'poverty alleviation migration' programme. Furthermore, the projects tend to be failures since other project parts, namely, educational and vocational training, are not well managed, if supplied at all. Problems of corruption on local administration levels sometimes further deprive resettlers of some of their benefits.

The framework of political and administrative conditions determining the transformation processes that pastoralists in Yushu – and elsewhere in the Tibetan plateau – undergo are considerably stimulated and led by processes related to changes of technical (media, communications), environmental (global warming) and demographic nature (population growth). Together with the integration of Chinese economy into the world market, they severely affect the socio-economic structure in the Tibetan plateau and the livelihood of nomadic societies. Households of herders make their choices: many stay in the countryside to continue their habitual life, adapting from modern life what they deem useful; others take their chance and settle in newly developing urban areas. By virtue of leaving their villages migrants help those staying back to dispose of, more or less, enough resources. Some of them may fail in towns, whilst others might be successful. Unfortunately, so far research has been done on state programme resettlers, but little is known about how those migrating on their own initiative deal with the transformation of their society.

15.4 Conclusion: Problems Remaining Unresolved and Future Prospects

The 're-nomadisation' of the 1980s in Tibetan pastoral areas has clearly exceeded its climax. The shrinking resource base of a growing number of pastoralist households has to be held responsible for this. Together with higher overall stocking rates and an increasing menace by ecological factors, rangelands gradually lose their capability for regeneration. Grassland is in higher demand since due to the growing human population, the overall number of the animals has grown even though more and more households have smaller herds. The disproportional change in the livestock–population ratio elucidates the decreasing importance of livestock for pastoral livelihoods. Simultaneously the surface of utilisable pastures has declined.

Until the mid-twentieth century, the economy of Tibetan nomads was a subsistence-oriented mobile pastoralism with yaks and sheep as most important livestock. Herders met their grain requirements by exchange with settled farmers. As livestock keepers, they need to produce surplus animal products in order to exchange them against goods like grains, oil, salt, etc. Still in the early 1980s, the situation of animal husbandry in Yushu allowed for equilibrium. With the growth of the human population, the situation has changed. A rising number of pastoral households find themselves forced to develop new economic opportunities. The increasing pressure on pastoral resources even resuscitated grassland conflicts that had been settled in the 1950s. To mediate such conflicts, the unloved state plays an increasingly important role – as he does in other problem fields. This explains the growing number of state interventions since animal husbandry was handed back into private responsibility.

At national and provincial levels, the Chinese central government has introduced policies that in their way of decision-making are still guided by Sino-Marxist ideology. However, due to changes in administration enabling a certain degree of decentralisation, the measures for implementation are adapted to local conditions. Not only are such conditions extremely varied, but also the engagement and the expertise of cadres as well as their willingness and ability to consider requests of the herders are highly variable. Hence, such policies are, on the local level, often diverse at best, if not conflicting, confusing or even counterproductive at worst. Whilst state interventions in China definitely aim at solving problems reported by both pastoralists and academics, successes tend to be limited and new problems arise.

In the long run, there will only be a chance to improve results if herders are not just meant to accept policies decided, but rather if they are included in the process of decision-making. Participation in planning and implementation still sounds like a long way to go, although there are cases in which local administrators and pastoralists, often under involvement of NGOs, have managed to do so – like in cases of co-management as advocated by ICIMOD (Banks et al. 2003). Local adaptations demonstrate that herders have expertise enabling them to play a part in the process, if they are allowed to. State interventions in the form of projects are accepted by pastoralists more than it is commonly believed. Major complaints are about measures that are not consistently implemented, conditions that have changed and local administrators that do not stick to what was promised:

> Since the government implements the project 'Pastures to Grassland' we thought we can get financial support from the government. The central government promised that we can get financial support if we give up our animals and return the pastures. (...) Even though we have lots of difficulties I do not want to complain to the central government. We need a house to live in and a regular income. The central government invests a lot of money, yet it is running to the pockets of officials. We received very, very little. This caused the suffering of the people in our village. (...) My hope is that the central government gives the money directly to us. This is not only my wish, it is the wish of all the villagers. If it will be like this, I can die one day in peace. I trust in the central government and I do believe in the PR China.[18]

One of the major issues of all projects in China is met with here: monitoring and evaluation. The same people who implement projects also give their feedback to higher-level authorities: This leaves little space for critical analysis. Official statements on projects in public media let the projects appear as a success, if huge amounts of allocated funds were fully spent. This provides good opportunities for misappropriation. Corruption amongst local Tibetan administrators is as big a problem as in the rest of China, and this creates tensions, even if projects are acceptable to local people. This becomes more serious for pastoralist households, since due to the remoteness of their places of living, they have less possibilities to access enough information about how policies are and can be interpreted, what kind of rights they have or should have, and by which means they could organise themselves to insist on them (Gruschke 2009).

It is true, as Melvyn Goldstein[19] expressed, that there is enough evidence that from 1949 until the end of the CR, there were less changes in the pastoral management in Tibetan areas than after the economic liberalisation that started off in the early 1980s. We cannot be sure, however, that all the changes were only induced by policies. Coincidences alone are no evidence, since they also coincide with global processes of modernisation, market integration, demographic change, etc. Those may play their part in the story. Policies and interventions in Tibetan pastoral areas cannot be understood without regard to the ideology-guided decision framework of China's general economic policy, whilst successes and failures need to include an analysis of the structural problems and shortcomings in its implementation, monitoring and evaluation. So far, it has to be worried that changes in policies will not help much as long as the monitoring and evaluation system cannot be improved.

Acknowledgement Data presented in this chapter were collected in the research project titled 'Nomads without Pastures?' carried out in the framework of the Collaborative Research Centre SFB586 'Difference and Integration' and financed by the Deutsche Forschungsgemeinschaft (DFG).

Notes

1. 'Nomadic' refers to the term drokpa ('brog pa) representing the self-image of people whose identity is built on being mobile pastoralists in marked contrast to the people practising farming. This reflects, until today, the Tibetan understanding of a way of life based on economic

activities related to animal husbandry, mobility and specific traditional values different from other groups of the region (cp. Gruschke 2009).

2. This concise term for 'Socialism with Chinese characteristics' (you zhongguo tese shehuizhuyi) was introduced by von Senger (1994, 207–209, 230–232; 2008).
3. For detailed references see Gruschke (2009, 65–70).
4. Cited according to Watson (1980, 24), based on an edition of Mao's Collected Works (1949).
5. Goldstein (1991, 95) and (1994, 110–112).
6. Between 1937 and 1945 the principal contradiction was the 'Chinese people against Japan'; thereafter until 1949 it consisted of the party confrontation between CCP and KMT (Chinese National Party). From the foundation of the PRC until 1978, the contradiction between the proletariat/peasantry and the bourgeoisie/gentry led the party to focus on class struggle.
7. Op. cit. Constitution of the Communist Party of China, amended and adopted at the 17th National Congress of the Communist Party of China on Oct. 21, 2007.
8. Statute of the CCP, cp. von Senger (2008, 111–116).
9. For a period of time, dues for the lease of pastures were levied as well as local taxes for productive livestock and school fees. Those were ultimately abolished in 2005.
10. The livelihood situation can be vaguely assessed by calculating the so-called 'sheep units' (SU) per person. It is a reference unit to make different livestock on the pastureland and herder's livelihoods comparable. Mostly one yak is calculated like five sheep. According to Miller, a person would need at least 25 sheep or five yaks to meet her basic needs (Gruschke 2008, 11).
11. Cf. the synopsis in Gruschke (2009, 91).
12. It should be noted that the pasture-rich western part of Yushu is underrepresented in the survey. The study reflects, however, the situation of the most densely populated pastoral areas and thus of the larger proportion of the Yushu pastoralists.
13. Cf., for example, Banks et al. (2003), Bauer (2005), Liu Yimin (2002), Goldstein (1996), Goldstein and Beall (1989), Horlemann (2002), Miller (2000), Sheehy et al. (2006), Wu and Richard (1999) and Yeh (2004, 2005).
14. Lobsang is the head of a nomad household in Gyiza, Zadoi County (interviewed in January 2007).
15. No specific months were mentioned for the 'summer' and 'winter' periods, but apparently the period of staying in the house was extended – at least for the elders.
16. China State Council, http://www.China-west.gov.cn/english (downloaded 10.11.2003).
17. The 11th Five-year plan, http://german.China.org.cn/china/archive/china2006/txt/2007-01/19/content_7681420.htm (downloaded on 6.8.2008).
18. Tsering Topgyal, 57-year-old nomad from Shang Laxiu, resettled to Gyêgu and interviewed in January 2007.
19. Acc. to Goldstein's presentation 'Changing Patterns of Resource Use and Pastoral Management in Western Tibet over the Past Half Century: A Case Study of the Nomads of Phala', Wittenberg, Nov. 30, 2007 (Symposium 'Paradigms of a Nomadic Mode of Living. Tenets and Perils of Coexistence', SFB 586 'Difference and Integration', 30.11.–2.12.2007).

References

Banks T, Richard C, Ping L, Zhaoli Y (2003) Community-based grassland management in western China: rationale, pilot project experience, and policy implications. Mount Res Dev 23(2):132–140

Bauer K (2005) Development and the enclosure movement in Pastoral Tibet since the 1980s. Nomad Peoples 9(1/2):53–81

Ekvall R (1972) Demographic aspects of Tibetan nomadic pastoralism. In: Spooner B (ed) Population growth. Anthropological implications. MIT Press, Cambridge, MA, pp 265–285

Foggin M (2005) Promoting biodiversity conservation and community development in the Sanjiangyuan region. In: Proceedings of a conservation planning meeting co-hosted by Plateau

Perspectives, Upper Yangtze Organization, Government of Zhiduo County and the Sanjiangyuan National Nature Reserve Management Bureau, 30 Sept–13 Oct 2005. Plateau Perspectives, Xining
Gertel J, Gruschke A, Breuer I (2009) Regionalisierung und Urbanisierung in nomadischen Gesellschaften – am Beispiel Osttibets. Internationales Asienforum 40(1–2):119–141
Goldstein M (1991) Change and continuity in nomadic pastoralism on the western Tibetan plateau. Nomad Peoples 28:105–122
Goldstein M (1994) Change, conflict and continuity among a community of nomadic Pastoralists: a case study from western Tibet, 1950–1990. In: Barnett R, Akiner S (eds) Resistance and reform in Tibet. Hurst & Co, London, pp 76–111
Goldstein M (1996) Nomads of Golog: a report. Manuscript
Goldstein M, Beall C (1989) The impacts of China's reform policy on the nomads of western Tibet. Asian Surv 29:619–641
Gruschke A (2008) Nomads without Pastures? Globalization, regionalization and livelihood security of nomads and former nomads in Kham. In: Bauer K et al (eds) In the shadow of the leaping dragon: demography, development, and the environment in Tibetan areas. JIATS, 4. http://www.thlib.org/static/reprints/jiats/04/pdfs/gruschkeJIATS_04_2008.pdf
Gruschke A (2009) Ressourcennutzung und nomadische Existenzsicherung im Umbruch. Die osttibetische Region Yushu (Qinghai, VR China). PhD dissertation, Leipzig
Harris R (2010) Rangeland degradation on the Qinghai-Tibetan plateau. J Arid Environ 74:1–12
Horlemann B (2002) Modernization efforts in Golog: a chronicle, 1970–2000. In: Huber T (ed) Amdo Tibetans in transition: society and culture in the post-Mao era. Brill, Leiden, pp 241–267
Liu Y (2002) A new strategy and the old land: the impact of China's regional development policies on a Drokba community in Tibet. In: Gelek L (ed) Nomads of Tibet: anthropological fieldwork reports from China. M.E. Sharpe, Armonk, Manuscripts
Miller D (2000) Tough times for Tibetan nomads in western China: snowstorms, settling down, fences, and the demise of traditional nomadic pastoralism. Nomad Peoples 4(1):83–109
Paul C, Cheng L (2011) China verstehen. St. Ulrich, Augsburg
Sheehy D, Miller D, Johnson DA (2006) Transformation of traditional pastoral livestock systems on the Tibetan steppe. Sécheresse 17(1–2):142–151
von Senger H (1982) Partei, Ideologie und Gesetz in der Volksrepublik China (= Schweizer Asiatische Studien 5). Peter Lang, Bern
von Senger H (1994) Einführung in das chinesische Recht (= Schriftenreihe der Juristischen Schulung, 124: Ausländisches Recht). Beck, München
von Senger H (2008) Moulüe – Supraplanung. Unerkannte Denkhorizonte aus dem Reich der Mitte. Hanser, München
Watson A (1980) Mao Zedong and the political economy of the Border region a translation of Mao's economic and financial problems. Cambridge UP, Cambridge
Winkler D (2008) The Mushrooming Fungi market in Tibet. Exemplified by Cordyceps sinensis and Tricholoma matsutake. In: Bauer K et al (eds) In the shadow of the Leaping Dragon: demography, development, and the environment in Tibetan areas. JIATS, 4. http://www.thlib.org/static/reprints/jiats/04/pdfs/winklerJIATS_04_2008.pdf
Wu N, Richard C (1999) The privatization process of rangeland and its impacts on the pastoral dynamics in the Hindu-Kush Himalaya: the case of western Sichuan. In: Eldridge D, Freudenberger D (eds) People and rangelands: building the future. VI International Rangeland Congress, Aitkenvale, pp 14–21
Yeh E (2004) Property relations in Tibet since decollectivization and the question of 'fuzziness'. Conserv Soc 2(1):108–131
Yeh E (2005) Green governmentality and pastoralism in western China. 'Converting pastures to grasslands'. Nomad Peoples 9(1–2):9–29
Yeoh K-K (2008) China's interregional imbalance: development, ethnicity and socioeconomic Malaise, ICS working paper no. 2008–6. Institute of China Studies, University of Malaya, Kuala Lumpur. http://ics.um.edu.my/umweb/ics/workingpaper/2008–6.pdf

YSB = Yushu TAP Statistical Bureau (ed) (2000) Yushu Zangzu Zizhizhou Tongji Nianjian 1950–1999 (Statistical yearbook of Yushu TAP 1950–1999). Yushu TAP Statistical Bureau, Yushu

YSB = Yushu TAP Statistical Bureau (ed) (2006) Yushu Tongji Nianjian 2005 (Statistical yearbook of Yushu 2005). Yushu TAP Statistical Bureau, Yushu

YZZ = Yushu TAP Local Archives Compilation Committee (ed) (2005) Yushu Zangzu Zizhizhou Zhi (Gazetteer of the Yushu Tibetan Autonomous Prefecture), 2 vols. Yushu TAP Local Archives Compilation Committee, Xi'an

Chapter 16
Enclosure and Resettlement in the Eastern Tibetan Plateau: Dilemma of Pastoral Development During the Last Three Decades

Wu Ning, Yan Zhaoli, and Lu Tao

Abstract The eastern Tibetan Plateau (ETP) in China is one of the most productive pastoral areas in highland Asia, which has been always placed on the crest of policy-driven changes for pastoral development and rangeland management in the last few decades. Profound changes from traditional pastoralism in the eastern Tibetan Plateau included the collectivization of livestock and rangeland from the late 1950s, privatization of livestock from the early 1980s, enclosing rangelands for private utilization from the 1990s and massive settlement and resettlement of herders in 'Pastoral New Villages' from the twenty-first century. The policy-driven changes initially aimed only to maximize livestock production but later on added in a deliberate intention to modernize pastoralism and prevent rangeland degradation. What in some places of experimental site seemed to work well in rangeland privatization or pastoralists' settlement, nevertheless, brought in many other situations more pressures to both people and their environments. China is spending significantly increasing amount of resources in rangeland management, but the effectiveness of the programmes in achieving dual targets of fair pastoral development and environmental conservation are questionable. The challenge is becoming even greater and has given the impacts of global development and climatic change to the ecologically fragile and economically vulnerable eastern Tibetan Plateau. New policies for ecosystem compensation or payment for ecosystem services might be an alternative if wisely formulated and implemented.

Keywords Tibetan Plateau • Herder settlement • Pastoral development • Rangeland enclosure • Ecosystem compensation

W. Ning (✉) • Y. Zhaoli • L. Tao
Chengdu Institute of Biology, Chinese Academy of Sciences,
P.O. Box 416, 610041 Chengdu, Sichuan, China
e-mail: wuning@cib.ac.cn; zhaoliy@cib.ac.cn; lutao@cib.ac.cn

H. Kreutzmann (ed.), *Pastoral practices in High Asia*, Advances in Asian Human-Environmental Research, DOI 10.1007/978-94-007-3846-1_16,

16.1 Introduction

The eastern Tibetan Plateau (ETP) covers northwestern Sichuan, southeastern Gansu, southeastern Qinghai and northwestern Yunnan Provinces in China. Most of the ETP is topographically mild with small hills interspersed with rivers, with an exception of high mountains and deep valleys of the Hengduan Mountain Range that cuts off the Tibetan Plateau in the east. ETP is generally pure pastoral, though small patches of barley or beans cultivation can be found in catchments. The whole ETP is the water source area of the Yangtze (Changjiang), Yellow (Huanghe) and Mekong (Lancangjian) Rivers. The average elevation in ETP is about 3,600 m above sea level, the annual precipitation is some 600–800 mm, and the annual mean temperature is about 1°C. Due to the winding rivers and changing riverbeds, much of ETP rangeland has extensive development of marshland and peatland. Various types of subalpine meadows are dominant plant communities in ETP. ETP rangeland is some of the most productive land on the Tibetan Plateau, with often 30–40 vascular plant species per square metre that produces an annual dry matter of 80–150 g.

Grazing of cold tolerant livestock mainly including yak, sheep and horse in ETP has been the dominant and environmentally sustainable land use pattern for thousands of years. However, due to a multitude of factors, rangeland resources in the Tibetan Plateau and in the eastern part in particular are increasingly under pressure to sustain increasing livestock production and to meet the increasing demand of local dwellers and an ever-growing market. Studies show about 50% biomass reduction in ETP from the 1950s to 2000s, companied by the increase of unpalatable species in plant community and the expanding desertification (Ze 2005). The key factors causing rangeland degradation on the Tibetan Plateau include population growth, climatic warming and policy changes. Dynamic policies have modified land use characteristics and have resulted in increased rangeland degradation and livelihood vulnerability (Dong et al. 2011).

As one of China's five major pastoral production areas, ETP is often chosen as a demonstration site for implementing new rangeland policies and pastoral development programmes. This section summarizes the authors' long-term study in China's rangeland policies and pastoral programmes, as well as field research findings from mainly Hongyuan and Ruoergai (also as Zoige) counties of Sichuan Province and Maqu County of Gansu Province in ETP (see site No. 15 in Fig. 1.3). It focuses on a time frame of the last three decades since the privatization of livestock in 1983 but covers a wider time span for backgrounds.

16.2 Policy-Induced Rangeland Enclosure

16.2.1 Major Policy-Driven Changes in Rangeland Management

Before 1958, the Democratic Reform of China, local people in ETP practised tribe-based communal tenure and highly nomadic traditional livestock grazing systems. With low population and livestock density, the traditional system had been able to retain its mobility and feasibility in adapting to the harsh environment without major changes for hundreds to thousands of years. Rangeland was however not open and accessible to all, but was controlled by different tribal leaders, rich families and monasteries with rough boundary demarcation by natural distinctive landmarks such as mountain ridges and water bodies. The average survival rate of young livestock was about 50%, and it was not uncommon for a family to lose all their livestock in a natural disaster, especially heavy snow in the early spring. The tribal leaders had generally unique decision-making privileges pertaining to rangeland use, and their herds were shifted to new rangelands ahead of others in the same community. Each year, they reallocated rangelands to different sub-tribes and determined grazing routes. They used to move their tents and herds in search for water and grass from late April to late October. When it got very cold in late October, they returned to their winter pastures and stayed there till next April.

Between 1958 and 1982, both rangelands and livestock were collectivized into two types of production units. One was 'Mutual Aid Agricultural Cooperation' (1958–1973) followed by 'People's Commune' (1973–1983) managing 'Production Brigades' and 'Production Teams', and another was the State-Owned Farm. During the period, livestock number in Hongyuan County alone increased from 130,120 to 486,500 heads (Hongyuan County History Editorial Committee 1996). Livestock owned by collectives was usually divided into different groups, like milking cows and calves, young yaks (2–3 years old), bulls, castrated and ready to sell yak, sheep, etc. Labour forces were accordingly divided into different groups, each having their own target livestock group, to accomplish their assigned tasks to earn work-points or wages. The ownership of pastures belonged to the state, but the actual use right was authorized to individual production units. The leadership of a production unit decided the division of tasks and grazing routes by listening to experienced herdsmen. People could only follow given routes to graze their assigned group of livestock within production unit boundary.

Starting in 1983, all livestock of production units were equally allocated to individual households according to the number of families regardless of their age and sex, and the pastureland was still kept for common use until the mid-1990s. Herders no longer needed to work together in groups as they used to in the last decades, and they gained full authority for managing their own livestock. The introduction of household livestock contracts greatly stimulated pastoral productivity, which also resulted in most households attempting to raise as many livestock as they could on common pasture within their production unit. Pasture boundary fights between administrative communities were very common. It was difficult to get an exact number of livestock from individual households. Rangeland degradation, however, was becoming more rapid and serious than ever before as illustrated by the increasing desertification areas or piles of exposing black soil caused by rodents, due to over grazing. At the beginning of the twenty-first century, about 90% of China's rangeland was degraded of which half was seriously degraded. For example, the desertification area in Maqu County in ETP was 1,140 ha in 1985 but 53,400 ha in 2003, and in the meantime, the biomass of Maqu rangeland reduced from 5,860 kg/ha in 1981 to 4,000 kg/ha in 2004.

In 1995, as one of the 25 selected pastoral demonstration counties in China, Hongyuan County initiated an 'Integrated Pastoral Socio-Economic Development Programme'. Rangeland was divided amongst individual households or household groups (from 10 to 20 households) on a long-term contract base. Each household was required to build a house, an animal shed and a hay storage barn. Under the new programme, pastoralists were supposed to have responsibility to their contracted pastureland and to invest in pasture improvement. Ever since then, rangeland lease policy was made effective throughout pastoral China. By the end of 2003, about 70% of China's rangeland was leased through long-term contracts, of which 68% was contracted to individual households and the rest to groups of households or to villages (Yan et al. 2005). Those rangelands which were not leased out were either identified as unusable rangeland (not suitable for livestock grazing), protected areas or kept for special uses such as seed production and disaster defensive base. By the end of 2008, a total of 22.8 million ha or 71.5% of China's usable rangeland had been contracted (MOA 2009).

In Aba Prefecture in ETP (including Hongyuan, Ruoergai and some adjacent counties in Sichuan Province), 54.36 million *mu* (a Chinese land measure, 1 ha is equivalent to 15 *mu*) or 94% of the total usable rangelands was contracted by 2000. Of this, 29.90 million *mu* was winter pasture and 24.46 million *mu* summer pasture. A total of 57% of contracted pasture or 31.00 million *mu* went to 31,000 households; 20% or 10.77 million *mu* went to 674 household groups (a household group contracting shared rangeland consisted of 6–12 households, which might be formed at voluntary base or Village Committee assignment); 22% or 11.70 million *mu* went to 1,500 villages; and the rest was kept for emergency use or was used by work units (Sichuan AHB 2001). These figures show that 29.90 million out of 31.00 million *mu* of rangeland contracted to individual households was cold season pasture, and this indicates that most pastures for warm season use were allocated to household groups or villages. About 390 households, or one-twentieth of the rural households in Aba County, Sichuan Province, have been allocated one plot of rangeland each for year-round grazing. The ratio of traditional winter pasture was around 30–40% of total useable grazing lands. However, in the Aba Prefecture in ETP records for rangeland contracts, about 55% of allocated rangeland was cold season pasture. This means that some traditional summer pasture has been leased to herders in the name of winter pasture.

16.2.2 Rangeland Enclosure: Main Policy Items and Local Practice

The Rangeland Law of the People's Republic of China (1985, revision in 2002) articulates that all China's rangelands belong to the state, except for those having legally assured collective ownership. It has also regulated for the first time that state or collectively owned rangeland can be authorized to all-people-owned units or collective economic bodies for protection, construction and proper uses, which can

Table 16.1 Rangeland grades and the productivity of standard *mu*

Sites	No. of rangeland grades	Equivalence to standard *mu*, starting with grade I	Green forage per standard *mu* (kg)	Percentage of standard *mu* in the county's useable rangeland (%)	Carrying capacity per sheep unit (*mu*)	Average rangeland allocated to per person (*mu*)
Hongyuan, Sichuan	6	1.4, 1.2, 1.0, 0.8, 0.6, 0.4	360	67	5	270–350
Ruoergai, Sichuan	4	1.0, 0.85, 0.65, 0.5	360	64	5	220–250
Maqu, Gansu	3	1.0, 0.9, 0.75	400	3	7.6	230–300

Source: Adapted from Yan et al. (2005)
mu is Chinese land measure and 15 *mu* equals 1 ha

then further lease the rangeland to individual households or household groups within its managerial territory. It is firmly believed by the decision-makers that privatization of pastureland should make herders more responsible for limiting their herd sizes and investing in rangeland improvement for sustainable use. The law has legalized rangeland enclosure, but did not give any instructions on how. Rangeland policy actually followed the great success of cropland privatization in China, without conducting in depth research on the differences of traditional management patterns and uses of these two types of land, or the heterogeneous nature of rangeland.

People living in pastoral areas nevertheless did their best to allocate rangeland within a village, which is the legal unit in dividing rangeland. Hongyuan County was in the first batch of 25 experimental counties to allocate rangeland to individuals in China. The pilot experiment allocation was carried out in two villages in Longrang Township near Hongyuan County Town, where there was good pasture and abundant water resources. Experimental rangeland allocation was then quickly extended in Hongyuan, Ruoergai and Maqu Counties, and massive rangeland distribution was done throughout ETP in 3 years time from 1996 to 1999. People applied standard *mu* to balance rangeland quality and quantity in the distribution process. One standard *mu* rangeland produces certain amount of forage, which means one actual *mu* of good rangeland with higher productivity is calculated as more than one standard *mu*, and vice versa. Grade levels and green forage productivity per standard *mu* were not the same in those three counties (Table 16.1).

The process of rangeland allocation in ETP mostly entailed, first, classifying and calculating the amount of village rangeland on a topographic or sketch map; second, calculating how much area each household could get according to their own agreed allocation criteria; third, having one representative from each household draw a numbered lot that matched a locality on map; fourth, allocating rangeland on map or in the field based on the locality and calculated area; and lastly, certifying each family's contracted area of rangeland by local government. The average rangeland area allocated per person ranged from 15 to 20 ha in ETP in different villages.

Rangeland allocation criteria included the number of family members at the time, number of livestock in 1982 when it was privatized (or occasionally that of the rangeland allocation time), the productivity of rangeland and the location of pastureland based on traditionally seasonal uses, but ratios of the weight varied between the different villages. Whilst many complaints and disputes emerged during the allocation process, it just went on, and most of the distribution results were non-changeable.

16.2.3 Local People's Perception About Rangeland Enclosure

Pastoralists' complaints about the rangeland allocation process in ETP were mainly with regard to the unfairness due to the rush without careful planning or assessment. Mr. Chike, a 65-year-old herder (in 2002) in No. 3 Village of Longrang Township, Hongyuan County, related the story of rangeland allocation in his village. Some experienced herders were firstly invited to look over all their available rangeland resources on horseback and give rough evaluation by classifying pasturelands into three grades. Much of pasture (or say usable rangeland) was categorized by them as the third class, but supervising leaders from the county town said later that there was far too much third class and revised a lot of the land into the second class. They used only 3 months from rough pasture evaluation by experienced herders in May to completing village rangeland allocation in August 1996. People responsible for rangeland allocation were not experienced, yet mistakes made in the process were not corrected. In the process, not enough consideration was given to the elevation or traditional uses in rangeland division, which resulted in some family contracting traditional summer or autumn pasture for year-round use – Huaerdeng family in the village, for instance, contracted over 2000 *mu* of pasture all of traditional winter pasture and they had to escape their own heavily snow covered land to rent pasture in lower area of 3,500–3,700 m every winter. In the same village, three *mu* of shrub land was calculated as one standard *mu* for allocation at the beginning, but it changed to two *mu* shrub land for one standard *mu* later on in the process due to insufficient pastureland being available for allocation. In No. 2 village of the same township, people also complained about the rush in rangeland allocation, going from a rough sketch map to unclear boundaries in the field. In Xiaman and Axi Townships of Ruoergai County, herders' complaints focused on water resources availability, where 80% people and livestock faced drinking water shortage after rangeland allocation.

Local herders' reaction to the rangeland privatization policy itself has been diverse and dynamic. Generally speaking, a rich family with large number of livestock hardly appreciates the policy, whilst a poor family either hates about being more marginalized in remote areas through unfair rangeland allocation or embraces it welcomingly if they contract good size and location of pasture. After the rangelands had been allocated, rich families with big herds started renting other's pasture for livestock grazing. So it actually would benefit poor families and would narrow down the wealth gap, if the distribution itself was open and fair. Most of the herders interviewed, rich or poor, mentioned increasing boundary conflicts after rangeland allocation, especially when boundary fences were not available. Many herders told about increases in cases of stolen livestock after rangeland allocation because of separating herder families by living far from neighbours. However, the majority of herders still seemed to support rangeland enclosure/privatization for the following reasons: (1) it promotes equitable access to rangeland, (2) poor families can rent out their land as an alternative source of livelihood, (3) it enhances herders' ownership and responsibility towards managing rangeland, (4) it is easier to manage a smaller area and (5) it reduces labour requirements for pastoral production and restricts livestock from running about and losing weight.

Herders who did not like rangeland allocation grouped themselves together by close kinship or neighbourhood to share their pasture, but lots of such voluntary small household groups eventually split up in the last 15 years except in the places where fencing is not available or feasible. For instance, people still practice community-based rangeland management in Upper and Lower Reer Villages of Tangke Township, Ruoergai County where fair rangeland allocation is practically not feasible. People do not acknowledge that rangeland allocation *per se* can improve ecosystem, since, they argue, the same number of livestock is still in the same area after rangeland allocation, only rich families have choices of over grazing on own or rented pastures if not reducing their herd size.

16.2.4 Post-Rangeland Allocation Policies and Programmes

Even after the legalized distribution of rangeland in ETP, the ownership of rangeland still belongs to the state, and, therefore, the state government supervises and can readjust the use patterns. With improved understanding of the multiple functions and services of rangeland ecosystem, the Chinese government initiated a number of big programmes in ETP not long after the allocation of the rangelands to promote integrated pastoral development and proper rangeland management. In 2000, the State Council initiated a national project to force the return of non-productive sloping croplands to forests and rangelands in head water areas in 2000, for which semi-pastoralists received compensation of grain, cash and seedlings or seeds annually. This was followed by another national project on convert degraded grazing lands into grasslands from 2003 to 2005. For implementing the second project, all severely degraded rangeland was supposed to be fenced to restrict or ban livestock grazing. In return, pastoralists get compensation in the form of cash and/or food. By the end of 2008, rangeland periodically or permanently banned from livestock grazing in China reached 9.9 million hectares or a quarter of the total.

The Chinese government spent a huge amount of money on controlling livestock grazing in rangelands, but the results of this outlay are questionable. The government subsidized herders with an amount of 173.25 Yuan in 2002–2004 and 210 Yuan in 2005–2006 for fencing 1 ha of rangeland, and annual fodder compensation of 82.5 kg grain per hectare or equivalent cash. From 2005 onwards, the state council has further increased restoration compensation to 300 Yuan per hectare rangeland, of which 70% is provided by the central government and the rest 30% is to be covered by local government and individuals. Meanwhile, herders could also receive 150 Yuan seeding fee per hectare for top sowing. In an extensive field survey on herders' perceptions about the programme in the summer of 2007, most of herders interviewed in ETP were not clear about the programme, and many of them thought the government subsidy was for poverty alleviation. As for the grazing ban either for year round or for the grass growing season, it did not seem to be very effective either. For instance, even around newly developed moving sand dunes in Hongyuan and Ruoergai Counties along Baihe (White River) and Heihe (Black River), we still found livestock grazing on supposedly grazing-free desertification rangelands. Local herders told us during interviews in summer 2007 that they were never strictly requested by anyone to ban grazing from sand dunes in their contracted rangelands.

16.3 Pastoral Settlement Programmes

16.3.1 Settlement on Individually Contracted Rangeland

Traditionally, most of herder families/households in ETP had their own simple winter shelters that were located on the warm and lee side of the foothills. The frame of the shelter was normally made of bush branches or sod layer blocks, which was covered with yak dung or clay to keep the heat in. The herders would stay in their winter camp from November to next April and then move to warm pastures from May to October (Richard et. al. 2006). Policy makers and scholars living in lowlands regard the traditional nomadic lifestyle as backward, something that needed improvement. As a result, massive programmes were initiated in ETP and throughout pastoral China to settle down herders in permanent houses.

The early pastoral resettlement programme in ETP encouraged herders to build permanent houses on their own long-term contracted rangeland. Integrated pastoral development programmes called '*san peitao*' (three constructions' package), '*wu peitao*' (five constructions' package) or alike were initiated in 1996 to couple with rangeland distribution. The major focus of '*san peitao*' programme implementation in Hongyuan County was that all the herder households were to build up their own facilities including house construction for people, barn construction for livestock and forage construction including a shed for hay storage and fences on their winter pasture for winter grazing. Local governments provided no-interest or low-interest loans and subsidies to encourage herder families completing construction within a short period of time. By 2001, over 90% herder families in Hongyuan County has completed the building of a house, barn and shed on their own contracted land, and the materials used ranged from locally available resources to steel cum concrete. The result was that for the first time in ETP, herders did not have to save their valuables only in the form of livestock or small and easy-to-carry items such as expensive clothes and ornaments, but bought heavy furniture and electronic goods to put in their own houses. They could watch television like other settled people do, and they appreciate the fact of being able to sleep in dry and warm beds which is good for the many rheumatic sufferers especially for the elderly generation. A noteworthy problem was that many herders had to build their individual houses on newly contracted rangeland even far from neighbours, which led to difficulties in services and market access and increased risk to livestock security.

16.3.2 Resettlement of Pastoralists in Local Rangelands

A completely new Pastoral New Village (PNV), consisting of 120 two-storied concrete houses and a Village Committee Centre, was built in Waqie Township of Hongyuan County in 2001–2002 as a provincial demonstration trial. The new houses had advanced facilities such as tap water, metal stove, flush toilet and television and telephone cables. During that time, only two hotels in Hongyuan County town had such facilities, whereas the county town residential apartments did not have toilets attached. Each building of the PNV cost about 90,000–200,000 RMB Yuan (some 11,000–24,200 USD) depending on size. Herders from different villages of Waqie Township were encouraged to buy the new houses by paying about one-fifth of original cost and taking about same amount of loan (interest subsidized by the government). This was not very attractive to all herders since the cost was tremendously high for them, yet many had just built self-contained constructions on their own contracted pastures and had spent most of their savings if not all. Besides, the herders liked or had to stay close to their herds on rangeland instead of clustered concrete buildings.

Before Waqie PNV was completely built up, Sichuan Provincial leaders pinpointed that pastoral life needed qualitative improvement and that the general pastoralists should live a modern civilized life. The main leader requested putting overall efforts in improving economic, social and infrastructure conditions of pastoral areas, through the construction of Pastoral New Villages (Photo 16.1). At the same time, it was planned to build up at least one high-standard demonstration PNV in eight pastoral counties in Sichuan Province within 2 years time, so that to extend the demonstration effects. Another PNV construction in Hongyuan County started in 2002 in Anqu Township. PNV buildings were soon becoming popular not only in western Sichuan but also other pastoral places in ETP (Lu et al. 2009). Many new PNVs have been built in the following years to provide herders better living condition (Photo 16.2), but people were not all settled, and the main labour force of each family still move in tents with their livestock especially during summer.

Photo 16.1 Linga Village, a new settlement that was built for Tibetan pastoralists in Ruoergei County, Sichuan Province, in 2009. In order to improve the living and productive conditions of Tibetan herders, many new settlement clusters were established in pastoral regions of western Sichuan since 2008 under the support of national and provincial governments (Photograph © Sun Geng)

Photo 16.2 The gate of Linga Village, Ruoergai County, Sichuan Province (Photograph © Sun Geng)

16.3.3 Resettlement of Pastoralists at Different Location

Some PNVs were also built under resettlement programmes to move herders away from badly threatened but ecologically vital rangeland ecosystems. In Gansu province, the construction of high-standard PNVs was to move pastoralists away from degraded rangeland or key conservation areas such as core zone of a Maqu National Wetland Preserve. From 2004 to 2009, 62 new PNVs for 6,375 households, costing 5.28 billion Yuan on over 27,000-square-metre house constructions were built in Maqu, Luqu and other four neighbouring counties in Gannan Prefecture of Gansu Province. Qinghai Province has also resettled thousands of herders from the water-head of Three Rivers (*Sanjiang yuan*), namely, the Yangtze, Yellow and Mekong Rivers. Qinghai provincial government subsidized 30,000–80,000 Yuan (some 4,615–12,308 USD) to each family for construction costs in town, whilst subsidizing a herder family 3,000–8,000 Yuan (462–1,231 USD) annual fodder and forage compensation fee for an initially determined 5 years (Chen and Luo 2009). Chen and Luo (2009) also reported a total of 30.16 million Yuan (4.64 million USD) to be spent in Duoma County of Qinghai Province to move out 1,800 people of 388 household that would ban livestock grazing from 372,000 ha rangeland (or a reduction of 110,000 sheep unit) in 2003–2004. Many of the migrants, resettled in a completely new area, had to abandon pastoralism, by receiving least income insurance as urban citizens do.

We have no statistics on the number of PNV sites built in ETP or how many herders have been affected in the last 10 years or so. In the nation's 11th Five-Year Plan (2006–2010), a number of new policies were made focusing on rural development, especially the improvement of agricultural production and rural infrastructures including housing. From 2006 onwards, China has exempted all agricultural taxes and spent an ever-growing amount of money (300 billion Yuan 2005) in supporting rural development. By the end of 2008, the average living area of Chinese rural residents reached 32.4 square metres per person, of which 34.3% is concrete and 58.8% wooden-brick buildings, and 17.5% houses had flush toilet (MOA 2009).

16.4 Dilemma on Pastoral Development and Discussion

Throughout history, pastoral and non-pastoral stakeholders all over the world often adopt extreme positions because they have been shaped by different cultural and environmental exigencies and therefore subscribe to different paradigms of development. China has long been dominated by agricultural societies and the pastoral sector has always been considered only as one small part in agricultural policies and programmes. As a result, rangeland policies are being made and implemented by people who do not necessarily recognize or appreciate indigenous pastoral systems, nor seek to understand the interaction between pastoralists and environment under dynamic climatic and socio-economic constraints. The intention of policies on rangeland enclosure and resettlement to convert nomadic pastoralism to a ranching system through rational use of rangeland resources, has caused an obvious paradox since policies in favour of individual usufruct rights and sedentarization do not really favour sustainable rangeland management in the Tibetan Plateau.

16.4.1 Economic Viability

Livestock production is the foremost important means and in many cases the only feasible way to sustain people's livelihoods in the Tibetan Plateau. Total income contribution of pastoralism to China's overall GDP is rather small. In Sichuan Province, for instance, pastoral and semi-pastoral western land produces only 5% of the provincial GDP. Before the end of 1990s the investment of central and local governments in rangelands mainly aimed at its animal products without any appreciating its crucial ecosystem functions. There were little government inputs in China in rangeland from 1950 to 1977. From 1978, when the country's opened up to a more market economy till 1999, central government invested annually less than 2% Yuan per *mu* of rangeland (less than 5 USD cents per hectare). Even today, with an increasing input since 2000, the investment has been far from enough to stop rangeland degradation and desertification.

Whilst there is no question that winter pasture allocation has contributed to improving income for some households in the region, the vast majority of inputs require heavy subsidies from the government. Demographic and economic changes are triggering adaptive responses in nomadic societies of the Plateau, such as towards market-driven livestock production; thus, traditional barter economies and associated systems of cooperation are declining in importance, especially those adjacent to population centres, increasing reliance on cash income and widening the gap between rich and poor (Wu 1997; Miller 1998).

The investment required for fencing and house construction is a big amount and often too costly to herders. The cost of a settlement programme requirement for building a house for people, a barn for livestock, a shed for hay storage and fence on their winter pasture ranges some 30,000–70,000 Yuan. Very few households, mostly those having contracted pasture along main roads, are selected as demonstration families, who get a significant part of their cost subsidized by the government and have relatively easy access to long-term credits. The majority of other herder families, on the contrary, can barely afford all the required construction costs. The cost for resettlement is much higher than settlement, which ranges from 70,000 to 200,000 Yuan including basic living facilities. Governmental investment in pastoral settlement, resettlement and grazing ban has been increasing in last one decade, but never was it enough to cover all herders' costs. For those people resettling in a completely different place, they need also to invest in starting a totally new life, which could be mentally and financially a lot more costly than physical resettlement facilities.

Meanwhile, the cost-effectiveness of facilities like fencing on the Tibetan Plateau is questionable. At some extent, fencing winter pasture and developing artificial pasture for supplementary foodstuff could help increase livestock off-take and pastoral incomes through more intensive management. Due to climatic and topographic constraints, however, the grass growing season in the Tibetan Plateau is from May to September, yet the livestock have to spend lot of calories to cope with winter cold. Therefore, both primary and secondary productivity of the Tibetan Plateau is low, with a carrying capacity of 1 ha pasture for two sheep unit even in the most productive ETP. Such natural restraints have decided that it is impossible to apply advanced ranching system technologies to enhance pastoral production on the plateau as lowland decision-makers wish. When the fencing scale is too large that becomes only boundary marks of allocated pastures, the investment is difficult to balance by any economic outputs of pastoral production. Moreover, the cost-effectiveness is not necessarily commercialization per se. Rangeland is rather a complex ecosystem with multiple functions that provide various ecosystem services to the wellbeing of people living locally and downstream.

16.4.2 Ecological Feasibility

For thousands of years, Tibetan pastoralists have kept their livestock and stayed in harmony with rangeland ecosystem, through flexible responses to short-term variations in climatic conditions particularly in the way of mobile livestock grazing. They evolved a system employing migratory, semi-sedentary and deferred grazing practices to produce ample quantities of animal products for their own households and for trade. Chinese government officials and researchers all confess that the nation's pastoral production enhancement oriented policy approaches from 1960s to 1970s, such as plough of rangeland, draining up wetland and blindly increasing livestock numbers, have been harmful to rangeland ecosystem. As for the ecological effects of rangeland enclosure and application of forage improvement technologies, there are debates. Wu and Du (2007) reviewed different authors' reports of 68.5–158.3% increase in the productivity of palatable species by fencing alone, but scholars (Wu and Richard 1999, Yan and Wu 2005; Yan et al. 2005) argued that rangeland enclosure cause ecosystem fragmentation and the constraints in wildlife movement and seeds disperse. Besides, depopulation areas and the reduction of livestock stocking levels might result in pronounced changes in ecosystem functions and dynamics in the Tibetan Plateau (Dong et al. 2007, 2011).

The Tibetan Plateau rangeland ecosystem has been increasingly recognized as a priority eco-region for conserving biodiversity because of its highly distinctive fauna and flora species, ecological processes and evolutionary phenomena. This eco-region is the headwater environment for many major rivers, and what takes place in these rangeland landscapes has important implications for billions of people living downstream. It also plays an important role in climate change. Main policy concern about rangeland enclosure and pastoralists settlement is to enhance rangeland ecosystem health and to impede rangeland degradation and desertification, in addition to improve pastoral livelihoods and living conditions. The importance of conserving Tibetan

Plateau rangelands is beyond all doubt, but the question is to find out how fencing and settlement actually affects rangeland ecosystem and its integrity in the long run, which has to be studied carefully in the context of global change.

16.4.3 People's Vulnerability

The pastoralists are generally vulnerable because their livelihood has to depend largely on highly variable climatic conditions. In traditional ways, livestock in the Tibetan Plateau grew in an annual cycle of 'replete in summer, fat in autumn, emaciated in winter, and shattered in spring'. Adult livestock (yak, Tibetan sheep and goats) lose about 20–40% of their body weight during an ordinary cold season, when many animals, particularly the weak ones, die in severely harsh winters especially when there is heavy snow disaster at the end of winter or in the early spring. Risk imposed by environmental conditions is always a factor in meeting animal demands and livelihood needs in a pastoral system. Pastoralists have to learn to cope with the changes and to mitigate risks and avert disasters, but even so it is not uncommon for a family to lose most or even all of their livestock during a heavy snow disaster (Wu and Yan 2002). The need of grazing animals to be productive and to survive well must continually be balanced with the availability of feed, water and shelter. In this aspect, fencing winter pasture, storing hay and building livestock shed for challenging times can help to reduce herder's vulnerability or exposure to risks.

In the meantime, the surest way of reducing risk in livestock production and ensuring sustainable livelihoods is to maintain flexibility of decision-making in animal production activities, mobility of adapted animals and access to a variety of spatially and temporally distributed resources. The policy-induced enclosing of rangeland has restricted livestock mobility from searching for grass and water in a reasonable scale, which indeed increases herder's exposure to uncertainties and risks in still climate-dependent livestock production systems in the Tibetan Plateau. Likewise, government-forced reduction of herd size is good for sustainable use of rangeland resources on the one hand, but it destroys herders' traditional way of insurance on the other – herders used to keep large herd sizes as a means of insuring more survivals from disasters (Wu 1997).

16.5 Suggestions and Way Forward

China's first priority for Tibetan and other rangelands has been clear in maintaining and restoring the ecological sustainability of the rangeland ecosystems for present and future generations, which means that the country is changing its rangeland policy focus from sustaining livestock outputs to sustaining ecological services and a wide variety of goods and values of rangelands. This should however not be at the cost of local dwellers' suffering, but requires well balancing the diverse economic and social needs of pastoralists along with conserving biodiversity and watershed values.

Rangeland in nature is heterogeneous that requires adaptive management, for which mobile livestock grazing has been proven to be the most effective scheme from African dry lands and more (Scoones 1994; Niamir-Fuller 1999). In this regard, fencing and settlement is not in favour of sustainable rangeland, but we cannot deny that the modernization programmes do help to improve access and services to remote pastoral areas in the Tibetan Plateau. As a matter of fact, mobility of livestock does not mean people cannot be settled. If pastoralists want, they can build winter houses where the old and young can live and easily go to hospital or school, but some people from a family or community should keep taking livestock to different seasonal pastures. Pastoralists should also have their own choices of living a way they like, be it mobile, settlement or resettlement.

There are always numerous uncertainties and risks that can hardly be controlled in accomplishing ecological sustainability into development plans for Tibetan rangelands, especially given notably greater than ever impacts of climate change and globalization. The following features of pastoral systems should thus be acknowledged when sustainability of Tibetan rangeland was planned: the dynamic nature of ecological systems, the significance of natural processes, the uncertainty and inherent variability of ecological systems, cumulative effects, preserved options, indigenous knowledge in natural resource management, conserved habitat for native species and productivity of ecological systems and reduced uncertainty through adaptive management and continuous learning.

Since the end of 1990s sustainable management of pastoral production system and rangeland ecosystem in the Tibetan Plateau has been adopted by Chinese government. From 2011 onwards, the central government has planned to spend an annually *ad hoc* rangeland ecosystem compensation fee of 13.4 billion Yuan for its eight western pastoral provinces and autonomous regions. Other than fencing and settlement programmes or simply giving money to herders, however, it is essential to involve pastoralists in the decision-making process and active actions for an integrated ecosystem. The ecosystem approach focuses on protecting and conserving entire ecosystems and well acknowledges positive roles of people living in it, which identifies specific objectives in relation to scale, social considerations and management.

Rangeland ecosystems of the Tibetan Plateau are not only the concern of local development but also ecological security of China and far beyond. Despite the ever increasing input from Chinese government in subsidizing rangeland ecosystem management in the Tibetan Plateau, it is still not enough to change rangeland degradation trends immediately or significantly. Therefore, more societal efforts and resources should be mobilized in conserving Tibetan rangeland and its multiple ecosystem services through means of carbon trade, voluntary environmental education and protection programmes and international aid.

Acknowledgements The work was partially supported by the National Natural Science Foundation of China. The authors also gratefully acknowledge the helpful comments and suggestions of the reviewers, which have improved the presentation.

References

Chen J, Luo D (2009) Survey on China's rangeland ecological healing. Shanghai Fareast Pressing, Shanghai

Dong S, Gao HW, Xu GC et al (2007) Farmer and professional attitudes to the large-scale ban on livestock grazing of grasslands in China. Environ Conserv 34(3):246–254

Dong S, Wen L, Liu S et al (2011) Vulnerability of worldwide pastoralism to global changes and interdisciplinary strategies for sustainable pastoralism. Ecol Soc 16(2):10. http://www.ecologyandsociety.org/vol16/iss2/art10/

Hongyuan County History Editorial Committee (1996) Hongyuan history. Sichuan Ethnic Publication House, Chengdu

Lu T, Wu N, Luo P (2009) Sedentarization of Tibetan Nomads. Conserv Biol 23(5):1074

Miller D (1998) Nomads of the Tibetan Plateau rangelands in western China, part one: pastoral history. Rangelands 20(6):24–29

Ministry of Agriculture of PR China (MOA) (2009) China's agriculture development report. Chinese Agriculture Publication House, Beijing

Niamir-Fuller M (ed) (1999) Managing mobility in African rangelands: the legitimization of transhumance. Intermediate Technology Publications, London

Richard C, Yan Z, Du G (2006) The paradox of the individual household responsibility system in the grasslands of the Tibetan Plateau, China. USDA Forest Service Proceedings RMRS-P-39, pp 83–91

Scoones I (ed) (1994) Living with uncertainty: new directions in pastoral development in Africa. Intermediate Technology Publications, London

Wu N (1997) Indigenous knowledge and sustainable approaches for biodiversity maintenance in nomadic society. Experience from Eastern Tibetan Plateau. Die Erde 128:67–80

Wu G, Du G (2007) Discussion on ecological construction and sustainable development of degraded alpine grassland ecosystem of the Qinghai-Tibetan Plateau. J Nat (Chin) 29(3):159–164

Wu N, Richard C (1999) The privatization process of rangeland and impacts on the pastoral dynamics in the Hindu-Kush Himalaya. In: Eldridge E, Freudenberger D (eds) People and rangelands building the future: proceedings of the VI International Rangeland Congress. VI International Rangeland Congress, Townsville, Australia, pp 14–21

Wu N, Yan Z (2002) Climate variability and social vulnerability on the Tibetan Plateau. Erdkunde 56:1–14

Yan Z, Wu N (2005) Rangeland privatization and its impacts on the Zoige Wetlands on the Eastern Tibetan Plateau. Mount Sci, 2(2): 105–115

Yan Z, Wu N, Yeshi D et al (2005) A review of rangeland privatization on the Tibetan Plateau of China and its implications. Nomad Peoples 9(1&2):31–51

Ze B (2005) Main ecological problems of Sichuan grasslands and relevant strategies. J Sichuan Grassland (Chin) 25(1):1–6

Chapter 17
Pastoral Communities' Perspectives on Climate Change and Their Adaptation Strategies in the Hindukush-Karakoram-Himalaya

Yi Shaoliang, Muhammad Ismail, and Yan Zhaoli

Abstract This chapter is based on four case studies carried out respectively in Afghanistan, Pakistan and Nepal by using semi-structured household interviews. Climate changes, especially rising temperature and more erratic precipitation, are strongly felt by pastoralists and are affecting their livelihoods strategies. Local communities have adapted to these changes, passively or proactively, by enhancing water resources management, changing the temporal and spatial pattern of seasonal migration, introducing drought-resistant crops or animal varieties or diversifying income-generating activities. However, the adaptive capacity of the pastoral communities to deal with the changes has been severely limited by multiple factors.

Keywords Climate change adaptation • Hindukush-Karakoram-Himalaya • Pastoral communities • Adaptability • Transformation • Rangelands

Y. Shaoliang (✉)
NRM (Land & Water), Aga Khan Foundation, P.O. Box 5753, House No.2, Qala-e-Fatullah, Kabul, Afghanistan
e-mail: Yi.Shaoliang@akdn.org

M. Ismail
Rangeland Resources Management, International Centre for Integrated Mountain Development, GPO Box 3226, Kathmandu, Nepal
e-mail: mismail@icimod.org

Y. Zhaoli
Chengdu Institute of Biology, Chinese Academy of Sciences, P.O. Box 416, 610041 Chengdu, Sichuan, China
e-mail: zhaoliy@cib.ac.cn

H. Kreutzmann (ed.), *Pastoral practices in High Asia*, Advances in Asian Human-Environmental Research, DOI 10.1007/978-94-007-3846-1_17,

17.1 Introduction

Pastoral communities in the Hindukush-Karakoram-Himalayan Region (HKH) that stretches over 3,000 km from Myanmar in the east to Afghanistan in the west mostly inhabit high-elevation temperate regions or arid or semi-arid zones that are likely to be most impacted by climate change. It is widely acknowledged that rangeland ecosystems and the related social and ecological systems in such areas often exhibit non-equilibrium nature (Vetter 2005), that is, their behavioural dynamics is mainly controlled by abiotic factors such as frequent droughts and extreme cold weather. Climate change is bound to enhance this non-equilibrium nature and increase the uncertainty pastoral communities have to face. Pastoral communities are known for their extraordinary capacity and knowledge in coping with and adapting to environmental uncertainty. In recent years, national governments, NGOs, development agencies and international organizations have started to increase their attention to assisting pastoral communities to adapt to climate change. Many national governments in the HKH region, for example, China, India and Nepal, have given special focus on the pastoral areas in their respective national or provincial government action plans or strategies for climate change adaptation and mitigation. It is foreseeable that this attention to climate change in the pastoral areas will be picked up by more and more organizations and government sectors.

To make external interventions and assistance more effective and specific to the needs of pastoral communities, it is necessary to understand how the local communities perceive climate change and its impacts and threats, how they respond to the changes and what factors are limiting their options for better adaptation and transformation. This chapter presents some findings from four case studies carried out in the pastoral areas of Afghanistan, Pakistan and Nepal.

17.2 Study Area and Methods

The case studies were carried out in Keshem District of Afghanistan, Upper Mustang and Langthang Districts of Nepal and Upper Hunza Valley in the Karakoram Range of Pakistan.

17.2.1 Keshem, Afghanistan

Keshem is located at the Keshem Valley on the western edge of Badakhshan Province, belonging to the Amu Darya river basin. The Keshem town (36° 48′ 23″ N, 70° 6′ 15″ E) is about 60 km to Faizabad, the capital of Badakhshan. Annual

Photo 17.1 Pastoralists on their way back from summer pastures between Faizabad and Keshem (Photograph © Hermann Kreutzmann September 1, 2006)

rainfall in the area is between 300 and 500 mm, mostly in the winter and spring seasons. It is cold in winter and hot in summer, with a very high evaporation rate. In the investigated villages, most of the inhabitants depend on both sedentary farming and migratory pastoralism as main livelihood means (Azarbaijani-Moghaddam 2006). The average household size is about 13 (range: 5–31). Major land resources include rangelands, irrigated farming lands (*abi*), orchards and rain-fed croplands (*lalmi*). Rangelands (including summer pastures) in the district were highly degraded with extremely low fodder productivity. Summer pastures and winter pastures are mostly used as commons, but contests and conflicts over the property of both are, however, common and have a long and complex historical background (Alden Wily 2004; Kreutzmann and Schütte 2011). Winter pastures were close to the villages, whilst summer pastures can be as far as 100 km away (Photo 17.1). Many farmers have no land of their own at all, and some others only have rain-fed lands. Irrigated lands were usually used for rice cultivation and horticulture and rain-fed lands for fodder plants, wheat and barley. Each household has about 60 sheep and goats and a few milking cows and bullocks. From April to October, the animals were free-ranged on the summer pastures or on the way between summer pastures and winter settlements. In winter months from November to March, the livestock has to be fed with hay and supplements, and thus fodder cultivation and hay collection are very important.

17.2.2 Upper Mustang and Langthang, Nepal

Upper Mustang (29° 05′ 00″ N, 83° 55′ 00″ E) lies in the northern part of Nepal, geographically and culturally very close to Tibet Autonomous Region of China. Upper Mustang is in the rain shadow area of the towering Himalayas and is extremely dry with annual rainfall being around 250 mm. The rangelands for pastoralism are mainly *Artemisia/Caragana* steppe and alpine meadows. People depend on diverse strategies for livelihoods. However, pastoralism is still the most important means of livelihood in terms of food security and the use of local resources. Major domestic animals include, in the order of population, goats, cattle, sheep, *chauri* (yak-cattle cross-breed), yaks, horses and mules (Photo 17.2). On average, each household has around 40–50 yaks, *chauri* or cattle. Those raising goats and sheep may have over 200–300 animals in each household. Crop cultivation is a very important part of the farming system but could only be developed on valley lands irrigated through man-made channels that allow only one crop (highland barley) per year. In winter months, almost the entire population moves south to escape the harsh winter months with the accompanying food shortage, and only the elderly remain behind to look after homesteads and animals. Increasing number of young population travels to the USA, Middle East, India and other parts of Nepal to take off-farm jobs (Chetri and Gurung 2004).

Langthang District is about 60 km north of Kathmandu. The study was carried out in the Langthang National Park (28° 12′ N, 85° 34′ E) which encompasses the partial catchments of two major river systems, the Trisuli River and the Sun Koshi River, and rises from 1,300 m at the valley's mouth to over 7,200 m at its highest point, Mt. Langtang-Lirung (Visit Nepal 2011). About 45 villages and 3,000 households (with an average household size of around seven people) are situated within the park. Most of them are agro-pastoralists who make their livelihoods through a combination of crop cultivation around the valley settlements and mobile pastoralism along mountain belts. Yak and *chauri* are the main livestock, which migrate seasonally between winter settlement areas and alpine summer pastures. On average, each household has 17 yaks (range: 11–33). Being on the southern slopes of the Himalayan Range, Langthang area is exposed to monsoonal climatic effects and gets its water in dry seasons mainly from the mountain glaciers and snow melting (Kharel 1997).

17.2.3 Upper Hunza Valley/Karakoram Range, Pakistan

The case study covered five villages in the Gojal region of Hunza-Nager District – Sost, Pasu, Ghulkin, Shimshal and Misgar – that span across the Karakoram Range (36° 10′ –37° 00′ N, 74° 06′ –76° 00 E). The region is characterized by high elevation (above 2,000 m asl), rugged terrain, cold winter, high solar radiation and low rainfall. Most of the settlements were located at the river valley. Local people depend on mixed farming (a combination of crop farming at lowland area and transhumant pastoralism) for livelihood. Yaks and goats were the main animals. Yaks migrate

Photo 17.2 Goats are of utmost importance for the pastoralists in Upper Mustang (Photograph © Renate Fliener August 2011)

seasonally between winter pasture at low elevation and summer pastures in alpine zone (Photo 17.3). Goats were usually kept around the valley belts all year round (Kreutzmann 2004, 2006). Altogether, 143 households were interviewed, and another 90 experienced herders and elders were consulted through group meetings.

In all case study areas, semi-structured interviews, participatory observation and group meetings were the main methods employed for field studies. Altogether, 244 households in Afghanistan, 143 households in Pakistan and 110 households in Nepal were interviewed with questionnaires. The researchers also visited the summer

Photo 17.3 Pasture settlement of Patundas (4,130 m) located at the top of a ridge between Pasu and Batura glaciers in Gojal, Gilgit-Baltistan (Photograph © Hermann Kreutzmann August 14, 1990)

pastures and the winter pastures of all the study sites and had meetings with the villagers to gather their opinions on the historical changes of climatic variables and their environments. The studies were carried out in 2009 and 2010.

17.3 Manifestations of Climate Change As Perceived by Pastoral Communities

Results from the three case studies indicate clearly that the environmental in general and climatic conditions in particular in the rangeland area have been changing, and local communities are fully aware of these changes. Some major indicators cited by the respondents as direct or indirect evidences of climate change include hotter weather, less rainfall/snow, more frequent and unpredictable extreme weather conditions, retreat or advancement of glaciers, changes in flora and fauna, longer summer periods, desiccation and compactness of soil, changes in the quality of forage, rivers that are no longer freezing in winter times and reduction of snow amount in the mountains.

In Afghanistan, almost all the herders interviewed mentioned the obvious reduction of rainfall and a rise in temperature in recent years compared with the past. They indicated that the number of rainy periods had decreased by four to eight times on monthly basis during the rainy season. The level of groundwater has gone down, and the number of dry days has increased. People also observed that the amount of snow in the mountains is decreasing.

The interviewees all felt that the weather was becoming hotter, albeit, there is a spatial difference as to the impact of rising temperature. Lowland areas (like Kunduz oasis in Afghanistan) have been suffering continuously 3 months of dry and hot weather, whilst in high-elevation areas, more people took temperature rise as an advantage to keep animals longer in the mountains. People pointed out that in the past 10–12 years, summers in the alpine regions have become longer and winter seasons shorter. Fruit trees now begin to flower 10–12 days earlier than in the past, and the fruit ripening time also advances accordingly.

> Rainfall has decreased and fodder grasses in the pasture land reduced; rainfall decreased by five to eight times per month; ... which caused water deficiency even in the pastures and also affected rain-fed cultivation. (Emam, Keshem, Afghanistan)
>
> Rainfall has been reduced and that causes a shortage of fodder in the pastures; water sources have decreased in the rangeland; irrigation water also has decreased. Rain spells have decreased by about three to six times per month during rainy season.... (Amallulah, Keshem, Afghanistan)

In Pakistan and Nepal, since the investigated villages are all located at high-elevation area and very close to major glaciers, people naturally cited the changes in the positions of glaciers and the amount of snow on the mountains as the most obvious evidences of warming weather. For example, in Pakistan, all the Ghulkin, Pasu and Batura Glaciers were quite away from the Karakoram Highway (KKH) in the early 1970s but have slided down gradually from their previous positions and are seen now very close to the KKH. People also mentioned that back in 1952 and till early 1960s, the rivers and streams used to stay frozen from September to January. People from certain villages such as Chupursan used to cross to other villages via the frozen Khunjerab River with horses and mules. However, this is no longer possible. Farmers in both Upper Mustang and Langtang of Nepal said that they were experiencing declining snowfall in the winter and rainfall in the rainy season. Snow used to be 5 ft deep 10 years ago, but there was no snow at all in recent years. White snowy mountains have become black rocky ones, leading to longer dry season, drier watershed areas, less water in the streams and increased areas with no vegetation.

A very interesting observation by the villagers in Pakistan study area is the change in wind regime. Local people used to use wind directions to predict weather and plan for various farm practices. According to the herders and farmers, the wind direction was quite predictable until 10 years ago, but it is not now. This has created ambiguity for the farmers to plan for the various practices they used to do.

In all the case study sites, people mentioned that the weather was become more unpredictable. For example, in Pakistan, people mentioned after a continuous dry spell for almost 10 years in the past, the area got an unusual heavy snowfall in 2009, the scale of which was never seen in the last 70 or more years. Afghanistan had witnessed unexceptionally good rain in 2009, but many people were not prepared for it and were thus not able to benefit from it.

17.4 Impacts, Threats and Opportunities

17.4.1 Impacts and Threats

Herder communities have felt the impacts of climate changes and the consequent threats to their traditional livelihood. Commonly mentioned impacts and threats in relation to climate change include drier environment, unhealthy rangelands, less fodder, increasing natural disasters, changing rangeland biodiversity, increasing animal and human diseases, water shortage and increasing conflicts amongst villagers.

In Afghanistan, people pointed out that decrease in rainfall affected mainly the fodder production of rangelands, leading to decrease of livestock numbers, but it also caused the reduction of irrigation water and the cultivation of rain-fed crops. Even in rainy season, water irrigation period reduced by 8–20 days each year. People feel that the reduction of rainfall is less stressful to agriculture than it is to grazing since high altitude reserve snow which is the source of water for the region can still provide enough water for irrigated agriculture. Due to continuous long drought months, spring sowing has become difficult. This decrease in rainfall especially consecutive drought has caused people to out-migrate to other places.

In Nepal, villagers mentioned that because of reduced rainfall, the grass growth on the rangelands is slow, and the fattening season for the *chauri* is getting shorter. As a result, most of the animals are lean and thin. Increase of wildfire occurrences was also mentioned as consequence of drier weather:

> The weather is so hot that fodder trees are dying-off and I have not seen rainfall for the last eight months. I did not see grass growth taking place, no moisture in the ground to support grass growth; instead, vegetation dried off… fire has caused substantial damage to forest and other vegetation, mainly because of excessively dry weather. There is fire everywhere. This year, almost half of the forest area and the pasture were gutted by fire and all the grasses were destroyed by fire. This is mainly because of long dry season without rain for eight months. Water sources are dried and no more water is coming. We have to go far away to get the water for the family and for the animals, which has added more labour, and thus we are having hard time. (Khamsung Sherpa, 45, Langthang, Nepal)

In Pakistan, unprecedented long dry spells caused the pasture to dry up, and the herders had to sell most of their livestock. Due to food shortage and low quality forage, with subsequent increase in livestock diseases, the herders suffered heavy economic losses. The prolonged summers caused earlier ripening of apricot, but the fruits were less tasty and heavily infected with the pests.

Villagers mentioned some changes in vegetation communities which can be traced to environmental changes driven by climate change. For example, the herders in Gircha, Pasu and Ghulkin noticed a prominent shift in vegetation on their pastures from willows (*Salix*) to emergence of profuse regeneration of the junipers (*Juniperus*) over the past 30–40 years. Changes in fauna are also quite obvious in recent years. There is a significant reduction in the number of migratory birds, frogs and other animals. Certain butterfly species have disappeared in the region in the past 10 years. People believed that change in climate has caused serious reduction in the availability of forage for the livestock. They feel that some of the pastures have been seriously depleted, producing only one fifth of what they had been producing about 10 years ago.

In all three countries, there is a significant increase of animal diseases and related livestock mortality. This may be due to reduced immunity to diseases as a result of poorer quality of fodder plants. Besides, rising temperature and decreased rainfall may have created favourable conditions to diseases and bacteria. Increased occurrence of livestock diseases has adversely impacted their income. In both Nepal and Afghanistan, there are increasing incidences of animal diseases and animal deaths attributed to hot weather. In Upper Mustang and Langtang of Nepal, the main animals are yaks or cross-breeds of yaks and cattle, both being very sensitive to temperature rise.

> ...crop harvest from the farm is just enough for the family for two to three months; for the rest, we make our living by selling *chauri*, milk and milk products. I have been herding *chauri* since the last 16 years. *Chauri* remain healthier during cold weather conditions. But the problem is that temperature is rising. Now we are moving *chauri* herds up into the high Himalaya pastures. *Chauri* wants to go up to find good grass and water. Many small water ponds and streams where *chauri* used to drink and where we used to collect water are already dried-off. A long dry season without rain has created multiple problems. (Nara B. Gauchan, 52, Upper Mustang, Nepal)

In Shimshal of Pakistan, exceptionally heavy snow in the winter of 2009 caused the death of over 3,000 goats and sheep and 500 yaks. Local people also exhibited a good understanding of the causal relations between climatic changes and their impacts. For example, in Misgar and Ghulkin of Pakistan, local people explained:

> the gradual decrease in natural precipitation, and increase in the duration of summers over the past several decades with sudden decrease in the last 15–20 years has changed, on average, the nature of the pasture soil. The soil, being soft and more fertile in the past has become drier and harder, with shorter and less nutritious forage, as compared to the past. This was because of less or no decomposition of the animal dung due to absence of required moisture in the soil. The dung used to decompose rather quickly in the past in the presence of moisture and adequate temperature. Fast decomposition of animal dung makes the pastures rather soft, which in turn makes it possible for the grasses and forbs to grow taller. Though the dung is still available, it doesn't decompose due to lack of enough moisture in the soil and is thus blown with the wind when dry. Even adding artificial fertilizers didn't help the pastures to regain their vegetation since it is too dry. The pasture vegetation has lost its past density and cover due to a less favourable environment for seeds and other sources of plants to germinate. (Ghulam Raheem, Misgar; Karim Khan, Ghulkin)

In Upper Mustang, a herder said:

> I have not seen the winter time of blizzards and snowfall in the upper Mustang and the surrounding mountains. Look at there! That White Mountain which I saw in my whole life has turned to black rocks. The white thick snow cover of the mountain which I saw all the seasons of the year in my whole life has disappeared since last three years. Not only that there are more mountains turning from white to black, but no snow in the mountains means no water in the streams and rivers. See, this stream used to irrigate all agricultural land of Lomanthang and had reduced by three quarters of its size and the land remained bared.

> No water, no crops, and no crops, no harvest. Instead of thick vegetation cover the rangelands are opened to bare land with no moisture and vegetation to feed the animals. Lomanthang used to be covered by one to two metres of thick snow in the winter for at least two months and used to get enough rain for cultivation and for crops plant in the summer, but for the last three years there was no snowfall in this land and very less rainfall. This has resulted to reduced crop-production by more than 50%. (Samduk Gurung, 63, Upper Mustang)

17.4.2 Opportunities Brought About by Climate Change

At the same time, people also see the positive impacts especially from this warming trend. In Afghanistan, people said that due to the rise in temperature, the animals can be in the rangelands (alpine pastures) for at least one more month, which has reduced the burden for winter stall-feeding. The growing season for plants has also become 1 month longer, making it possible for many plants and new crops to get mature. Some non-fruit trees like plane and willow trees can be planted in areas out of its previous distribution range.

> Weather became warmer, bringing about both positive and negative impacts. The positive impact is: grazing period has been elongated for about one month, opportunity for more grazing time…. (Amir, Keshem, Afghanistan)

17.5 Pastoral Communities' Adaptation to Changing Environments

Pastoral communities in the study areas responded to climate change and the subsequent environmental changes in a remarkably wide range of ways. According to the degree of activeness involved in the adjustment and whether such adjustment helps to enhance the resilience of the communities, we have categorized the responses into active adaption strategies and passive reactions (Table 17.1). Some strategies adopted by the pastoral communities are very good examples of adaptation. For example, to cope with the recurrent drought and take advantage of the rise in temperature, pastoral communities managed to improve infrastructure to increase water supply, introducing new crops that need less water, increasing agriculture input to guarantee output, increasing double cropping, starting animal migration earlier, shifting animal species (from sheep to goat), increasing hay collection and cultivation and many others. In Afghanistan, since most of the rainfall was shed in autumn and winter months, villagers had increased autumn sowing or double cropping to capture rain opportunities and reduce risks. Diversifying income sources by turning to business or off-farm jobs is another important step taken by many pastoral households to reduce the dependence on agriculture and livestock production. In Nepal, increasing number of young people went to India as wage earners, and many households turned to tourism activities.

Table 17.1 Responses of pastoral communities to changing environments

Active adaptation strategies
Increasing water supply/water use efficiency
Improving infrastructure for water management
Increasing autumn cultivation and reducing spring cultivation (in Afghanistan, most of the rain is shed in winter months)
Making better use of irrigated lands
Making efforts on cleaning water streams to reduce water losses
Making use of improved thermal conditions
Introducing new crops and trees (such as willow and plane which in the past could not grow well in the region) to take advantage of increased growing period
Taking animals to high pastures at earlier time
Keeping animals outside/rangelands for longer time
Early sowing
Double cropping in the field to make use of the longer growing season
Increasing orchard to get some fruits and non-fruit tree products
Adopting varieties/breeds better adapted to changed environment
Changing sheep to goat
Tried to cultivate drought-resistant crops
Reducing plants sensitive to hot weather
Double cropping of crops needing less water
Turning to crops needing less water
Taking pre-emptive measures to reduce risks
Reducing animal to avoid more losses
Using some improved seed and fertilizer to increase the production
Diversifying livelihoods
Seeking for other jobs and businesses
Putting more efforts on business
Migrating to other places to reduce cultivation and livestock numbers
Taking wage jobs
Innovating institutions
Restoring indigenous institutions for pasture management
Passive reactions
Introducing fodder from other sources
Reduction paddy cultivation (due to water shortage)
Purchasing more water for agriculture
Reducing rain-fed cultivation
Reducing rice cultivation and other crops that need more water
Migrating to other provinces or countries

People also resorted to institutional innovations in the process of adaptation to changing environments. In the Pakistani study area, the traditional practice of joint decision-making by village and community members on pasture uses has been eroding since the 1970s, but has been revived in recent years by villagers as a strategy to enhance their capacity to adapt to environmental change.

17.6 Adaptability and Transformability: From Constraints to Options

According to complex adaptive systems theory, the future trajectory of a social ecological system is determined by its resilience, adaptability and transformability (Walker et al. 2004). Adaptability is the capacity of the actors of the system to cope with changes and disruptions whilst maintaining system resilience. If a social ecological system can no longer maintain its resilience, it should seek system transformation, that is, changing itself into a fundamentally new system by introducing new parameters to the system. The capacity of the actors of a system to create a fundamentally new system is called transformability. For pastoral communities in the study area, new variables that could be introduced to help the transformation may include such measures as introducing new varieties or breeds, stopping animal migration and start stall-feeding, reducing animal numbers and finding urban employment or off-farm jobs.

As is shown in previous analysis, pastoral communities in the study areas responded to climate-driven environmental changes through multiple strategies. However, options available for the individual households or communities for successful adaptation or transformation or their abilities to react to and deal with climate change were frequently limited by a lot of factors. In the following sections, we will discuss the major difficulties the pastoral communities were facing in livestock production, what factors were limiting their options to change their farming methods to improve agriculture for better adaptation and what factors are preventing the pastoral communities from successful transformational changes in their life as a whole.

17.6.1 Difficulties Facing Pastoral Communities: Priority Concerns of the Herders

To understand major difficulties facing the pastoral communities and assess the weight of climate change impacts on the pastoralists, we asked the respondents to list up to five major difficulties they are facing in livestock production, and the results from Afghanistan were presented in Table 17.2.

The most frequently mentioned difficulties facing livestock production in the study area include, in the order of frequency, animal deaths due to lack of quality veterinary service, poor winter sheltering, shortage of winter fodder, lack of nearby markets for livestock products and limited fodder productivity. It is worth noting that most of the challenges mentioned by the respondents with the highest frequency were not directly related to climate change. This on one hand reflects that there were many issues rather than climate change that are of the priority of the herders, whilst on the other hand it may also indicate that pastoral communities in Afghanistan are more accustomed to high climatic variability.

Table 17.2 Major difficulties facing livestock production

Difficulties facing livestock production	Total (N=244)	%
Lack of veterinary service, quality medicines and health clinics	229	93.9
Severe long winter cold and lack of proper animal shelter	162	66.4
Shortage of winter fodder and nutritive elements	155	63.5
Lack of nearby market for selling livestock products	110	45.1
Shortage of fodder/pasture in general due to livestock increase	64	26.2
Limited knowledge about new technology (no training programme)	62	25.4
Loss of animals at birth and on way to pastures (by diseases and wolf predation)	57	23.4
Lack of product processing/storage plants	48	19.7
Long winter and stall-feeding period and severe cold	31	12.7
Shortage of labour to work on pastures	25	10.2
Shortage of feeding elements like salt and grain	21	8.6
Long distance from settlement to grazing areas, causing animal loss	19	7.8
No summer pasture of own	10	4.1
No breeding facilities/service	10	4.1
Increase of diseases	9	3.7
Conflicts with communities on the way	7	2.9
Unclear rangeland boundary	7	2.9
Water shortage	6	2.5
Difficult topographical condition and climatic constraints	5	2.0
Lack of good animal breeds	3	1.2
Low economy and short of human resources	3	1.2
Animals poisoned to death by poisonous plants	3	1.2
No development projects	2	0.8
Reduction of grazing area due to land conversion	2	0.8
High labour cost	2	0.8
Lack of animal husbandry equipment	2	0.8
Insufficient government support	2	0.8
Lack of agricultural tools and chemicals	1	0.4
High cost of agricultural tools and chemicals	1	0.4
Poor accessibility to summer pastures	1	0.4
Summer seasons too hot	1	0.4
Negative effects of warm weather on animal products	1	0.4
Drought causing fodder shortage	1	0.4

Source: All information displayed in the tables is derived from the interviews in the case study areas

17.6.2 Factors Affecting Farmers' Capacity to Adapt and Transform

As farming is an important part of the whole livelihood strategy, improving farming practices will surely enhance farmers' adaptive capacity. Table 17.3 shows the responses of respondents in Afghanistan to the questions on major factors preventing them from adjusting their farming systems for better adaptation to reduced precipitation and rising temperature.

Table 17.3 Limitations to farmers' capacity to change farming methods

Dimensions	Limitations
Physical conditions	Remoteness, difficult topography and poor accessibility
	Cold weather and short growing season
	Shortage of water, in particular irrigation water
	No fertile land
	Unpredictable weather
Skills and capacity	Lack of skills
	Limited knowledge about new technologies
	Limited manpower/labour shortage
Economic capacity	Limited agricultural diversification services
	Poverty/lack of economic capacity
Technology and production materials	Shortage/lack of agricultural materials and tools
	High cost of agricultural tools and materials
	Lack of improved seeds, fertilizers and pesticides
	Limited agricultural land
	No land for farmers (unfair land tenure)
Social service	Limited agricultural improvement services
	High cost of agricultural inputs if available
	Lack of capacity-building programmes
	No agro-technique services
	No product selling market
Security and conflicts	Security instability was another major problem for people
	Conflicts over water use

The major constraints can be roughly grouped into categories such as unfavourable physical conditions, lack of personal skills and capacity, low economic capacity, lack of technologies and inadequate social service such as capacity-building programme and market access. Security and conflicts also limited people's choices.

To further assess the capacity of the pastoral communities of the study area to adopt measures that could help successful transformation of their current status, we asked the villagers in Afghanistan why they did not adopt the following strategies: (1) changing crop varieties, (2) stopping migration and turning to stall-feeding, (3) reducing their animal number and (4) taking off-farm jobs. The responses are presented in Table 17.4.

The vast majority of respondents (95%) said that they did not shift to new crop varieties, because they did not have any information on new varieties. Lack of sufficient lands for farming or lack of money to buy lands makes it impossible for many people to stop migration and turn to staff-feeding and farming activities. The simple fact of no other means of living is stated by all the respondents as the reason for not reducing livestock number. Lack of skills or no job opportunities at all makes it difficult for pastoralists to move away from their current occupation.

An interesting point is that it is frequently mentioned by herders that prediction of climate is difficult; therefore, it is hard to pursue any strategies. A particular case in point is that in Afghanistan, people mentioned that the ample rainfall of 2009 did not bring any benefit to them since they never expected it.

Table 17.4 Major barriers affecting successful transformational changes of pastoral societies

Options/strategies	Most frequently mentioned reasons for not being able to pursue			
1. Changing crop varieties	Lack of information (95%)	Lack of money (55%)	Climatic limitations (25%)	Shortage of labour force (0)
2. Stopping seasonal migration and settle down	Lack of money to purchase agricultural land (82.5%)	Difficult for livestock (8%)	Shortage of labour (2.5%)	Others (2%)
3. Reducing number of livestock	No other means of living available (100%)			
4. Taking off-farm jobs	Lack of skills (100%)	Lack of job opportunities (92.5%)		

17.7 Discussions and Conclusions

Pastoral communities are fully aware of climate change and its impacts on their surroundings and life. Across the study area, people felt rising temperature, decreasing precipitation and more unpredictable weather pattern. These changes in climatic variables have further caused the rapid movement of glaciers, reduction of fodder production, drying up of rivers and water shortage and are threatening the very subsistence of the pastoralists. Pastoral communities are quick to respond to these changes through a wide range of strategies within their capacity, for example, improving infrastructure for water management, increasing agriculture input, changing cropping pattern and increasing autumn sowing and early migration of animals. Some demonstrate good examples of successful adaptation and transformation. However, the adaptability or transformability of the pastoral communities in response to changing environment is severely limited by factors such as harsh physical conditions, poor economic capacity and lack of adequate technology, skills, information and social service. Improving any of the above aspects will help to enhance the capacity of the pastoral communities to maintain its resilience and adapt to the changing environment.

There are many projections for the future scenarios of climate change in the HKH region (Tsering et al. 2010). One agreement with all the projections was that changes in the region will be greater than the global average. For example, Indian Institute of Tropical Meteorology (IITM) study suggests that there will be a decrease in monsoon precipitation of up to 20% by the end of the century in most parts of south-eastern Afghanistan, the southern and eastern Tibetan Plateau and the central Himalayan range and an increase in the range of 20–30% for the western Himalayas (Shrestha 2009). How to build the adaptive capacity and transformability of the pastoral communities so as to reduce their vulnerabilities to the even more drastic climate-driven environmental changes present a challenging task to all social sectors.

At the same time, it should also be noted that even though climate change does present real and serious challenges to pastoral areas, there are also a lot of other issues that are of priority concerns to the pastoralists. Quality veterinary service, proper winter sheds for animals, market accessibility and overall fodder supply are some of the long-standing and urgent issues limiting livestock development in the rangelands area and should be tackled with more earnestness and urgency by government departments and NGOs. Even though those issues are not directly related to climate change impacts, properly addressing them could definitely reduce the vulnerability of the pastoral communities.

Acknowledgement This chapter is based on case studies carried out by ICIMOD staff in Afghanistan, Nepal and Pakistan. We thank Ashiq Ahmad Khan, Mohammad Shafi Hafizi and Tara Pandey for their field work and resulting reports.

References

Alden Wily L (2004) Looking for peace on the pastures. Rural land relations in Afghanistan. Afghanistan Evaluation and Research Unit, Kabul

Azarbaijani-Moghaddam S (2006) Dastarkhan Weaving in Tagab-i Keshem, Badakhshan. Consultancy undertaken on behalf of Zardozi – Markets for Afghan Artisans. www.afghanartisans.com/PDFs/dasterkhwan_keshem_study.pdf. Accessed 19 Aug 2011

Chetri M, Gurung CR (2004) Vegetation composition, species performance and its relationship among livestock and wildlife in the grassland of Upper Mustang, Nepal. In: Jincheng Z, Xiangdong Z, Jianlin H, Zhihua C (eds) Yak production in central Asian highlands. Proceeding of the fourth international congress on Yak. Sichuan Publication House of Science and Technology, Chengdu, pp 235–244

Kharel FR (1997) Agriculture crops and livestock predation by wildlife in the Langtang National Park. Mt Res Dev 17(2):127–134

Kreutzmann H (2004) Pastoral practices and their transformation in the North-Western Karakoram. Nomad Peoples 8(2):54–88

Kreutzmann H (ed) (2006) Karakoram in transition: culture, development and ecology in the Hunza Valley. Oxford University Press, Karachi

Kreutzmann H, Schütte S (2011) Contested commons – multiple insecurities of pastoralists in North-Eastern Afghanistan. Erdkunde 65(2):99–119

Shrestha AB (2009) Climate change in the Himalayas. Icimod, Kathmandu

Tsering K, Sharma E, Chettri N, Shrestha A (2010) Climate change vulnerability of mountain ecosystems in the Eastern Himalayas. Icimod, Kathmandu

Vetter S (2005) Rangelands at equilibrium and non-equilibrium: recent development in the debate. J Arid Environ 62:321–341

Visit Nepal (2011) Nepal National Parks Information. http://www.visitnepal.com/nepal_information/nepalparks.php#LangtangNationalPark. Accessed 19 Aug 2011

Walker B, Holling CS, Carpenter SR, Kinzig A (2004) Resilience, adaptability and transformability in social-ecological systems. Ecol Soc 9(2):5. http://www.ecologyandsociety.org/vol9/iss2/art5/

Chapter 18
Pastoralism: A Way Forward or Back?

Hermann Kreutzmann

Abstract Perceptions and experiences differ widely in the world of pastoralism. The case studies presented in this volume provide fieldwork-based insights and evidence from a widespread area between the Pamirs, Tien Shan, Hindukush, Karakoram, Himalaya and the Tibetan Plateau. More important than the ecological breadth and spread of environmental properties and changes seem to be the societal embeddedness of pastoralism, the politico-economic framework and the understanding of 'modernisation'. The debate on the 'tragedy of the commons' seems to have developed through a supposed 'drama of the commons' to an institutional 'tragedy of responsibility' under similar pretexts as in the early stages. Norms and viewpoints govern judgements about actors and victims in relation to their pastoral practices.

Keywords Pastoral embeddedness • Future of pastoralism • Payment for ecosystem services • Tragedy of responsibility • Norms and values

18.1 Contemporary Perceptions

Pastoralism has often been classified as a backward way of living and surviving. Nevertheless, one observation drawn from our case studies in High Asia is the significant persistence of a utilisation strategy that has mastered severe challenges and major constraints posed by neighbours, rulers, states and their administrations and, not to be forgotten, by development agents worldwide. The transformation of pastoral practices is ubiquitous and a signifier of its adaptive capacity.

H. Kreutzmann (✉)
Human Geography, Department of Earth Sciences, Centre for Development Studies (ZELF) and Institute of Geographical Sciences, Freie Universität Berlin, Malteserstrasse 74-100, D-12249 Berlin, Germany
e-mail: h.kreutzmann@fu-berlin.de

H. Kreutzmann (ed.), *Pastoral practices in High Asia*, Advances in Asian Human-Environmental Research, DOI 10.1007/978-94-007-3846-1_18,

At a recent e-conference, a conceptual grounding of pastoralism was attempted by linking its definition to climate change and livelihood debates:

> Pastoralism is a complex form of natural resource management, which requires maintaining an ecological balance between pastures, livestock and people, and it is an adaptive strategy to a stressful environment. This adaptation faces a myriad of challenges, of which climatic change is but one. Indeed, the challenge of climate change seems insignificant to many pastoralists who are faced with extreme political, social and economic marginalisation: relax these constraints and pastoral adaptive strategies might enable pastoralists to manage climate change better than many other rural inhabitants. (Nori and Davies 2007, 7)

This description perceives pastoralism as a singular system that has been adversely affected by a number of constraining influences but is a successful adaptive strategy to cope with a multitude of challenges. Various kinds of socio-economic and politico-historical challenges caused pastoralists to adapt, evade or escape and/or abandon certain practices. Most of these constraints were much more powerful than the felt effects of climate change to date. Global warming and its effects are perceived as a gradual process. It has always been the practice of pastoralists to adapt to gradual changes of that calibre. Mitigation was beyond their scope anyhow. Sometimes pastoralists are even not mentioned when it comes to ecosystem services of mountains in times of climate and global change (Macchi and ICIMOD 2010). Marginalisation and neglect are one approach; modernisation and development discourses meet pastoralists with significant effects. Social experiments such as the Soviet collectivisation and the Chinese 'great leap forward' have affected pastoral livelihoods not gradually but in the manner of a pending threat to survival and causing severe disasters. Social organisation and pastoral practices were transformed in a short span of time by external und un-experienced planners resulting in human tragedies and casualties and huge losses of pastoral wealth and resources. Within the twentieth century, these interventions were more forceful than anything linked to climate change. Nevertheless, climate change is on today's global agenda and therefore referred to more often than social change and political transformations (Schröter et al. 2005). Development practitioners and consultants, decision-makers and regional planners have developed a tendency to neglect social and political effects when it seems to be easier to blame climate change for adverse occurrences (e.g. Kelkar et al. 2008; Kohler and Maselli 2009). Some day we might regret the confusion of ideas as the 'climate change dilemma' in a similar manner as Jack Ives commented on the Himalayan Dilemma:

> It was a development paradigm that confused cause and effect, resulted in the misdirection of large financial resources, and sidelined some of the real needs of the people. Attempts were made to solve a problem that did not exist, or at least, one that had been exaggerated beyond measure. Thus, 'development' was distorted and the identification and prioritization of circumstances that demanded attention was delayed. (Ives 2004, 229)

It would be worthwhile to consider both phenomena as the superimposition of gradual processes related to environmental changes with efficient socio-political interventions that sometimes come in the disguise of development, modernisation and resettlement. The resulting effects are meeting pastoralists' livelihoods and transforming them. A holistic approach that operates from the perspective of pastoralists might avoid the fallacies of confusing cause and effect as was said above.

Photo 18.1 Visitors from Changtang pastoral areas enjoying pilgrimage and urban life in Lhasa (Photograph © Hermann Kreutzmann July 30, 2004)

In the framework chapter, we addressed pastoral practices; here, the mosaic of presented case studies attempts to contribute some insights into significant changes in the livestock sector. Furthering the argument, land degradation and measures to counteract these developments become prominent features affecting pastoral practices. At the same time, the transformation of pastoral livelihoods mirrors societal developments that are often planned at urban centres and in capital cities but always have significant effects even in the remotest and mountainous peripheries (Photo 18.1).

In some cases, the adaptive potential might be surprising. In Afghanistan, an environment of insecurity and threat poses the major challenges to pastoralists. Nevertheless, they have shown that animal husbandry can be a profitable undertaking today, fulfilling the important role of supplying the bazaars with valuable livestock products. India and Pakistan have experienced continuous disputes about pasture access, legal rights and grazing fees in their mountainous regions of the North. Since colonial times, agricultural and forest departments have been challenging the space utilised by pastoralists, and revenue officials have been keen on dues from cross-border trade. Although pastoralism remained an important economic resource even after the closure of international borders (Photo 18.2) with China, it is quite surprising how decentralised the attitudes regulating pastoral migration and pasture use appear and how differently the respective administrations treat aspects such as animal health and marketing facilities (Dangwal 2009; Inam-ur Rahim and Amin Beg 2011).

The space left for pastoral activities has been shrinking further since the competition between combined mountain farmers and pastoralists increased the demand for grazing lands, whilst at the same time nature protection and privatisation of common properties are reducing their degrees of freedom. In both countries, a selective process can be observed: sedentarisation of former mobile communities and

Photo 18.2 The Irshad-e-win (4,979 m) resembles a pass separating the Little Pamir (Afghanistan) from the Chapursan Valley (Gilgit-Baltistan, Pakistan) and connecting both. Politically, both countries do not treat this pass as an official border-crossing point while at the same time Irshad-e-win is an important lifeline for Kirghiz to barter and exchange goods with neighbouring Wakhi who even have acquired the right to pasture their flocks across the border (Photograph © Hermann Kreutzmann June 15, 2000)

marginalisation of others. The latter are increasingly becoming more dependent on external suppliers and buyers. They continue to occupy a low social status or are transformed into hired shepherds in a transhumance system, but they remain a vital asset to meet a growing demand for livestock products. A higher degree of regulation has entered the Hindukush-Karakoram-Himalaya realm in the nexus of modernisation and nature protection. With more than 30% of its land surface under protected status (Bajracharya 2011, 128), Nepal is probably one of the most significant examples of the shrinking space in formerly accessible commons. As cattle slaughtering and meat production play only a minor part in the economic activities of many communities, the value attributed to animal husbandry is lower, making it easier to turn to alternative employment or additional income from tourism, outmigration and services (Photo 18.3). These patterns differ from neighbouring China where – as all case studies have significantly shown – a concentrated and centrally guided transformation process in the pastoral sector is under way. The rationale for this is inspired by modernisation theory and justified by green policies that are a response to perceived land degradation and the need for nature protection. The process of implementation seems to be controlled and steered by powerful and effective administrations. Inspired by a similar dual approach of nature protection and modernisation, the Indian model appears to have quite different effects. The public

Photo 18.3 The Lama of the Monastery in Beding (3,692 m; Rolwaling Valley, Gauri Shankar VDC, Nepal) is greeting yaks and its hybrids that are part of household herds. Only a few households practice combined mountain agriculture in which the contribution of livestock is fading. Yaks are still kept for transport purposes and as an investment that can be converted in cash money in times of need (Photograph © Hermann Kreutzmann September 29, 2011)

discourse operates along similar lines, the public administration's belief in the messages of modernisation coincides with that of their Chinese colleagues, but China's efficiency and drive as well as the budgetary allocation and implementation of resettlement are not obvious in India yet. The future will show whether the Indian model follows the Chinese blueprint.

The former Soviet Central Asian Republics bear a similar heritage of collectivisation and state-controlled operations in the livestock sector to that in the PR of China (Photo 18.4). Nevertheless, their transition from state to private ownership and their shifting pastoral practices follow a different model in which state interference has diminished except for legislation and allocation of pasture rights. In recent years, the gap has widened between successful pastoral entrepreneurs and subsistence herders. For some, the applicable regulations and enhanced liberties provided the framework to practise a viable form of animal husbandry by creating a value chain that supplies the newly established markets. For others, the summer spent on the pastures is a means to lower the cost of living in an environment where other job opportunities besides outmigration are rare.

All cases have shown the importance of the position of pastoralism within the legal framework and societal set-up of the respective nation states and their economies. In most cases, pastoralists are citizens without a strong political lobby; Afghanistan might be a partial exception (Tapper 2008).

Kyrgyzstan as the land of the Kirghiz incorporates a pastoral tradition in its name. The national symbol – the *tunduk* – refers to the central part of a pastoralist's yurt and is meant to root its society in pastoral traditions. However, the five case

Photo 18.4 The privatised flocks of shepherds from Rangkul are pastured close to the indicator near Chechekti that the Kolkhoz Lenin Zholu (the collective farm named 'way of Lenin') was here before (Photograph © Hermann Kreutzmann September 6, 2007)

studies on the Kirghiz in different societies – Afghanistan, Tajikistan, Kyrgyzstan and China – have explicated how varied can be the effects of state intervention and resource competition, their impacts on livelihoods and participation in decision-making, the secular and symbolic values of modernisation and development and how they can structure daily life experiences and pastoral practices. Pastoralists rarely get centre stage attention when global agendas such as desertification, nature protection, global warming and/or climate change trigger national politics to implement related policies on regional and local levels. Pastoralists have often become the addressees of ideology-driven modernisation strategies that have significantly affected their livelihoods and pastoral practices.

18.2 From the 'Tragedy of the Commons' to the 'Tragedy of Responsibility'

In a shrinking potential environment where growing external demands and powerful interests govern structural changes in the political arena, land-grabbing and encroachments into 'traditional' pastures and common properties, it is still surprising how pastoral practices have always adapted to new and threatening challenges and found an outlet to cope with mounting constraints. Societal and political changes have dominated over all kinds of climate and environmental changes. The case studies have revealed that in some regions, pastoralism as a solitary strategy to make a living might be abolished in the near future for some practitioners, but pastoral practices might prevail in other niches and offer ample opportunities to those who are connected and embedded in their way of accessing resource potential that others still cannot utilise in a meaningful way. The niches are shrinking, but there may well be a way forward.

Growing pressure on the commons has changed the attitude of policymakers and rangeland management planners who had treated rangelands and their inhabitants over long periods as 'marginalised people in regions of neglect'. The debate on the 'tragedy of the commons' triggered off by Garrett Hardin (1968) has developed and gained pace. In times of land-grabbing and expropriation of resources when customary rights can easily be breeched and community practices do not count, it could well be that the notion of a 'drama of the commons' (Ostrom et al. 2002) is much more appropriate. Even during the last decade, the pressure on land resources has grown further and led to an unequal positioning of interests (Fig. 18.1). Hardin's solution for alleviating the 'tragedy of the commons' was privatisation. The presently observable process of selling-off vast tracts of agricultural land resources to powerful multinational state and private investors in Africa and Asia is exactly stimulating the land-grabbing and expropriation of weak communities without lobby. The 'drama of the commons' gains pace and appears to be a 'drama of responsibility' where the vital interests of rural people and communities are at stake and grossly neglected.

In our case, neglect is meant to express the notion of inadequate policies for pastoral communities and their stakes. The subsequent information on policies and legislations will document that attempts at state evasion (Scott 2009) have been in vain since the 1950s at the latest. Administration and bureaucracies have penetrated pastoral areas with different degrees of efficiency. China has been mentioned in great detail already, and it has become obvious that man-made changes to the environment are treated in different ways. Sometimes they are explained as caused by natural hazards and climate change; in rare cases, ideology-based experiments and societal transformations are made responsible for adverse effects in the rangelands.[1] Whilst China has a legacy of top-down interventions accompanied by all kinds of legislation, incentive packages and modernisation programmes, other neighbours are beginning to rethink their attitudes. In India and Pakistan, rangeland management was inherited as a colonial legacy, and policymakers of today refer to early legislation such as the 'Cattle Trespassers Act' of 1871 and the 'Forest Policy' of 1894. In Pakistan, the 'National Forest Policy' of 1962 was the first step towards a rangeland management strategy after independence, this policy being extended to wildlife in 1980. To further the new 'Pakistan Forest Policy' of 1991, a 'National Rangeland Policy' has been announced; a decision about the draft is still pending. Nevertheless, the vested interests of pastoralists and tenure issues are mentioned only in passing.

As late as in 1988, India envisaged a paradigm shift with the 'National Forest Policy' in which rangelands played an important role – followed by the 2006 'National Environmental Policy' – that affects the four mountain provinces (Jammu and Kashmir, Himachal Pradesh, Uttarakhand and Sikkim) and to a minor degree Arunachal Pradesh and parts of West Bengal. The intention of Indian rangeland policies is to intensify livestock production in an arena of decreasing rangeland

Fig. 18.1 The position of High Asia and its rangelands within the wider context of Asian boundaries and political systems

availability. Inherited legislation has been adapted to current challenges and is probably failing to cope with them. A lack of coordination and adaptation to specific frame conditions is prevalent.

In Nepal, the creation and expansion of protected areas contributed to the exclusion of herders from their inherited pastures. Low production and low productivity can be held responsible for such an approach. The 'Pasture Nationalisation Act' of 1975 has transferred the right to provide access to rangeland to the authority of local communities. The present state of affairs can be linked to general observations and to the perception of inefficient traditional management, non-adaptation of scientific knowledge, lack of investment, confusion of ownership and conflicts resulting in a low national priority and the neglect of indigenous knowledge, skills and techniques. The 'Nepal Biodiversity Strategy' of 2002 has highlighted the need for a 'National Rangeland Policy'. The outcome is to be awaited, and it remains to be seen whether the envisaged improvement of herders' livelihoods based on an increased productivity will materialise.

Only in Bhutan were rangelands nationalised as part of the government forests. In the framework of the 2007 'New Land Act of Bhutan', the government recognises pastoralists as eligible users. Through a '*tsamdro* management plan', pastures can be leased for periods of up to 30 years and beyond. Pastoralists are assumed to constitute a tenth of Bhutan's population, and they are supported from the top in order to create an environment in which herders remain in high-altitude areas.

In Afghanistan, the 'Pasture Law' of 1970 codified the property rights of the government. The law was last amended under the Taliban in 2000 and is currently being re-drafted under the guidance of international agencies to incorporate community-based pasture management systems, but the provisions of 1970 remain the official policy to date with little effect on pastoral practices (Kreutzmann and Schütte 2011).

In the former Soviet Central Asian Republics, a process of transition from state-owned property rights to leasehold and private and/or community-based pasture rights is still in the making. Whilst Kyrgyzstan passed a new law on pastures in 2009 and Tajikistan has also already achieved some legislation, the implementation and practical value of an allocation of pastures to local communities and the decentralisation of responsibilities still have to be awaited.

All brief descriptions of policies and plans reveal a varied set of attitudes towards the management of the commons.[2] Still, a 'tragedy of responsibility' may be observed in countries such as India, Nepal and Pakistan – whilst others have recognised a new challenge that demands an answer. Whether the design of national policies might be an adequate answer to the challenges for the livelihoods of pastoralists remains unanswered here.

18.3 Actors and/or Victims: The Future of Pastoral Practices

Perceiving pastoralism solely from the organisational and strategic aspects of its adaptive potential, its 'direct and indirect values' (Davies and Hatfield 2007) and/or its appropriateness for utilising a widespread and extensive natural potential would omit a discussion about the people that are involved. The case studies presented here cover a wide range of experiences and experiments with changing attitudes, strategies and societal set-ups. In the following, a narrative is related about people of differing backgrounds and biographies who met at two international conferences on pastoralism in 2010.[3]

The story begins 20 years earlier when land disputes increased in Gojal, Upper Hunza Valley, Gilgit-Baltistan in Pakistan (Photo 18.5). Amongst the main arenas of disputes and conflicts were encroachments in pasture areas by neighbouring villages and communities, questioning of traditional pasture rights and outright refusal to provide passage and access to formerly used grazing grounds. In an area without land assessment and cadastral surveys, without written documents and maps, the disputes gave all sides some negotiation power and leverage:

> Never before have village funds been spent to such an extent in legal disputes in religious and civil courts… The village of Gulmit is the most severely affected of all and serves here as an extreme example. Gulmit's pastures lie scattered comparatively far away from the permanent settlement and are not located just above the homestead… During the 1990s different disputes arose with neighbours about the hereditary rights of pasture use. In 1990 a severe dispute began with Shishket across the Hunza river. The Bori kutor clan of Gulmit was to be deprived of its right to access Gaush, and the Ruzdor clan had similar experiences in Bulbulkeshk and Brondo Bar. Although kinship and marriage relationships exist between the inhabitants of Gulmit and Shishket, no solution could be reached through the local institutions and negotiations by mutually accepted and respected neutral persons. The whole conflict escalated and became a major affair of defending property rights that had not been laid down in written documents. Representatives of public and religious institutions were consulted in vain before the legal proceedings started. Up to the present day [2003] more than 0.5 million Pakistani Rupees have been spent on lawyers and court fees alone by the people of Gulmit. Similar or even higher contributions were invested by the opponents, not counting all travel expenses and secret meetings of representatives. No solution is in sight, despite 'stay orders' issued by the courts permitting both sides to use the pastures. The funds spent exceed by far the commercial value from animal husbandry in these pastures for the next decade. (Kreutzmann 2004, 70)

The case described here has been only one of several disputes in addition to quarrels about the exclusion of pastoralists from the Khunjerab National Park (Knudsen 1999). In sum, the Gojali people spent significant amounts of money on finding a legal solution for their disputes and a basis for a future understanding. Looking back at the Gaush case, the villagers from Gulmit and Shishket have since made sure that always at least one representative shepherd from their respective villages is present in Gaush during the summer season. Their huts and corrals are located next to each other (Photo 18.6).

The arguments brought forward in the dispute were informed less from a pastoral point of view than from a perspective highlighting honour and land entitlements. It would be a loss of community honour if a piece of land inherited from their ancestors were to be surrendered to neighbours even if – as in this case – they are close blood relations. The second argument brought forward was directed towards entitlements to land rights, mining and water. Nobody knew at this point whether the pastures would be future settlement grounds, as has happened in many other cases nearby. Neither was there any evidence whether

Photo 18.5 Wakhi mountain farmers have left their village of Gulmit to fight with their relatives from the neighbouring village of Shishket about the pasture rights on the mountain slopes in the left background. For more than two decades, monetary funds and other resources were invested in this and similar disputes about pastures in Gilgit-Baltistan. Nevertheless, a lasting solution has not been reached yet (photograph © Hermann Kreutzmann April 29, 1990)

mineral wealth was located here and whether extractive industries might become interested in future.

Somehow the villagers have proved to be right. The pasture of Gaush lies above the village of Goshben (lower Gaush). Goshben was flooded and completely inundated due to a major landslide that occurred on January 4, 2010 (Kreutzmann 2010);

Photo 18.6 One of the disputed pasture settlements in Gaush (3,600 m) is now divided between the two villages of Gulmit and Shishket (cf. Photo 18.5) from where the permanent settlements are clearly visible. Each village needs to be present here with at least one shepherd in order to claim the persistence of their 'traditional' pasture rights (photograph © Hermann Kreutzmann July 7, 1991)

all village lands were lost. The pasture has remained dry and high above the lake level. Nevertheless, the common notion in both contesting villages is that it remains a burden to send shepherds up to Gaush every year and that the economic returns do not justify the effort.

A son of the contesting combined mountain farmers from Gulmit, who has become a development practitioner and consultant, participated in the above-mentioned two conferences and in the related fieldtrips. When the participants were crossing Kulma Pass (4,363 m) from Gorno-Badakhshan in Tajikistan into the Kizil Su Kirghiz Autonomous Prefecture of Xinjiang, PR of China, delegates from Nepal, Pakistan, Tajikistan and China had the opportunity to see with their own eyes the effects of transformation. In Tajikistan, two major transitions had occurred in the twentieth century, with collectivisation and privatisation as their ideological markers. Similar transformations had taken place within the PR of China, albeit at different times and in varying contexts. Whilst pastoralists in Tajikistan are struggling – with few exceptions – to make a meagre living from animal husbandry, the situation observed in Xinjiang was quite different for all delegates. Here the state was present at all instances: infrastructure development, provision of planning and extension, veterinary services, marketing facilities and now resettlement. It needs to be noted that all delegates from Nepal, Pakistan and Tajikistan admired a 'caring government' that supports pastoralists in becoming 'modern'. The observation in Bulunkul (see Photo 6.3) was inspired by admiration and by criticism for their own respective governments. Delegates familiar with the Tibetan situation mentioned that the backward pastoralists of Xinjiang would now enjoy the same developments that are well known from the Tibetan Plateau: fencing, housing and resettlement. A further package is being experimented with already. Payment of ecosystem services could be an alternative, thus enhancing pastoral lifestyles whilst at the same time contributing to nature protection (Wilkes et al. 2010). It might be worthwhile to consider the advantages of having pastoralists as active landscape managers instead of removing them in great style from pastures that have been utilised for centuries. The indigenous knowledge accumulated by pastoralists over many generations seems too valuable to be just neglected or omitted. The framework of these two conferences provided a forum where 'experts' and 'practitioners' could meet and be exposed to the experiences made in other societies. The aim of the book presented here has been to provide further insights into background, circumstances and prospects of pastoral practices in High Asia. Whether the route is always straight or whether it involves backward and forward turns has to be judged from the respective viewpoints and norms. Changing practices are the result of the application of norms and their implementation. A sound measure would be to listen to the voices of the pastoralists whose lifestyles and economic prospects are being addressed and considered.

Acknowledgements Insights presented in this chapter were collected during the last 28 years in several research projects that received generous funding provided by the *Deutsche Forschungsgemeinschaft* (DFG) and the *Volkswagen Foundation*; their support is gratefully acknowledged. In recent years, *InWEnt – Capacity Building International* (renamed in *Deutsche Gesellschaft fuer Internationale Zusammenarbeit* (GIZ) GmbH in 2011) – and the *International Centre for Integrated Mountain Development* (ICIMOD) have put more emphasis on pastoralism and rangeland management. Their timely efforts have initiated a number of conferences, stimulated the debate and produced a number of valuable publications in the field of pastoralism from which this chapter could make ample use.

Notes

1. The explanation of causes and effects is surprisingly weak as Harris (2010, 8) observed for the Qinghai-Tibetan Plateau: '… there has been very little rigorous Chinese research into the reasons for overgrazing and rangeland degradation. Most Chinese biological research has not asked, much less answered, questions regarding human motivations among the pastoralists using the rangelands…'.
2. This information on recent policies in the Hindukush-Karakoram-Himalaya was derived from the presentations of country papers during the ICIMOD workshop on 'Regional Rangeland Management Programme (RRMP). Development and policy review for the Hindukush-Himalayas' held in Kathmandu August 22–23, 2011. I am indebted to Imtiaz Ahmad (Pakistan), Ruchi Badola (India), Shikui Dong (China), Tsering Gyeltshen (Bhutan), M. Arif Hossini (Afghanistan) and Devendra Kumar Yadav (Nepal) for sharing their insights during the workshop. For Kyrgyzstan and Tajikistan, evidence was presented by Ermek Baibagushev (2011), Bernd Steimann (2011), Andrei Dörre and Tobias Kraudzun (cf. Chaps. 5 and 7 in this volume).
3. Both conferences were organised by InWEnt – Capacity Building International (renamed in Deutsche Gesellschaft fuer Internationale Zusammenarbeit (GIZ) GmbH in 2011), aiming to bring together academics, decision-makers and development practitioners. The proceedings were published by Kreutzmann et al. (2011a, b).

References

Baibagushev E (2011) Recent changes in pastoral systems. Case study on Kyrgyzstan. In: Kreutzmann H, Abdulalishoev K, Zhaohui L, Ritcher J (eds) Pastoralism and rangeland management in mountain areas in the context of climate and global change. GIZ, Bonn, pp 102–118

Bajracharya SB (2011) Tourism development in annapurna conservation area. In: Kruk E, Kreutzmann H, Richter J (eds) Integrated tourism concepts to contribute to sustainable mountain development in Nepal. GIZ, Bonn, pp 126–142

Dangwal DD (2009) The lost mobility. Pastoralism and modernity in Uttarakhand Himalaya (India). Nomad Peoples 13(2):84–101

Davies J, Hatfield R (2007) The economics of mobile pastoralism: a global summary. Nomad Peoples 11(1):91–116

Hardin G (1968) The tragedy of the commons. Science 162(13):1243–1248

Harris RB (2010) Rangeland degradation on the Qinghai-Tibetan plateau: a review of the evidence of its magnitude and causes. J Arid Environ 74:1–12

Inam-ur-Rahim, Amin Beg (2011) Production and marketing of livestock products in the Hindu Kush-Karakoram-Himalaya. In: Kreutzmann H, Yong Y, Richter J (eds) Regional workshop in Lhasa 2010 – pastoralism and rangeland management on the Tibetan plateau in the context of climate and global change. GIZ, Bonn, pp 177–199

Ives JD (2004) Himalayan perceptions. Environmental change and the well-being of mountain peoples. Routledge, London/New York

Kelkar U, Narula KK, Sharma VP, Chandna U (2008) Vulnerability and adaptation to climate variability and water stress in Uttarakhand State, India. Glob Environ Chang 18:564–574

Knudsen A (1999) Conservation and controversy in the Karakoram: Khunjerab National Park, Pakistan. J Polit Ecol 56:1–29

Kohler T, Maselli D (eds) (2009) Mountains and climate change – from understanding to action. Geographica Bernensia and Swiss Agency for Development and Cooperation, Berne

Kreutzmann H (2004) Pastoral practices and their transformation in the North-Western Karakoram. Nomad Peoples 8(2):54–88

Kreutzmann H (2010) An inevitable disaster. Himal Southasian 23(6):24–25

Kreutzmann H, Schütte S (2011) Contested commons – multiple insecurities of pastoralists in North-Eastern Afghanistan. Erdkunde 65(2):99–119

Kreutzmann H, Abdulalishoev K, Zhaohui L, Richter J (eds) (2011a) Regional workshop in Khorog and Kashgar. Pastoralism and rangeland management in mountain areas in the context of climate and global change. GIZ, Bonn

Kreutzmann H, Yong Y, Richter J (eds) (2011b) Regional workshop in Lhasa 2010. Pastoralism and rangeland management on the Tibetan plateau in the context of climate and global change. GIZ, Bonn

Macchi M, The ICIMOD team (2010) Mountains of the world – ecosystem services in a time of global and climate change. Seezing opportunities – meeting challenges. ICIMOD, Kathmandu

Nori M, Davies J (2007) Change of wind or wind of change? Climate change, adaptation and pastoralism. The world initiative for sustainable pastoralism. IUCN, Nairobi

Ostrom E, Dietz T, Dolšak N, Stern PC, Stonich S, Weber EU (eds) (2002) The drama of the commons. National Academy Press, Washington, DC

Schröter D, Polsky C, Patt AG (2005) Assessing vulnerabilities to the effects of global change: an eight step approach. Mitig Adapt Strat Glob Chang 10:573–596

Scott JC (2009) The art of not being governed. An anarchist history of upland Southeast Asia. Yale University Press, New Haven/London

Steimann B (2011) Making a living in uncertainty. Agro-pastoral livelihoods and institutional transformations in post-socialist rural Kyrgyzstan, vol 26, Human geography series. Department of Geography, University of Zurich, Bishkek/Zurich

Tapper R (2008) Who are the Kuchi? Nomad self-identities in Afghanistan. J R Anthropol Inst 14:97–116

Wilkes A, Tan J, Mandula (2010) The myth of community and sustainable grassland management in China. Front Earth Sci China 4(1):59–66

About the Authors

Christoph Bergmann is a research assistant and Ph.D. student at the Department of Social Anthropology at the South Asia Institute, University of Heidelberg, Germany. His research interests are mountain pastoralism, ethnicity and ritual practice with a regional focus on the Central Himalayas.

Ted Callahan is a social anthropologist specializing in Afghanistan, Central Asia, pastoral nomadism and political anthropology. His Ph.D. thesis, 'To Rule the Roof of the World: Power and Patronage in Afghan Kyrgyz Society', looked at post-Taliban tribal politics in Afghanistan through the lens of political clientelism. He has spent two years conducting research in Afghanistan, including 11 months living with the Kirghiz in the Afghan Pamir.

Andrei Dörre is a research associate at the Centre for Development Studies at the Freie Universität Berlin, Germany, and has dealt with different aspects of change in post-Soviet transformation societies. In connection with his doctorate, he carried out research on changes of pastoral practices in Kyrgyzstan. Other fields are the interrelationship between international intervention, security promotion and development activities in Afghanistan. His research interest focuses on social change and human-environmental interactions in Central and South Asia.

Eckart Ehlers is professor emeritus (geography) and former chairman of the German National Committee on Global Change Research and IHDP. His regional research interest is focused on the Middle East, thematically on rural and urban research, and nature-society interrelationships in the Anthropocene.

Fazlur-Rahman holds a M.Sc. and M.Phil. from the University of Peshawar, Pakistan, and a doctoral degree from the University of Bonn, Germany. He is currently associate professor at the Institute of Geography, Urban and Regional Planning, University of Peshawar. His research interests include mountain geography, development research and resource management, especially in the Hindukush and Karakoram.

H. Kreutzmann (ed.), *Pastoral practices in High Asia*, Advances in Asian Human-Environmental Research, DOI 10.1007/978-94-007-3846-1,

Martin Gerwin is research assistant and Ph.D. student at the Department of Geography at the South Asia Institute, University of Heidelberg, Germany. He has specialized in development studies, mountain geography and natural resource use in the Central Himalayas.

Melvyn Goldstein is a member of the US National Academy of Sciences and has conducted extensive research in Tibet over the past 24 years on a diverse range of topics including pastoral nomadism, monasticism, modern Tibetan history, oral history and modernization in rural Tibet. He is author or coauthor of over 100 articles and books on Tibet.

Andreas Gruschke is a researcher at the Special Collaborative Research Centre 'Difference and Integration', Institute of Oriental Studies, University of Leipzig, Germany. His main research interest focuses on transformation processes, development geography of the Tibetan Plateau, cultural history of East and Central Asia.

Arnd Holdschlag studied geography, geology, economics and urban planning at the Universities of Bochum and Bonn, Germany. He holds a doctoral degree in geography from the University of Bonn and is currently assistant professor at the Institute of Geography, University of Hamburg, Germany. His research interests include human-environment interactions, development studies, agricultural geography, mountain geography, island studies, complexity theory, South and Southeast Asia and the Caribbean.

Toni Huber is a professor for the anthropology and cultural history of Tibet and Himalayan areas at the Humboldt University, Berlin. His long-term research interests are environment and society, popular ritual and pastoralism on the Tibetan Plateau and highland societies of the far eastern Himalaya.

Muhammad Ismail has more than ten years of working experiences with different international organizations in rural development and natural resources management. His academic interest is in community-based natural resources management and culturally selective technology adoption in climate change adaptation in the pastoral communities. He is a project coordinator in the Rangeland Resources Management Department of the International Centre for Integrated Mountain Development (ICIMOD).

Tobias Kraudzun is a research associate in the Centre for Development Studies, Institute for Geographical Sciences at Freie Universität Berlin, Germany. He has been conducting livelihood-related fieldwork with pastoralists in the Tajik Pamirs since 2003. Development research focusing on livelihoods, human-environment interactions and exchange processes in Central Asian borderlands are his research interests.

Hermann Kreutzmann is chair of human geography at the Institute of Geographic Sciences and Director of the Centre for Development Studies at Freie Universität Berlin, Germany. His research focuses on issues such as migration, conflict, development, water utilization, pastoralism and political geography in Central and South Asian contexts.

Wu Ning is the director general and a professor at the Chengdu Institute of Biology, Chinese Academy of Sciences. He received his Ph.D. from Free University Berlin, Germany, in 1995, and finished his postdoctoral research funded by Alexander von Humboldt Fellowship from 1999 to 2001. His research interests are mainly on the ecology and geography of alpine vegetation and pastoral development on the Tibetan Plateau.

Marcus Nüsser studied geography, geology and development sociology at the Freie Universität Berlin, Germany, leading to a Doctoral degree in the Natural Sciences in 1996. Currently he is a Professor of Geography at the South Asia Institute (SAI), University of Heidelberg (since 2006). His research focuses on issues of human-environment interaction, environmental dynamics and development processes in mountain regions of South Asia and sub-Saharan Africa.

Jarmila Ptackova is a Ph.D. candidate at the Central Asian Seminar of the Institute of Asian and African Studies, Humboldt University, Berlin. Her research concentrates on the influence of the current development policy of the Chinese government on the livelihoods of Tibetan pastoralists in China.

William S. Sax is professor and head of the department of social anthropology at the South Asia Institute, University of Heidelberg, Germany. He has been conducting research in Uttarakhand (India) since 1977 and focuses on issues such as religious practice, local social structure and politics, territoriality, forms of theatre and performance as well as ritual healing.

Stefan Schütte is an assistant professor at the Centre for Development Studies, Geographic Sciences at Freie Universität Berlin, Germany. He has extensive research experience in Afghanistan, India and Pakistan and worked on issues of rural and urban development with a special focus on human security and livelihoods.

Yi Shaoliang has over 20 years of work/research experiences in alpine ecology, nature conservation and mountain pastoralism. He worked for many years in the International Centre for Integrated Mountain Development (ICIMOD) as a project coordinator and team leader in rangeland resources management and in Chengdu Institute of Biology (Chinese Academy of Sciences) as a researcher. His academic interest is in common pool resources management and human-nature interactions in mountain regions. He is currently the manager of the Natural Resources Management (Land & Water) Department of Aga Khan Foundation, Afghanistan.

Chetan Singh is a professor of medieval Indian history at Himachal Pradesh University, Shimla (India). He has several publications on seventeenth-century Punjab as well as on the environmental and socioeconomic history of the Western (Himachal) Himalaya during the colonial period. He is currently working on the historical transformation of popular culture in Himachal.

Bernd Steimann is a research fellow at the Department of Geography, University of Zurich, Switzerland. His research interests include development studies in Central and South Asia, with a particular focus on the governance of natural resources, rural livelihoods and processes of institutional change.

Emilia Róża Sułek received her master degrees in social anthropology and in Tibetan studies from the University of Warsaw, Poland. In 2007–2010, she conducted research on the emergence and social consequences of the caterpillar fungus economy in Tibetan pastoral areas of Qinghai Province, China. She is completing her doctoral dissertation at the Central Asian Seminar of the Institute of Asian and African Studies at Humboldt University, Berlin.

Lu Tao is an associate professor at the Chengdu Institute of Biology, Chinese Academy of Sciences. He received his Ph.D. from the Research Centre for Eco-Environmental Sciences, Chinese Academy of Sciences, in 2007. His major research interests are in landscape ecology and grassland ecology.

Yan Zhaoli is an associate professor at the Chengdu Institute of Biology, Chinese Academy of Sciences. She received her Ph.D. from the Chinese Academy of Sciences in 2003 and henceforth worked as rangelands specialist at the International Centre for Integrated Mountain Development for over six years. Her research interests are mainly in rangeland ecology and management.

Index

H. Kreutzmann (ed.), *Pastoral practices in High Asia*, Advances in Asian Human-Environmental Research, DOI 10.1007/978-94-007-3846-1,

Q

R

MIX
Papier aus verantwortungsvollen Quellen
Paper from responsible sources
FSC® C105338

If you have any concerns about our products,
you can contact us on
ProductSafety@springernature.com

In case Publisher is established outside the EU,
the EU authorized representative is:
Springer Nature Customer Service Center GmbH
Europaplatz 3, 69115 Heidelberg, Germany

Printed by Libri Plureos GmbH
in Hamburg, Germany